CW00594528

Adult Crying

BIOBEHAVIORAL PERSPECTIVES ON HEALTH AND DISEASE PREVENTION

From the perspective of behavioral science, the series examines current research, including clinical and policy implications, on health, illness prevention, and biomedical issues. The series is international in scope, and aims to address the culturally specific, as well as the universally applicable.

Series Editor: Lydia R. Temoshok, PhD, Institute of Human Virology, Division of Clinical Research, University of Maryland Biotechnology Center, 725 West Lombard Street, Room N548, Baltimore, Maryland 21201, USA

Volume 1
Stress and Health: Research and Clinical Applications
edited by Dianna T. Kenny, John G. Carlson, F. Joseph McGuigan and John L. Sheppard

Volume 2
Psychosocial and Biomedical Interactions in HIV Infection
edited by Kenneth H. Nott and Kav Vedhara

Volume 3
Adult Crying: A Biopsychosocial Approach
edited by Ad J.J.M. Vingerhoets and Randolph R. Cornelius

Adult Crying
A Biopsychosocial Approach

Edited by

Ad J.J.M. Vingerhoets
Tilburg University, The Netherlands

and

Randolph R. Cornelius
Vassar College, Poughkeepsie, NY, USA

First published 2001 by Brunner-Routledge
27 Church Road, Hove, East Sussex BN3 2FA

Simultaneously published in the USA and Canada
by Taylor & Francis Inc.
325 Chestnut Street, Suite 800, Philadelphia PA 19106

Brunner-Routledge is an imprint of the Taylor & Francis Group

Printed and bound in Great Britain by Biddles Ltd, Guildford and King's Lynn

British Library Cataloguing in Publication Data
A catalogue record for this book is available from the British Library

Library of Congress Cataloging-in-Publication Data
A catalogue record for this book is available from the Library of Congress

ISBN 1-58391-225-8

For Bregje (AJJMV)
For my late father, Earnest M. Cornelius (RRC)

Contents

List of Figures

List of Tables

Foreword

Nico H. Frijda

Crying is a very common everyday phenomenon. It is one of the behavioral manifestations that define what it is to be a human. And yet it is a puzzling behavior. Why do we do it? What is it that elicits crying? Why has crying the appearance that it has, the attendant noise, the facial contractions, the convulsions, and the tears? What, if any, are its functions?

The psychology of emotion and emotional expressions has made great strides forward during the last decades. Yet, there have emerged few answers to these questions concerning crying. There is relatively little insight into the nature of this response that has obvious biological origins. Until recently, there were mainly acute observations, but there was also much speculative reflection, and little empirical research. This book departs from that line. It manifests a renewed effort to find answers by the use of empirical research, as well as by systematic clinical observations. It provides an overview of much of the major effort in the domain, by major investigators in the field.

Why the slow progress? To a large extent, this is due to the absence of animal predecessors of human crying. Such predecessors have done much to help find plausible hypotheses on the functions of other expressions. Crying is uniquely human, at least as far as tears are concerned; and tears are the most prominent sign of crying in adults, the subject of the present volume.

But there are other reasons for the slow progress: most prominent among these is the complexity of the subject. Traditionally, crying is linked to distress, and notably to infant distress, pain, and sadness. However, it is obvious that there are many more conditions leading to crying, and that the variety of emotions in which tears appear is considerable. As will be made clear in many of the chapters in this volume, crying may appear in the context of almost any negative emotion. It occurs in anger and in shame, there are tears of remorse, there are tears of frustration. In addition, crying occurs in the context of positive emotions: in joy, happiness, hope, relief, orgasm. And crying is prominent in that ambiguous class of emotions called 'being moved.' Being moved includes joy with a melancholy overtone, and sadness with a not unpleasant tone, as well as responses to what is appraised as beautiful, endearing, or cute. There can also be satisfaction in tears themselves. Sweet tears were mentioned by Homer, and were discussed more frequently in the literature of past centuries (e.g. by Montaigne, 1993). They have, of course, not disappeared in the meantime. The Czech language even has a word for grief that one wallows in, namely, *lítost*, which may be one of the few subjects connected with crying that is not specifically discussed somewhere in the present volume.

The diversity of conditions for crying is well recognized and discussed in many of the chapters here. One will also encounter the many ways in which researchers try to deal with this diversity. One can hold onto the notion that crying is properly an expression of distress. In positive emotions, it can still be an expression of past distress that is allowed to come out. One can also hold that crying corresponds to a psychological entity that is common to the various forms of distress under which crying appears, and to the other emotional conditions where this is the case. The psychological entity may, for instance, be the attitude of helplessness and dependence. Or one can hypothesize that crying, in fact, embraces several types of responses, each with a different function. These three approaches, obviously, lead to quite different specific hypotheses about the functions of crying.

They each have to deal in their own ways with an additional complexity. None of the emotions mentioned as a condition for crying is invariably accompanied by crying. Not all distress, all sadness, nor all joy or happiness are. This points to the need to specify much more closely the conditions under which a particular emotional source actually leads to crying. In the language of cognitive emotion theory, one will have to seek to specify the specific appraisals that lead to crying, in addition to those that lead to the emotion concerned. Appraisal of helplessness, or of accepted or recognized helplessness, is an obvious candidate for such a specification. In sadness one may cry when truly recognizing that nothing can be done about the event; in anger, crying may occur when the anger is impotent; and so forth.

It is important to remember the variety of emotional conditions under which crying occurs when interpreting the data on crying frequencies, reported in different chapters of this volume. Individual and group differences in the frequency of crying may reflect differences in occasions for crying, and also in predominance of the emotions that give rise to crying. Certain social groups, for instance, may be more prone to impotent anger than others, and some may be more vulnerable to conditions of helpless sadness. Future research may well be inclined to seek more information on these matters.

In line with this complexity of conditions, there is the complexity of possible functions of crying. The contributions in this volume suggest and discuss a wide array of such functions. The most obvious, and frequently mentioned, is its communicative function, to communicate the need for help and to call for help. The imperative quality of hearing someone cry is mentioned in several places; the innate disposition to cry seems to be matched by an innate disposition to be disarmed by crying, or to have one's capacities for caring awakened. There is indeed some older research that supports such a surmise, at least with regard to infant crying. Murray (1979) has observed that five seconds go by between an infant's onset of crying and the mother's movement to pick it up. Some observations along the same lines are mentioned in some of the chapters here (see Chapters 3 and 8, this volume), but the topic would seem to lend itself perfectly to both research via questionnaire and experimental research.

But there is an abundance of other possible functions. Crying is discussed as a discharge phenomenon, a means of tension release, and also as a chemical or symbolic purification. It is viewed as a medium for cognitive reorientation or affective reintegration and as a mode of defensive regression. It is also interpreted as an aggression inhibitor or elicitor of compassion. It has also been recognized as a mechanism for social bonding, whether in infant-parent attachment or in adult encounters, along the road of compassion, comforting, and emotion sharing; Rimé *et al.* (1997) mention it in that context. Crying may even be given an explicitly social function, in crying together. It can be a truly communal activity, in public ceremonies of mourning as well as of veneration, and be a direct medium of communion.

One may doubt that the social functions of crying might be the fundamental ones, because people often cry when alone, and sometimes prefer being alone. The facts are interesting, but their meaning is unclear. Fridlund (1994) made the assumption of "implicit sociality", but that assumption would seem unnecessary: evolution may have installed an automatic help-calling provision that needs no presence of help-givers to elicit it; a clam need not detect predators to close tightly.

Analysis of the possible functions of crying should proceed carefully, and with sophistication. The notion of function is polysemous. It may refer to why a biological function might have come into existence (that is, the functional benefit by which it was retained in evolution); it might refer to the purpose to which it was put, once it existed for other reasons; and it might refer to how it happens to function, again it is there. Voluntary manipulation of others would be an instance of the second kind of function; social bonding might be a function of the third kind. The distinctions are important because they show that quite different functions are not mutually exclusive, depending upon how their origin is conceptualized.

The multiplicity of functions of crying is recognized transculturally. In many cultures, crying is seen as a sign of weakness or unmanliness. This is so particularly in macho cultures, like Muslim cultures. And yet, in several of those cultures, that condemnation knows one major exception: crying in response to 'sentimental' stimuli, like sad love songs. It is completely acceptable that men wallow in them, and sob openly and loudly during the performance of such songs; this is so in Turkey (Mesquita, 1993) and among the Ahwad'Ali Bedouins of Egypt (Abu-Lughod, 1986), and it explains in part the tremendous popularity of singers like Oum Kalthoum in Egypt who specialized in such songs.

One aspect of crying does have an animal predecessor, namely in distress or alarm calls. As mentioned, they are releasers of care and considerateness, mainly but not merely with regard to infants of the species (De Waal, 1996). But this does not hold true for tears. Tears are an almost complete mystery with regard to their biological origin as well as to their present function. That does not merely mean that it is unclear what they communicate, but also how they do it, that is, the capacity for receptivity in those who perceive them. Darwin's (1872) hypothesis—protecting the eyes during shrieking—is clearly inappropriate, because animals do shriek but do not shed tears.

Perhaps one indeed has to recognize that crying involves more than one function, and perhaps is only one name for what in fact are different types of responses, belonging to different behavioral systems. It may be a worthwhile approach to examine that possibility. Perhaps there is a basic difference between noise production, the crying proper or shrieking, and the shedding of tears. The loud noise is meant to bridge distances; the tears can be seen only when close by. Tears thus should have a function in influencing others in proximal social situations. What could they be? Perhaps the functions of loud crying and weeping tears might be those of help-requesting versus submission, or the displaying of humility. The latter obviously is relevant mostly in close contact.

There are several arguments for this view of tears, at least as one of their functional contexts. One is that weeping can function as an aggression inhibitor. I think that it can be shown that weeping is seen as endearing and a sign of vulnerability and harmlessness. Another, more important one is that weeping is frequent in reverential situations. One often weeps when confronting superiors, and particularly those superiors whose superiority is recognized and accepted to the full. The public tends to weep when cheering the pope or, at the time, Khomeini, the Iranian head of Islam and of the state. The weeping occurs in a humility context, that is, of other submission displays like kneeling, kissing the hem of the target person's dress, or prostrating. In a less reverential context, many people weep when seeing and hearing their media stars, their top singers or actors. Perhaps this humility or submission interpretation also fits the conditions of sentimental emotions and aesthetic admiration (Tan and Frijda, 1999). These conditions can all be interpreted as conditions in which one faces someone or something that is recognized as great, pure, absolute, a condition in front of which one can put one's weapons down and to which one can submit. This applies to suggestions of the possible perfections of human affection, as when ET wants to go home, the Miracle Worker achieves her miracle of teaching the deaf-mute Helen Keller to speak (Efran and Spangler, 1979), or when witnessing the fullness of human acceptance in a rough world, as when the grandfather of Oliver Twist finally receives his grandson. They are all conditions for sentimental tears.

The present volume brings together research results about crying from a large number of methodological approaches and theoretical orientations. It surveys biological bases, questionnaire data on occurrence of crying and individual and cultural differences in that regard, and information from therapy and pathology. It ranges over the variety of possible interpretations, some of which I have just broadly outlined. It crystallizes a large amount of research effort from the last decade or so. We trust that the present volume will make a valuable contribution to this new research domain.

References

Abu-Lughod, L. (1986). *Veiled sentiments*. Berkeley, CA: University of California Press.

Darwin, C. (1872). *The expression of emotions in man and animals*. London: John Murray (1965, Chicago, IL: University of Chicago Press).

De Waal, F.B.M. (1996). *Good natured: The origins of right and wrong in humans and other animals*. Cambridge, MA: Harvard University Press.

Efran, J.S., and Spangler, T.J. (1979). Why grown-ups cry. *Motivation & Emotion, 3*, 63–72.

Fridlund, A.J. (1994). *Human facial expression: An evolutionary view*. New York: Academic Press.

Mesquita, B. Gomes de (1993). *Cultural variations in emotions. A comparative study of Dutch, Surinamese and Turkish people in the Netherlands.* Ph.D. Thesis, University of Amsterdam.

Montaigne, M. de (1993). *The complete essays.* M.A. Screech (transl.). Harmondsworth: Penguin.

Murray, A.D. (1979). Infant crying as an elicitor of parental behavior: An examination of two models. *Psychological Bulletin, 86*, 191–215.

Rimé, B., Finkenauer, C., Luminet, O., Zech, E., and Philippot, P. (1997). Social sharing of emotions: New evidence and new questions. In: W. Stroebe and M. Hewstone (Eds.), *European Review of Social Psychology, Vol. 7.* Chichester: Wiley.

Tan, E.S.H., and Frijda, N.H. (1999). Sentiment in film viewing. In: C. Plantinga and G. Smith (Eds.), *Passionate views: Film, cognition, and emotion* (pp. 48–64). Baltimore, MD: Johns Hopkins University Press.

Acknowledgment

We would like to thank Rinus Verkooyen for his fine text-editing and reference checking, Sue Ann Tasselmyer and Gail Garrison for the many things they did to make this book possible, Robin Nussbaum for ferreting out obscure facts in the psychological literature, and Leah Warner for the wonderful job she did indexing the US contributor's chapters. We would like to thank James Gross for the permission to reprint the figure depicting his model of emotion regulation and Tineke K'tinka Storteboom for her perfect cover painting. We would like also to thank the contributors to this volume for their hard work and dedication to this project. Finally, we would like to offer our many thanks to Harwood Academic Publishers and Lydia Temoshok, the editor of the series of which this volume is a part, without whom this book would not have been possible.

List of Contributors

Marleen C. Becht
Department of Psychology, Tilburg University, Tilburg, The Netherlands

Marrie H.J. Bekker
Department of Psychology and Department of Women Studies, Tilburg University, Tilburg, The Netherlands

A. Jan W. Boelhouwer
Department of Psychology, Tilburg University, Tilburg, The Netherlands

Robert F. Bornstein
Department of Psychology, Gettysburg College, Gettysburg, Pennsylvania, USA

Randolph R. Cornelius
Department of Psychology, Vassar College, Poughkeepsie, New York, USA

Antje Eugster
Department of Psychology, Tilburg University, Tilburg, The Netherlands

Nico H. Frijda
Department of Psychology, University of Amsterdam, Amsterdam, The Netherlands

Janice L. Hastrup
Department of Psychology, State University of New York at Buffalo, Buffalo, New York, USA

Myriam Horsten
Statistics Netherlands, Voorburg, The Netherlands

Jeffrey A. Kottler
Department of Psychology, University of New England, Armidale, Australia

Ban N. Khan
Houston Neurocare, Houston, Texas, USA

Deborah T. Kraemer
Department of Psychology, Southern Connecticut State University, New Haven, Connecticut, USA

Susan M. Labott
Department of Psychiatry, University of Illinois—Chicago, Chicago, Illinois, USA

Marilyn J. Montgomery
Department of Psychology, Florida International University, Miami, Florida, USA

Vikram Patel
Sangath Centre for Child Development and Family Guidance, Goa, India

Ype H. Poortinga
Department of Psychology, Tilburg University, Tilburg, The Netherlands

Marwan N. Sabbagh
Department of Neurosciences, University of California San Diego and Department of Neurology, San Diego VAMC, San Diego, California, USA

Jan G.M. Scheirs
Methodology Department, Faculty of Social Sciences, Tilburg University, Tilburg, The Netherlands

Aziz T. Shaibani
Houston Neurocare, Houston, Texas, USA

Klaas Sijtsma
Methodology Department, Faculty of Social Sciences, Tilburg University, Tilburg, The Netherlands

Glenn R. Trezza
Boston VA Medical Center; Boston University School of Medicine and Tufts University School of Medicine, Medford, Massachusetts, USA

Nico J. Van Haeringen
The Netherlands Ophthalmic Research Institute, Amsterdam, The Netherlands

Guus L. Van Heck
Department of Psychology, Tilburg University, Tilburg, The Netherlands

Miranda A.L. Van Tilburg
Department of Psychiatry and Behavioral Sciences, Duke University Medical Center, Durham, North Carolina, USA

Ad J.J.M. Vingerhoets
Department of Psychology, Tilburg University, Tilburg, The Netherlands

Debra M. Zeifman
Department of Psychology, Vassar College, Poughkeepsie, New York, USA

Introduction

Everyone cries. Our lives begin with a cry that brings relief to our mothers because it is a sign that we are in good shape after our entrance into the world. As babies, most of us go on to cry on a regular basis—even if we are not cry babies—because this is the one means we have to attract the attention of our parents or caretakers and to indicate that we are in need of food or comfort, or that we are in pain. As Darwin (1872) noted in his observations of his infant son William, a baby's cries of hunger, frustration, and distress are soon accompanied by tears, and thus a mystery is presented. While many animals utter distress cries, humans appear to be the only animals that shed emotional tears. If this is indeed the case, then we must ask, Why tears? What functions do tears serve? How did emotional tearing evolve? What is the relationship between crying considered as an acoustic distress call and the shedding of emotional tears?

The mystery of the origin of tears is complemented by others: As we grow up and acquire the verbal skills to communicate our wants, needs, pains, and frustrations, we begin to cry less. But in some conditions, even for the most emotionally reserved of us, we nevertheless find ourselves with tears in our eyes. Why, if we can communicate our distress in other ways, do we still find it necessary to cry? Why does our crying change form from primarily acoustic to visual/acoustic? What does crying communicate to others? To ourselves? What do we know about the situations that make us cry? About the emotions that elicit and accompany crying? People differ considerably in why, when, and how much they cry. What do we know about the underlying causes of these differences? What are the contributions of nature and nurture to such differences? Are there cultural differences in the frequency and experience of crying? Is crying healthy? These are some of the questions that inspired us to assemble the contributions to this book.

When we began our respective research programs on crying, we were surprised to learn that so little was known about crying and that what there was consisted of much more speculation than theory driven research. Adult crying in many ways still is '*terra incognita*' in the behavioral sciences. Therefore, we felt that it was most appropriate and timely to assemble a book that not only summarizes the state-of-the-science of crying, but that also would inspire scholars to start their own research programs on crying and to formulate specific hypotheses about the functions of crying and how crying relates to and is influenced by both personal and environmental factors.

In putting together this book, we wanted to pay attention to all relevant aspects of adult crying (this includes taking a brief look at infant crying) and sought contributions from researchers all over the world who are currently studying adult crying. We challenged the contributors to discuss all the phenomena that they felt might play a role in adult crying. The ultimate aim of this volume is to come to the formulation of a preliminary model of crying that might be helpful in generating testable hypotheses and directing further research in this area. Of course, it is up to the reader to decide to what extent we have been successful in achieving these goals.

References

Darwin, C. (1872/1965). *The expression of the emotions in man and animals.* Chicago: University of Chicago Press.

1 THEORIES OF CRYING

Jeffrey A. Kottler and Marilyn J. Montgomery

It was dark outside, and cold, so cold the man's eyes stung from the moisture on his lids. He was hunched over, trying to make himself as small a target as possible against the frigid wind. He shuffled along at an unsteady gait, trying to keep his balance on the slick sidewalk. Occasionally, he would stop, peek out from under his hood to reorient himself in the blowing snow, and then step cautiously but purposefully onward toward his destination.

The man was late for a dinner engagement, partly because he hadn't anticipated the slow driving conditions, but also because he had to park quite a way from the restaurant. He had been thinking about some problems at work during this agonizing walk in fierce weather, partly to distract himself but also to work out some solutions to the challenges that he would face the next day. Suddenly, without warning and with frightening intensity, he could feel the heat of tears running down his face; they seemed scalding when compared to the frozen skin of his cheeks. By the time the tears reached his upper lip, they tasked like salty ice particles.

"Now, what on earth could be going on?" he wondered to himself, half amused but also alarmed. "One minute I'm on my way to a restaurant to meet some friends, trying desperately not to fall on my behind, and the next minute I'm standing here bawling like a baby."

Actually, that wasn't quite right. Far from the sobs of a full fledged, dramatic crying episode, he was shedding tears quietly, with dignity. Standing there in the swirling snow, he was like Ed Muskie, the presidential candidate who saw his campaign go down the drain once he let tears show on his face during a speech. Muskie claimed, just as this man would had he been asked, that the moisture on his cheek was only snowflakes.

But the man knew he was crying. He admitted it to himself, however disturbing, even though he couldn't figure out why he was upset, nor even what he was worked up about. It wasn't something at his job, he quickly surmised. He also sensed it had little to do with where he was headed. But something was going on inside him, even if he couldn't put it into words.

In developing a theory to account for this tearful episode, the man considered, then rejected, a number of explanations. He wondered if it had something to do with his hormones, or even his recent sleep patterns. He thought perhaps the weather was triggering some involuntary reaction, or maybe even some memory that was being accessed through associations with the present conditions. Hadn't he read somewhere that crying is really a matter of excreting toxic chemicals from the system? Or that it involved a release of tension. Or maybe it was a way to bring his attention to some issue he was ignoring?

In his search for causes to account for his behavior, this man experienced what most any scholar might feel in trying to unravel one of the greatest mysteries of human functioning since the earliest human occupants on this planet first noticed the weird phenomenon of water dripping out of their eyeballs during times of emotional upheaval. Quite a number of investigators, from Darwin (1872) and Montagu (1959) to contemporary scholars have been similarly stumped as to what causes crying, what purposes it serves and how it evolved. Various emotion researchers such as Collier (1985), Denzin (1984), Damasio (1994), Izard (1991), and Lazarus and Lazarus (1994), who are otherwise quite informed about the mechanisms of emotional expression, admit that they are perplexed by the mystery of crying. Indeed, most texts and books on the subject of emotion ignore the subject altogether (Kottler, 1996).

Problems in Theorizing About Crying

In this chapter, we review the major theories of crying, beginning with ancient conceptions, continuing to historical paradigms, then moving to contemporary models. Finally, we present an integrated theory of crying that in our opinion combines the best features of the theories reviewed.

We acknowledge that much of the material presented in this chapter is based on speculation rather than on hard empirical science. The reality is, that at this time, most theories to account for crying behavior are relatively primitive, and often contradictory. Adult crying has been hypothesized by some to be a cry for help, by others as a sign of surrender. It is conceptualized by some theorists as a form of healthy self-expression while others view it as a symptom of psychopathology. It is described as the ultimate in human authenticity, or in manipulation. It is a sign of utter, abject helplessness, or of the courage to reveal one's inner most feelings. There is even considerable disagreement as to whether it represents emotional arousal, or emotional recovery from arousal.

One of the reasons that crying may be so difficult to research and explain is because we may not really be looking at a single phenomenon but rather a complex mix of several distinctly different behaviors that are influenced by a host of biochemical, endocrine, social, systemic, and cognitive processes (Cornelius, 1988). Perhaps the tears that are shed during endogenous depression are elicited by different physiological mechanisms than those that accompany an acute grief reaction. Even more confusing, it is likely there are different factors and processes that influence tears of shame compared to those that accompany a profound experience of spiritual transcendence.

These competing explanations are only part of the reason why it is so challenging to construct a comprehensive theory of crying. There are few aspects of human behavior that seem to involve so many different disciplines, each with their own language and focal areas. Certainly psychology has a large investment in this subject, not only as a social phenomenon but also because clinicians are so befuddled as to the most helpful way to respond to various crying episodes that occur in therapy. There is considerable disagreement among therapists as to when crying should be encouraged in session versus when it should be stifled (Mills & Wooster, 1987; Chapter 12, this volume).

Sociology has added much to the body of literature as well, especially the work that examines crying from a symbolic interactionist perspective (Mead, 1934). Cultural theory has offered much to the body of literature as well, especially the work that examines crying from a social cultural perspective looking at how crying behavior is shaped through roles, beliefs, values, and customs (Betancourt & Lopez, 1993; Lutz, 1999; Triandis, 1989; Wellenkamp, 1992). Anthropologists also have constructed theories to account for cultural differences in this behavior. While in some parts of the world, crying is ritualized into an honorable form of communication (tuneful weeping in India, for example), in other regions, it is seen as the ultimate in shameful acts.

In addition to the social sciences, biologically based disciplines have also offered much to the subject. Ophthalmologists, pediatricians, and biochemists have developed their own theories to account for weeping. Evolutionary psychologists have tried to account for how and why this strange behavior developed in the first place, and why it communicates so many different messages.

Although the array of explanations offered by different disciplines does complicate matters, it also enriches the landscape. A far more difficult challenge is the problem of measuring the behavior. How do we know when someone is crying, especially those of the male gender who may not actually be shedding tears or sobbing, but who are clearly crying inside? Must tears drip from the eyelid in order for it to count as an "official" tearful episode?

While one of us (J.K.) was leading a group therapy class recently, a student in the class was talking about his frustration and pain from not yet being able to conceive a baby. While talking about his feelings, his lip trembled visibly. Moisture pooled in his eyes. He had that peculiar, scrunched up look to his face that is so unique to "brave" men showing they are too strong to cry. Somehow, with furious effort, he talked through the pain in a trembling voice, without allowing the actual tears to fall (although several times he cheated by casually wiping his arm across his face).

After the episode was over, one of the other class members referred to the incident by saying: "When you almost cried earlier..." There was no doubt that he *did* cry, in the unique way that many men weep—with little noise and almost no visible tears. Still, it is unclear whether this incident would have qualified in many studies of crying episodes. (In an exceptionally generous categorical scheme, Williams and Morris (1996) differentiate between "close to tears", "tears-in-eyes", "watery eyes", "tears plus sobbing", and full fledged "weeping"; in their study, this man's behavior would have made the cut.) Attempts to measure crying quantitatively by noting the length of the tearful episode or the amount of tears expelled omit important instances and information that should receive our attention.

Another problem in constructing theories about crying is that this single act expresses so many different emotion states. We can see tears in a person's eyes, but without other contextual cues, we find it difficult to determine whether this person is happy or sad, frustrated or angry, profoundly moved or utterly miserable. Moreover, crying is hardly a single universal language, but rather a human form of communication with self and others that has hundreds of different dialects (Kottler, 1996). There is weeping associated with despair and sadness, but also tears of shame, loss, emotional release, joy, and spiritual transcendence. Furthermore, there is clearly a cultural and gender context for the behavior that make it nearly impossible for a single theory to account for the behavior.

All of these methodological and hermeneutic challenges make it especially difficult to explain what crying means with any sort of precision and comprehensiveness. Nevertheless, a number of philosophers and scientists throughout the ages have attempted to explain what is going on.

Historical Perspectives

Human beings have been trying to understand the meaning of their tears since the first time they were observed as an accompaniment to emotional upheaval. In ancient times, tears themselves were regarded as potent. According to an ancient Egyptian text, humans were born from the tears of the Gods. Others also made cosmic associations; in the Dark Ages, the shooting stars often observed in Europe during late summer were called "Tears of Saint Lawrence," in honor of the Roman patron saint of beggars.

In Western literary epics and folk tales, the weeping of the protagonist often symbolizes a crucial psychological turning point. Gilgamesh, the hero of the ancient Mesopotamian epic, weeps when he accepts the futility of his quest for immortality and surrenders to an existential acceptance of his humanity. Ulysses counters the spell of the immortal nymph, Calypso, by allowing himself to tearfully yearn for his home and his mortal wife. In countless narratives, crying seems to help loosen and wash away existing rigid patterns and prepare the way for new possibilities (Clarus, 1991).

Crying has also been used, both formally and informally, in the determination of guilt or innocence. In 17th Century America, a notion prevailed that future marital happiness was doomed if the bride did not weep profusely at the wedding. The belief behind this superstition was that witches could only cry three tears, and those from the left eye only, so a copious flow of tears was proof that a bride had not "plighted her troth" to Satan, and was therefore not a witch (Evans, 1981). In this social context, the ability to cry literally had survival value. Even today, juries are swayed by the guilty party's ability to show remorse, and a flow of tears is often seen as the primary evidence of genuineness.

Philosophers, too, have tried to explain tears. Medieval doctrines of the emotions were proffered by thinkers who wrestled with the contrasts between human and divine nature (Gardiner et al., 1937). Gregory of Nyssa (361/1864) criticized the common doctrine that the soul is seated in the heart, and was the first to address the subject of tears. Weeping happens, he thought, because blood vessels are contracted and their functions in the viscera impeded. Tears evaporate from the blood, therefore, and rise to the head, accumulate as moisture and descend to the eyes where they are pressed out by the eyelids as tears.

During the Renaissance, all aspects of human inner life, including emotion, were regarded with new interest and much speculation. De la Chamber (1658) devoted an entire chapter to the subject of weeping in his book, *Les caractères des passions*. A common opinion during his lifetime was that weeping with grief was an expression that procured aid for the sufferer by arousing the sympathy of others. De la Chamber recognized, however, that there are tears of joy, shame, anger, and pity, as well as those of grief. He saw the tears themselves as humors distilled from the moist brain. Vives, another writer of that time, argued that tears flow profusely when the brain is filled with moisture, as when a man is drunk, or when it is soft and tender, as it is for boys, young women, and the sick. When for any reason the brain is heated or the moisture dried up (as in the hot anger of men, melancholia, and prolonged grief), there are no tears to flow (Vives, 1555).

Descartes' *Les passions de l'âme* (1647/1985) relates his theory that tears are formed of vapors akin to sweat; they are changed into water when the normal evaporation is interrupted by a process like that of the formation of rain. Weeping is not continuous, and it does not occur in extreme grief, Descartes reasoned, so it arises from the alteration of grief with love or joy (grief contracting the pores of the eyes; love and joy sending more blood to the heart and so increasing the quantity of vapors). Descartes thought that emotions themselves arise in the soul, which is completely distinct from the body, influencing the body through the brain's release of animal spirits into visceral systems.

Centuries later, William James argued that there is nothing mental (or metaphysical) about emotions, they simply are a type of awareness that results from subtle changes in the blood vessels and internal organs. We undergo physiological changes in reaction to an event, and we interpret those physical changes as having emotional meaning. "We feel sorry because we cry," James wrote, and not the other way around (James, 1894, pp. 527).

Current Theories: Levels of Explanation

Current theories of crying could be organized in a number of ways, for example, according to their history of development or popularity. We have chosen to discuss the various explanations according to the levels of analysis at which answers are sought to the question, "Why do we cry?" In the following discussion, we will review a range of theories that vary from "micro" levels of analysis to "macro" levels of analysis.

BIOCHEMICAL AND PHYSIOLOGICAL EXPLANATIONS

Darwin's work on emotional expression (1872) left him confused when it came to crying; he could think of no clear way that crying might be adaptive. He concluded that tears are simply a by-product of muscles contracting around the eye. This contraction, he reasoned, serves a protective function by preventing the facial muscles near the eyes from becoming too engorged with blood. For Darwin, the resulting tears seemed to pose an inexplicable exception to the rule of the evolutionary functionality of all behavior and body structure.

Almost a century later, the evolutionary psychologist Montagu (1959) proposed a biological explanation for how tears could be adaptive. He hypothesized that crying originated as a protective mechanism that prevented the rapid drying out of the mucous membranes of the nose and throat when people were agitated and sobbing. Tears, which contain the antibacterial enzyme lysozyme, also reduce the risk of contracting ocular and upper respiratory diseases. In this way, he thought, tearing and crying make a physiological contribution to the survival of the human species.

The American biochemist Frey (1983; 1985) conducted the first landmark research on the biochemical aspects of crying. Frey discovered that the chemical composition of tears springing from emotion is different from the tears produced continuously to lubricate the eye or tears produced in response to an irritant: they contain higher concentrations of proteins, manganese, and prolactin, a hormone produced during stress-induced danger or arousal. Therefore, he reasoned, perhaps the function of crying is the removal of potentially toxic substances that are released when people are distressed. When these waste products are removed from the body, Frey believed, one's mental state is improved, so crying helps to physically restore the biochemical conditions for emotional well-being.

While Frey's theory is compelling, it does not appear to hold the conclusive answer. His initial observation that increased prolactin production might be responsible for crying output has not yet been empirically confirmed. In one study testing his hypothesis, elevated plasma prolactin did not affect the threshold for weeping (Vingerhoets et al., 1992).

Still, medical and psychological authorities have been saying for decades that crying and other forms of emotional release are good for one's health (see Chapters 11 and 13, this volume). Holding in tears has been believed to contribute to a host of ailments from hives and acne to cancer and colitis. Crepeau (1981) put this theory to the test by observing a number of physiological benefits enjoyed by fluent criers. Like Frey's theory, however, subsequent research has not necessarily supported her claim that crying is physiologically beneficial. In some cases, quite the opposite was found, in that people who cried profusely were more likely to end up feeling worse, at least in the short run (Gross et al., 1994).

PSYCHOLOGICAL EXPLANATIONS

Psychodynamic theories. In his early work, Freud (1895/1953) saw emotions as psychic energy, and the repression of psychic energy as the main cause of hysteria. In this view, crying is evidence of a healthy catharsis of aroused emotions. In later work Freud (1910/1957) revised this view and theorized that when the quantity of energy associated with instinctual processes becomes excessive, it is discharged in the form of emotions. Some modern psychoanalysts, including Rapaport (1960), have used this 'drive discharge' model to argue that the failure to express feelings represses drives and results in neurotic behavior. Crying is alleged to "wash away pain", functioning as a kind of safety valve that seeps tears to release negative emotional energy (Sadoff, 1966).

Psychoanalysts Wood and Wood (1984) also believe that instinctual, cathartic drives are at the heart of crying, but they employ an equilibrium model in which tears are connected to unresolved issues of the past, and are essentially triggered as a form of emotional regression by some stimulus in the present. One such example of this phenomenon in action is when someone cries at a testimonial dinner in his honor, not because he is happy at getting an award, but because of a release of pent-up feelings that have been simmering for years (Weiss, 1952).

The notion of crying as a release of pent-up psychic energy has been a favorite theory in the literature, mostly because there seems to be a grass-roots consensus that people report feeling better after a good cry. Sloboda (1991) applied this theory to account for the kind of transcendent crying that sometimes accompanies a moving passage of music. Whether someone cries in response to the love theme of Tchaikovsky's *Romeo and Juliet* or Puccini's *La Bohème*, he noted that certain melodic constructions most reliably produced a tearful result. A similar result has been observed during those times in which people weep at movies, not during the sad scenes, but *after* the tension is resolved—when the lovers are finally reunited or when the protagonist's achievements are finally recognized (Efran & Spangler, 1979; Labott & Martin, 1988).

Cognitive approaches. While the theory that crying exists to leach toxic levels of emotional states is attractive if not intuitively compelling, there is not a lot of empirical evidence to support this belief (Labott & Martin, 1988; Chapter 13, this volume). Other scholars have explored crying from a cognitive or informational perspective. Cornelius (1997), for example, doesn't quibble with the reality that crying *can* bring relief, but points out that much depends on the particular context and internal cognitive states.

Ellis (1962) and quite a number of other cognitive therapists have operated from the assumption that all emotional responses, crying included, are the result of cognitive interpretations of events rather than the spontaneous result of any automatic feeling. Thus according to this model, individuals actually "decide" to cry based on their perception and labeling of their experience as sad, joyful, or depressing. A whole group of people might be subjected to the same circumstances, whether watching a sunset or missing an airline flight, and yet some individuals would be moved to tears while others would sit stony-faced, depending on how they chose to think about the phenomenon. The thought, "This is no big deal," would create a very different reaction than, "This reminds me of my dead father," or "This sort of thing always happens to me. I deserve nothing better."

Rather than recognizing the primacy of biochemical influences on the tendency to weep, cognitive and constructivist models explore the ways that people create their own emotional realities based on their internal thinking and prior experience. Behavioral theorists, as well, emphasize the learned aspects of crying behavior, and the environmental responses that maintain it. From this theoretical perspective, the maintenance of crying in a person's behavioral repertoire will depend upon the contingent responses the person receives from those nearby. Indeed, whether a person decides to cry in the first place, much less how passionately, will depend to a great extent on how others respond to the show of emotion.

Other cognitive theories focus more specifically on adaptive information and how it is used internally to produce particular emotional responses like crying. With their research on when and how viewers respond tearfully to sad films, Efran and Spangler (1979) proposed a model that focused not so much on emotional arousal but rather on recovery from tension. Their two-factor theory looked at the ways crying functions as a tension reducer, mostly as a result of how the person thinks about what is going on. They argued it is the cognitive shift that produces the intense release, not the catharsis itself. This is consistent with changes in the ways that clinicians deal with emotional arousal. Whereas it was once considered desirable to stir up feelings and express them openly and fluently, contemporary approaches (Beck, 1976; Ellis, 1962; Greenberg & Safran, 1989; Kennedy-Moore & Watson, 1999; Lazarus & Lazarus, 1994) emphasize not so much expressing emotion as processing it constructively, and using them to orient the individual toward more satisfactory beliefs and behaviors. Labott and Martin (1987) built on this early work to propose a theory that also focuses on the resolution of tension and reduction in arousal.

Personality traits and temperament. There is some debate among social scientists and psychotherapists alike about whether fluent criers are more likely to be relatively healthy or unhealthy individuals. Of course, much depends on the context of the crying behavior, whether it is indicative of chronic, intractable depression or an intense appreciation for life. For every person who cries himself to sleep each night in despair, there is another who weeps for joy at the spectacle of a sunset.

There are indeed tremendous individual differences in what provokes crying (Kraemer & Hastrup, 1986; Williams & Morris, 1996). A number of personality variables, including levels of self-esteem, affect the inclination to cry (Vingerhoets et al., 1993; Chapter 7, this volume). Other personality variables, such as neuroticism and the tendency toward excitability and inhibition have also been found to be influential (De Fruyt, 1997; Vingerhoets et al., 1993).

Anderson (1996) examined a dozen instances of involuntary weeping in which women felt transformed by the experience. The women in the study reported the following increased awareness of self (including mind, body, and spirit): an altered sense of reality, an appreciation and honoring of the tragic dimensions of life, and a deeper glimpse into some facet of existence. In all cases of her investigation, this type of transformative crying was associated with self-actualized rather than psychopathological individuals.

INTERPERSONAL EXPLANATIONS

Symbolic interaction. Symbolic interaction theory emphasizes the social meaning of behavior. Early ethologists like Bowlby (1958) looked at social interactions with an eye toward the capacities and behavior patterns that have been "selected" by evolution, contributing to long-term survival. Crying is thus viewed as part of the attachment system whereby innate capacities for crying draw attention and evoke sympathetic responses (Bowlby, 1988).

Ethologists have also observed that among mammals, various forms of displaying surrender are employed during dominance rituals. This ensures that needless injuries do not result from tests that are designed to establish power hierarchies among social creatures. Chimpanzees and other primates, for example, will display their hindquarters to a more powerful foe as a show of deference. Hyenas, elk, chickens, and many other species have their own means of telling others to: "Back off! I've had enough!" Among humans, one of the most effective ways to stop excessive intimidation in its tracks is a show of tears. Thus, crying can serve the same symbolic function as other "white flags", but works far more dramatically than simply saying, "you win."

One such scenario, described by Kottler (1996), illustrates this interaction sequence vividly. A male physician had been verbally abusing a female hospital administrator. The more she apologized, the more he berated her. It was clear he was not accepting her signal of surrender.

"All of a sudden, a tear welled up in her eye, just a single tear, and ran down her cheek. He stopped cold. This guy, big time surgeon and all, used to having his way and blustering onward, just stopped dead. This tiny spot of wetness communicated to him very clearly what he otherwise had not seen. "He started backpedaling so fast, apologizing like crazy. That single tear had meaning for him a way that nothing else did" (Kottler, 1996, p. 68–69).

In addition to its role as a form of surrender, crying serves as an indirect means of asking for help, without the same assurance of reciprocal favors that would normally be expected during a direct request.

Anthropologists have noted that in various cultures, crying has a number of symbolic functions. Grief ceremonies such as the "tangi" among the Maoris of New Zealand or the tearful enactments by the Bosavi of New Guinea are used to communicate heartfelt feelings in the most authentic way possible. One Maori friend explained that it would be rude to tell someone you are sorry for their loss at a funeral. "If you truly feel their pain," he said, "then show it with your tears." Anything less would appear rude and insensitive. Thus, sharing tears communicates and strengthens the solidarity of the community.

Communication theory. Among its interactive causes and effects, crying can be treated as a language system, or rather an evocative *para*-language that is designed to communicate condensed information through nonverbal cues (Kottler, 1996). Complete with its own syntax, grammar, vocabulary, and varied dialects, crying may have evolved, or at least been adapted, to communicate intense emotional states to oneself or others. In fact, there are few other ways to say so much in such a brief interval, as testified by the number of academy awards given to actors who cry during dramatic scenes (Gullo, 1992). From Tom Hanks in *Philadelphia* and William Hurt in *Kiss of the Spider Woman,* to Robert DeNiro in *Raging Bull*, profuse tears are guaranteed to command attention in a way that words could never touch.

Whether employed to communicate messages such as: "I'm sad," "I'm distressed," "Help me!," "Leave me alone!", there is little doubt that crying is among the most powerful means to bring people closer or push them away. In addition, it functions as a sort of emotional marker for oneself, highlighting emotionally laden experiences so that they may later be retrieved.

When examined in an interpersonal context, crying appears to communicate a number of distinct messages that either invite people to offer support or to back off. As such, it is a distance regulator in relationships. For example, one woman reported a time when she was meeting a new colleague for the first time at lunch. The pair were sitting in a public restaurant making polite conversation when, all of a sudden, the woman spontaneously broke out in tears as her mind flashed to her mother, who was sick and in the hospital. Immediately, the woman's new colleague reached across the table and held her hand, asking what was wrong.

"In exactly five minutes," the woman reported, "we moved to a deep level relationship that normally would have taken months to develop." Each person in the relationship felt closer to the other because one shared her vulnerability.

Because crying works so well to influence and affect others' behavior, it also has tremendous potential for manipulation. Tears can lie just as words can, a phenomenon that works well to get needs met when other direct appeals are not effective. Tears can evoke sympathy, empathy, comfort, assistance, or irritation in the part of those who witness them (Frijda, 1986). But very often, they do strengthen the social bond between people.

Although crying as a communication tool may be part of its role and function in our lives, it is hardly the only one. Vingerhoets et al. (2000) have concluded that it is a coping mechanism of last resort: when words fail, then crying begins.

Social and cultural explanations. Emotions, including crying, are viewed by some scholars as "socially constructed" (Harre & Parrott, 1996). From this theoretical perspective, the social function of crying is emphasized, and this social function is expected to vary across cultures. In one analysis of the social construction of grieving, Stearns and Knapp (1996) traced several transformations of what they call our "grief culture." In pre-Victorian times, they argued, grief and its expression was firmly constrained. The religiously-influenced thinking of the times saw excessive crying as a sign of too great an interest in this world and too little faith in the next one. Victorian views, however, cast death as a defeat of medical science, and the disruption of loving family ties as a tragedy. In this climate, death-bed and funerary rituals were elaborated; intense and life-long grieving was encouraged. Hundreds of songs about death were published, often depicting emotion-laden visits to the cemetery—"All night I sat upon her grave, and sorely I did cry." The religious contradictions within this emphasis re-emerged, however, and practical objections to the excesses of the Victorian grief culture were raised. Following the First World War, emphasis shifted toward impatience with the indulgence of any backward-looking emotion, including the tears of grief, as a waste of time and energy. Stearns and Knapp thus argue that, rather than thinking of emotions as unitary and universal phenomena, we must realize how our emotions have been shaped by our culture, and that a large degree of plasticity exists in human emotional nature. Thus, crying has no inherent meaning or explanation outside its direct social context, and therefore studying it as a universal human phenomenon would not be useful.

Toward a Multi-Dimensional Theory of Crying

After reviewing the different theories of crying presented in this chapter we are inclined to agree with Cornelius (1988) that no single explanation can account for the complex phenomena of crying. Adult crying seems to serve a number of functions—as a state of emotional arousal and a discharge of energy, but also as a kind of distance regulator in relationships. Crying either brings people closer or pushes them away from one another. Like other forms of dramatic emotional expression, its meaning is lodged in the interactional context; it can be both constructive and counterproductive. Certainly, we must also remember that we are dealing with a variety of behaviors rather than a single human act. It may very well be that any one of the preceding theories is useful in explaining the causes and processes of a particular tearful event but is rather useless in examining another situation.

Returning to the story that began this chapter of the man who began crying as he walked to a dinner party, it is clear there is *something* going on within him that provoked the tears that began falling down his cheeks. Even if he is not consciously aware of the cognitive activity that may have precipitated that emotional upheaval, there may be some thoughts lying beneath the surface of his awareness that are making him feel upset. There is certainly a social and environmental context to his behavior as well, not only where he is going (to a party with certain people), but where he is located at that moment in time. The situational environment may be triggering reminiscences from the past of which he has not yet captured, or at least been able to put into words. The social meaning he attaches to crying influences his regard of his own tears—whether he sees them as silly, needless, embarrassing, cleansing, or informative. Certainly, different physiological processes may be involved, especially those that increased his fatigue level and lowered his threshold to weep.

Since we are clinicians who try to help people make sense of their tears, as well as teach other therapists to do so, we would like to suggest that any weeping episode should be understood within multiple contexts and dimensions. Any truly comprehensive theory of weeping must take into account the following factors:

(1) Biochemical, endocrine, and neurological mechanisms of the body that affect the threshold for weeping.
(2) Individual differences in the temperament, personality, and psychological health and maturity.
(3) Situational precipitants and environmental stimuli that tap into issues from the past or trigger sentimental/tragic/joyful reminiscences.
(4) Cognitive interpretations or personal narratives that lead one to interpret reality steeped in emotional activation.
(5) Social and interactional responses to the first weeping cues, reinforcing or discouraging continued crying. (How do others respond? Does it draw people closer or push them away?)
(6) Setting in time, place, circumstances, and others present. (What are the norms for the particular setting? Is the person alone or with others?)
(7) Functional benefits or secondary gains that accrue as a result of crying behavior. (What impact does it have on others? To what extent does it meet needs and gain desired outcomes?)
(8) Culturally-influenced communication messages, both symbolic and overt. (If the tears could speak, what would they say?)
(9) Gender scripts and contextual cues that predispose people to cry or not to cry.
(10) Unconscious influences, whether dynamic or systemic, that might be regulating the crying behavior.

It is obvious that no single theory or model can take all of these processes into consideration, nor can it explain the multiple levels of influence that come to bear at the same time. It is our recommendation, therefore, that research efforts be explicit about the types of explanation that are being offered. We can all acknowledge that understanding tears at the biochemical level does not negate the fact that tears have a symbolic meaning as well.

References

Anderson, R. (1996). Nine psycho-spiritual characteristics of spontaneous and involuntary weeping. *Journal of Transpersonal Psychology*, *28*, 167–173.

Beck, A.T. (1976). *Cognitive therapy and emotional disorders*. New York: International Universities Press.

Betancourt, H., & Lopez, S.R. (1993). The study of culture, ethnicity, and race in American psychology. *American Psychologist*, *48*, 629–637.

Bowlby, J. (1958). The nature of the child's tie to his mother. *International Journal of Psychoanalysis, 39,* 350–373.

Bowlby, J. (1988). *A secure base: Parent-child attachment and healthy human development*. New York: Basic Books.

Clarus, I. (1991). Die Tränen, ihre süssere und innere Wirklichkeit. [Tears: Their outward and inner reality]. *Analytische-Psychologie*, *22*, 295–314.

Collier, G. (1985). *Emotional expression*. Hillsdale, NJ: Lawrence Erlbaum.

Cornelius, R.R. (1988, April). *Toward an ecological theory of weeping*. Paper presented at the Eastern Psychological Association Conference, Buffalo, NY.

Cornelius, R.R. (1997). Toward a new understanding of weeping and catharsis? In: A.J.J.M. Vingerhoets, F.J. Van Bussel, & A.J.W. Boelhouwer (Eds.), *The (non)expression of emotions in health and disease* (pp. 303–321). Tilburg, The Netherlands: Tilburg University Press.

Crepeau, M.T. (1981). A comparison of the behavior patterns and meanings of weeping among adult men and women across three health conditions. *Dissertation Abstracts International, 42*, 137B–138B.

Damasio, A.R. (1994). *Descartes' error*. New York: Putnam.

Darwin, C. (1872/1955). *Expression of the emotions in man and animals*. New York: Philosophical Library.

De Fruyt, F. (1997). Gender and individual differences in adult crying. *Personality and Individual Differences, 22*, 937–940.

De la Chambre (1658). *Les caractères des passions*. Paris.

Denzin, K.K. (1984). *On understanding emotion*. San Francisco, CA: Jossey-Bass.

Descartes, R. (1647/1985). Les passions de l'âme [The passions of the soul]. In J. Cottingham, R. Stoothoff, & D. Murdock (Eds.), *The philosophical writings of Descartes (Vol. 1)*. Cambridge: Cambridge University Press.z

Efran, J.S., & Spangler, T.J. (1979). Why grown-ups cry: A two-factor theory and evidence from The Miracle Worker. *Motivation and Emotion, 3,* 63–72.

Ellis, A. (1962). *Reason and emotion in psychotherapy*. New York: Lyle Stuart.

Evans, I.H. (1981). *Brewer's dictionary of phrase and fable: Centenary edition, Revised*. New York: Harper & Row.

Freud, S. (1895/1953). *Studies in hysteria*. Standard Edition (Vol. 1). London: Hogarth.
Freud, S. (1910/1957). *Five lectures on psychoanalysis*. Standard Edition (Vol. 11). London: Hogarth.
Frey, W.H. (1985). *Crying: The mystery of tears*. Minneapolis: Winston Press.
Frey, W.H., Hoffman-Ahern, C., Johnson, R.A., Lykken, D.T., & Tuason, V.B. (1983). Crying behavior in the human adult. *Integrative Psychiatry, 1*, 94–98.
Frijda, N.H. (1986). *The emotions*. Cambridge: Cambridge University Press.
Gardiner, H.M., Metcalf, R.C., & Beebe-Center, J.G. (1937). *Feeling and emotion: A history of theories*. New York: American Book Company.
Greenberg, L.S., & Safran, J.D. (1989). Emotion in psychotherapy. *American Psychologist, 44*, 19–29.
Gregory of Nyssa (361/1859). De hominis opificio. In: J.P. Migne (Ed.), *Patrologiae cursus completus, series graeca*. Paris.
Gross, J.J., Fredrickson, B.L., & Levinson, R.W. (1994). The psychophysiology of crying. *Psychophysiology, 31*, 460–468.
Gullo, J. (1992). When grown men weep. *Premiere, 4*, 19.
Harre, R., & Parrott, W.G. (1996) (Eds.). *The emotions: Social, cultural, and biological dimensions*. London: Sage.
Izard, C.E. (1991). *The psychology of emotions*. New York: Plenum Press.
James, W. (1984). The physical basis of emotion. *Psychological Review, 1*, 516–529.
Kennedy-Moore, E., & Watson, J.C. (1999). *Expressing emotion. Myths, realities, and therapeutic strategies*. New York: The Guilford Press.
Kottler, J.A. (1996). *The language of tears*. San Francisco, CA: Jossey-Bass.
Kraemer, D.L., & Hastrup, J.L. (1986). Crying in natural settings: Global estimates, self-monitored frequencies, depression and sex differences in an undergraduate population. *Behaviour Research and Therapy, 24*, 371–373.
Labott, S.M., & Martin, R.B. (1987). The stress-moderating effects of weeping and humor. *Journal of Human Stress, 13*, 159–164.
Labott, S.M., & Martin, R.B. (1988). Weeping: Evidence for a cognitive theory. *Motivation and Emotion, 12*, 205–216.
Lazarus, R.S., & Lazarus, B.N. (1994). *Passion and reason: Making sense of our emotions*. New York: Oxford University Press.
Lutz, T. (1999). *Crying. The natural and cultural history of tears*. New York: Norton.
Mead, G.H. (1934). *Mind, self, and society: From the standpoint of a social behaviorist*. Chicago: Chicago University Press.
Mills, C.K., & Wooster, A.D. (1987). Crying in the counseling situation. *British Journal of Guidance and Counseling, 15*, 125–131.
Montagu, A. (1959). Natural selection and the origin and evolution of weeping in man. *Science, 130*, 1572–1573.

Okada, F. (1991). Is the tendency to weep one of the most useful indicators of depressed mood? *Journal of Clinical Psychiatry, 52,* 351–352.
Rapaport, D. (1960). The structure of psychoanalytic theory: A systematizing attempt. *Psychological Issues,* 2, (Monograph 6).
Rogers, C.R. (1959). A theory of therapy, personality, and interpersonal relationships, as developed in the client-centered framework. In: S. Koch (Ed.), *Psychology: A study of a science* (Vol. 3). New York: McGraw-Hill.
Sadoff, R.L. (1966). On the nature of crying and weeping. *Psychiatric Quarterly, 40,* 490–503.
Sloboda, J.A. (1991). Music structure and emotional response: Some empirical findings. *Psychology of Music, 19,* 110–120.
Stearns, P.N., & Knapp, M. (1996). Historical perspectives on grief. In: R. Harre & W.G. Parrott (Eds.), *The emotions: Social, cultural, and biological dimensions*. London, UK: Sage.
Triandis, H. (1989). The self and social behavior in different cultural contexts. *Psychology Review, 96,* 506–520.
Vingerhoets, A.J.J.M., Assies, J., & Poppelaars, K. (1992). Prolactin and weeping. *International Journal of Psychosomatics, 39,* 81–82.
Vingerhoets, A.J.J.M., Cornelius, R.R., & Van Heck, G.L. (submitted). Crying: To cope or not to cope?
Vingerhoets, A.J.J.M., Cornelius, R.R., Van Heck, G.L., & Becht, M.C. (2000). Adult crying: A model and a review of the literature. *Review of General Psychology, 4,* 354–377.
Vingerhoets, A.J.J.M., Van den Berg, M., Kortekaas, R., & Van Heck, G.L. (1993). Weeping: Associations with personality, coping, and subjective health states. *Personality and Individual Differences, 14,* 185–190.
Vives, J. (1555). *De anima et vita.* Basel.
Weiss, J. (1952). Crying at the happy ending. *Psychoanalytic Review, 39,* 338.
Wellenkamp, J.C. (1992). Variation in the social and cultural organization of emotions: The meaning of crying and the importance of compassion in Toroja, Indonesia. In: D.D. Frank & V. Gecas (Eds.), *Social perspectives on emotion.* Vol. 1 (pp. 189–216). Greenwich, CT: JAI Press.
Williams, D.G., & Morris, G.H. (1996). Crying, weeping, or tearfulness in British and Israeli adults. *British Journal of Psychology, 87,* 479–505.
Wood, E.C., & Wood, C.D. (1984). Tearfulness: A psychoanalytic interpretation. *Journal of the American Psychoanalytic Association, 32,* 117–136.

2 THE (NEURO)ANATOMY OF THE LACRIMAL SYSTEM AND THE BIOLOGICAL ASPECTS OF CRYING

Nico J. van Haeringen

All terrestrial animals, amphibians, reptiles, birds and mammals produce tears to keep the surface of their eyes in wet condition. The formation of the lacrimal apparatus was one of the changes that took place during the evolution of fishes into amphibians, which were the first vertebrates living outside the water. Fishes, whose eyes are bathed by the fluid medium in which they live, do not possess a lacrimal apparatus. In antique science it was believed that tears came from the heart (Egyptians, about 1500 B.C.), the brain (Hippocrates, 5th, 4th centuries B.C.), or glands at the puncta lacrimalis (Galen, 2nd century A.D.). However, after Stensen (1662/1992) described the tear ducts and the main lacrimal gland in 1662 it was accepted that tears originated there.

The Lacrimal System

In humans, the lacrimal system consists of the lacrimal glands, the preocular tear film in contact with the conjunctival and corneal surface, and the lacrimal drainage system, which brings the tears through the lacrimal puncta, canaliculi and lacrimal sac via the nasolacrimal duct to the mucosa of the nasal cavity.

TEAR FILM

The preocular tear film is composed of a trilaminar structure consisting of:

(1) A thin innermost mucus layer overlying the cornea and the conjunctiva, about 1 μm thick. The mucus layer has two main functions. It is essential in maintaining tear film stability, establishing a tenuous layer which converts an otherwise hydrophobic surface into a hydrophilic one, and enabling the overlaying aqueous tear to spread. This mucus

coating is, however, tenuous and tends towards breakdown unless resurfacing via blinking occurs periodically; (2) an intermediate aqueous layer is the thickest portion of the tear film (5–10 μm), which is the product of the main and accessory lacrimal glands, and; (3) a thin outermost lipid layer of about 0.2 μm, produced by the meibomian glands, that prevents the escape of lacrimal fluid over the eye lid margins and reduces the rate of evaporation of water from the open eye.

LACRIMAL GLANDS

The secretory system may be divided into *the basic secretors* and *the reflex secretors* (Jones, 1966). The basic secretors are made up of three sets of glands (see Figure 1):

(1) The mucin secretors consisting of (a) the conjunctival goblet cells, secreting mucus into the tears by apocrine secretion, where the greater part of the cell content is secreted, (b) the crypts of Henle found along the full length at one third of the upper and of the lower tarsus, and (c) the glands of Manz, found in a circumcorneal ring of limbal conjunctiva. These not only contribute most to the lubrication of the lids, but form the inner or "fixed" mucin layer of the precorneal film.
(2) The lacrimal secretors consisting of (a) the accessory lacrimal glands of Krause in the upper and lower conjunctival fornix, and (b) the accessory lacrimal glands of Wolfring embedded in the upper eye lid. These are exocrine glands, where no protoplasma is lost from the secreting cells, lying in the subconjunctival tissue and contributing to the intermediate layer of the precorneal film.
(3) The oil secretors are made up of (a) the meibomian glands in the upper and lower tarsus, (b) the glands of Zeis at the palpebral margin of each eyelid, and (c) the glands of Moll, found at the roots of the eye-lashes. The meibomian glands are the most important in the production of the outer layer of the tear film.

The reflex secretors are represented by the main lacrimal gland, located in the orbit at the temporal upper side of the eye and divided into two parts: an orbital and a palpebral portion. It is also an exocrine gland but differs from the glands of Krause and Wofring in that it has an efferent nerve supply. Usually 10–12 excretory ducts leave the gland and empty into the superior fornix just above the tarsal margin.

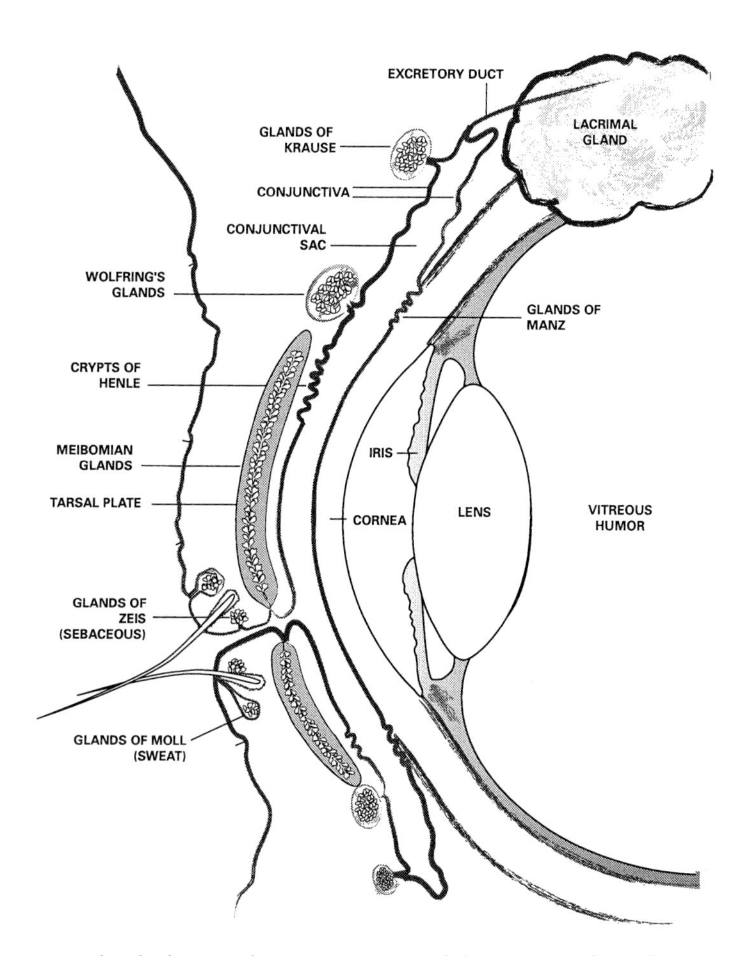

Figure 1. Glands that produce tears surround the outer surface of the eye (based on Botelho, 1964). The lacrimal gland helps to form the watery substance in tears, along with the glands of Krause and Wolfring. The crypts of Henle and the glands of Manz produce a mucoid layer. The tarsal and Meibomian glands and the glands of Moll and of Zeis help to make the oily layer.

THE BASIC SECRETION OF TEARS

Basic secretion is the fundamental, indispensable part of the secretory system. It alone can produce all three layers of the precorneal film. All terrestrial animals as well as the aquatic mammals possess this type of secretion for the lubrication of the surface of the eye.

Tears are secreted continuously during waking hours at a basal rate of about 1 μl per minute. Since tear fluid does not normally accumulate, its rate of secretion is presumably adjusted to compensate exactly for the rate of loss by evaporation and drainage of the fluid through the small orifices (lacrimal puncta) at the medial margin of each lid to the lacrimal sac and from here through the nasolacrimal duct to the mucosa of the nose.

A simple test to quantify lacrimal flow, proposed by Schirmer (1903), involves measuring the length of wetting by tears of filter paper abutting the lower conjunctival sac. Precut standardized filter paper strips (35 × 5 mm, bent 5 mm at end) are hooked over the lower eyelid border. The room where the test is performed should be moderately lighted with the subject facing away from any direct light. The strips are removed after a 5 minute absorption period and the extent of wetting, excluding the bent part of the strip, is recorded in mm. A measurement of 15 mm is regarded as standard for normal tear production. The Schirmer test represents a relatively gross measure and the results of the test were found so variable and inconsistent by some investigators (Feldman et al., 1978; Frankel & Ellis, 1978; Pinschmidt, 1970), showing large inter- and intra-subject variations, that they concluded the test itself to be an unreliable method of assessing tear production. A "basic" Schirmer tear test (Jones, 1966) even has been devised, in which topical anesthetic is applied to the eye, the conjunctiva gently dried with a cotton swab, and a Schirmer test performed, in an attempt to control sensory reflex secretion caused by the irritant of placing filter paper on the lower eye lid. Less than 10 mm of wetting of the Schirmer strip is considered abnormal.

Upon eye closure during sleep, a time period when irritative and light-induced stimulation is significantly reduced, the basic tear production decreases or ceases. This gives rise to a relative stagnant tear layer that has properties different from that of the open state of the eye (Sack et al., 1992). Also during sleep the cornea increases in thickness by edematous swelling and becomes thinner after awakening.

INNERVATION OF THE LACRIMAL GLAND AND REFLEX LACRIMATION

Innervation. Neuronal regulation of tear secretion may be through parasympathetic and sympathetic innervation. The main lacrimal gland receives an abundant efferent secremotor nerve supply and is parasympathically innervated via the pterygopalatine (sphenopalatine) ganglion (Duke-Elder & Wybar, 1961; Mitchell, 1953) and sympathetically via the superial cervical ganglion (Ruskell, 1975). Lately (see Figure 2), this has been confirmed in a study on cynomolgus monkeys, using modern retrograde tracing techniques, showing in addition a dual parasympathetic innervation of the lacrimal gland by the ciliary and

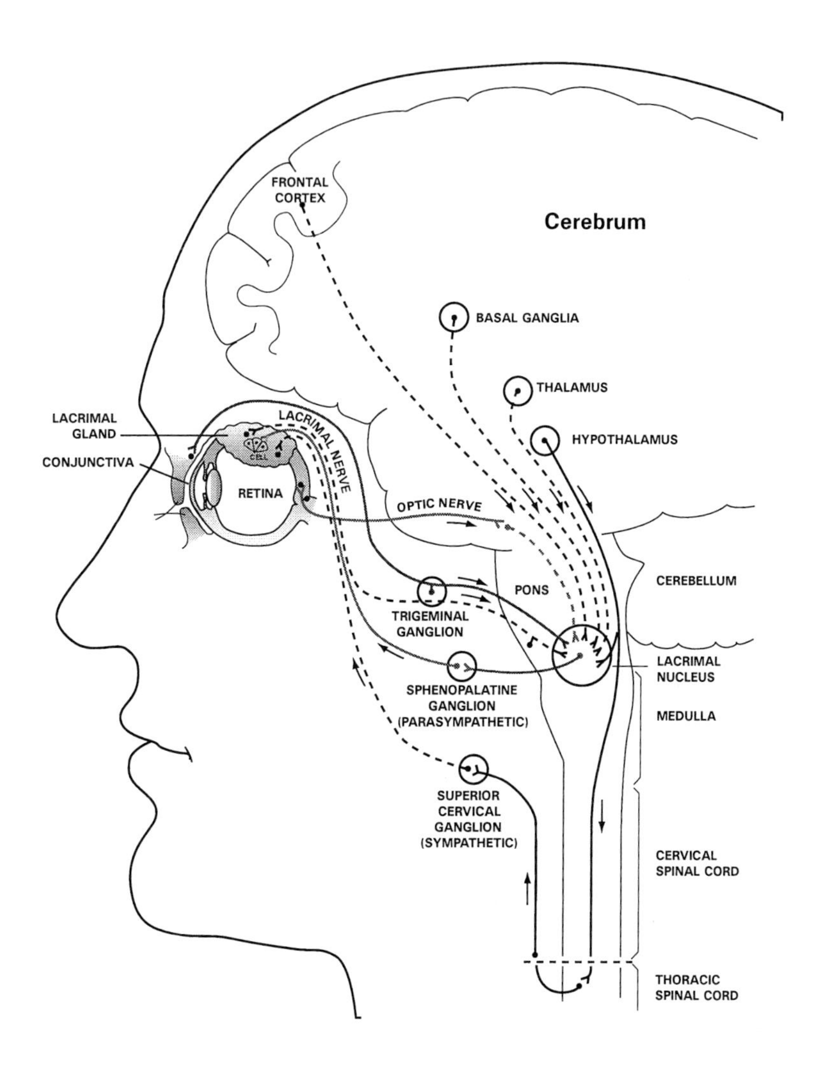

Figure 2. Schematic representation of the nerve connections between the lacrimal gland and retina of the eye and brain that contribute to the formation of tears (based on Botelho, 1964). Sensory innervation to the lacrimal nucleus arises from the conjunctiva, the cornea, the anterior uvea and the nasal mucosa through the fifth nerve. The lacrimal nucleus receives neuronal input from the frontal cortex, the basal ganglia, the thalamus, and the hypothalamus, as well as from the retina.

Parasympathetic secretory fibres from the lacrimal nucleus pass through the geniculate ganglion, synapse in the superior cervical ganglion and then follow the course of the carotid, the ophthalmic artery, and its lacrimal branch to provide sympathetic stimulation of the small arteries within the lacrimal gland.

pterygopalatine ganglion (Van der Werf et al., 1996). Within the lacrimal gland, parasympathetic fibres terminate on the surface of the secretory cells and about the myoepithelial cells of the ducts. The sympathetic supply terminates about the arteries and arterioles serving the gland. When parasympathetic fibers are stimulated, an evident increase in tear secretion can be observed. Stimulation of sympathetic fibers appears to have little effect on tear secretion but does act through the regulation of the blood supply of the main lacrimal gland. Besides the nerve fibers containing the classical neurotransmitters acetylcholine (parasympathetic) and norepinephrine (sympathetic), fibers are present that contain neuropeptides such as Vasoactive Intestinal Polypeptide (VIP), Met- and Leu-Enkephalin (M- and L-Enk), Neuropeptide Y (NPY) and Substance P (SP) (Dartt, 1994). The VIP and the M-and L-Enk nerves in the lacrimal gland are mostly of parasympathetic origin, where VIP and M-and L-Enk coexist presumably with acetylcholine. NPY in the periphery in most cases coexists in postganglionic sympathetic neurons with norepinephrine. SP is of primary sensory origin, differentiating from the trigeminal ganglion. The colocalisation in the close association of the peptidergic fibers with the secretory structures of the gland suggests that the neuropeptides are important neuromodulators of lacrimal secretion. This complex innervation of the lacrimal gland may reflect different populations of acinar cells that are activated separately thus producing a different secretory mix of fluids or proteins in the tears. Another view is that it represents a necessary redundancy, a safety factor, in the control of tear production.

The cornea is densely innervated by trigeminal sensory nerve fibers. Thus, even slight stimulation of the corneal epithelium is intensely painful. The stimulus required to elicit a sensation of pain at various parts of the body is in fact the lowest at the cornea. Recent ultrastructural investigation has confirmed that sympathetic, parasympathic, as well as peptidergic nerves innervate the human cornea (Müller et al., 1996).

Reflex lacrimation. Reflex tears are formed discontinuously as overflow tears in response to nerve impulses and secretogogues. Production of reflex tears occurs "de novo", there is no storage of tears that can be shed suddenly during the act of crying.

The main lacrimal gland has only one efferent nerve supply. But because of the many afferent sources from which this gland may be stimulated, reflex secretion is subdivided into peripheral sensory and central sensory (retinal and psychogenic) types.

The *peripheral sensory type* of reflex secretion occurs in all terrestrial vertebrates except snakes and most amphibians. It occurs whenever the sensory nerve endings of the conjunctiva, cornea, uvea, nasal mucosa and surrounding skin are abnormally stimulated by mechanical, thermal or chemical stimuli that, apart from a sensation of pain, activate the afferent trigeminal supply to the lacrimal nucleus and trigger the efferent parasympathetic nerves reaching the main lacrimal gland, which then produces the tears. Whenever the basic secretion becomes inadequate, conjunctival sensory stimulation may activate the reflex system.

Reflex trigeminal irritation, caused by a foreign body in the eye, is a common cause of tearing. On the other hand, tears produced in response to the burning sensation of onion vapor on the eye are well-known. These are caused by propanethial-S-oxide, the volatile lacrimatory factor of onions (*Allium cepa*) (Brodnitz & Pascale, 1971; Bloch et al., 1979). More plants contain pungent principles, such as the peppers belonging to the genus *Capsicum* (Andrews, 1980), containing capsaicin and the black pepper (*Piper nigrum*) (Szolcsanyi & Jansco, 1976) containing piperinoyl-piperine. Some animals, such as bombardier beetles (*Brachynus crepitans*) from Kenya, make use of tear gas as repellent or to incapacitate opponents, ejecting a spray of irritating benzoquinones (Dean et al., 1996). The carrier beetle (*Silpha obscura*) uses ammonia vapors (Habermehl, 1983) and soldiers of some species of lower termites such as *Prorhinotermes simplex* secrete toxic nitroalkene (Spanton & Prestwich, 1981) and those of *Schedorhinotermes lamianus* produce vinyl ketones in response to disturbance of their colony (Spanton & Prestwich, 1981). Methyl-crotyl-thiosulfinate is the substance responsible for the lacrimatory action of the secretion of the North America common striped skunk (*Mephitis*) (Maugh, 1974). Man uses synthetic tear gas in warfare, police action or self defense. The most widely used weapons do not produce a real tear gas but are of the solvent spray type, containing chloracetophenone (CN-gas) or chlorobenzylidene malonitrile (CS-gas). Pollution of the air by smoke of tobacco also gives rise to tearing as a consequence of the presence of irritant chemicals, such as ammonia, acrolein, collidine, and furfural. Reflex lacrimation as a symptom of ophthalmic disease occurs in allergic, bacterial or viral conjunctivitis, together with a burning sensation, edema and hyperemia of the conjunctiva. In photophobia, lacrimation is caused by trigeminal irritation, e.g., in cerebral hemorrhage. Obstruction of the canaliculi by inflammation may cause overflow of the tears over the eyelid margins.

The nasal reflex by mechanical stimulation of the mucosa of the nose and the gag reflex by depressing the tongue are examples of reflex secretion of peripheral sensory origin occurring without stimulation at the surface of the eye. Reflex tearing may also occur to a variable degree during the acts of yawning, sneezing, vomiting, and coughing.

A paradoxical gustatory lacrimal reflex, called "crocodile tears," occurs in patients recovering from a facial palsy. These patients lacrimate during meals or in response to stimuli that normally elicit salivation. It is assumed that it is due to antidromic stimulation of nerve endings in the acini of the main lacrimal gland or to misdirection of the regenerating nerve fibres so that some that should be destined for the salivary glands actually arrive at the main lacrimal gland. The labeling of this condition by Bogorad (1928) was based on the belief in former times, described by Plinius the Old (1st century A.D.) in his "Naturalis Historia" that crocodiles are weeping when they mangle their victims.

Retinal stimulation of the reflex system by light entering the eye furnishes the component of lacrimal fluid which together with basic secretion make up the normal flow of tears. Retinal adaptation to light keeps this amount fairly constant. This component is usually increased whenever intense light enters the eye or when suffering from photophobia. It ceases in complete darkness and when the eyes are closed in sleep.

Of all the vertebrates, including the primates, humans alone possess the *psychogenic type* of reflex secretion, designated as crying or weeping. This affective lacrimation is controlled in the frontal cortex and in the anterior portion of the limbic lobe of the brain. There is no evidence of any animal other than humans shedding tears due to emotion rather than stress or irritation, despite many anecdotal reports about pets and other animals. Asian elephants (*Elephas maximus*) may show tears, wetting the surrounding lids, because a groove in the skin, continuous with the medial canthus of the lids, drains the tears onto the face (Murphy et al., 1992). Lacrimal puncta for normal drainage of tears are not visible. The aquatic mammals such as seals, dolphins and whales secrete a watery mucus to protect their eyes from sea water. The overflow of these tears due to lack of a drainage system may have been misconstrued as emotional tears.

Patients with a proven decrease or absence of conjunctival sensory nerve impulses in the Schirmer test will give a history of having copious tears during emotional stress. The Schirmer test was applied for the first time as an objective test for psychogenic reflex tearing by Delp and Sackeim (1987). In their study on the impact of psychological manipulations of mood on tearing, lacrimal flow was assessed before and after mood manipulations intended to produce states of happiness and sadness. Lacrimal flow, at least among women, appeared to be responsive to manipulations of mood and may be an index of aspects of affective experience that are incompletely or poorly assessed by self report techniques.

BIOLOGICAL ASPECTS OF LACRIMAL SECRETION

The first chemical analysis of tears was recorded by Fourcroy and Vauquelin (1791), demonstrating the presence of salt, mucus and water. In 1928 Ridley (1928) demonstrated large concentrations of lysozyme in tears, the bacteriolytic enzyme also found in other secretions and tissues by Fleming & Allison (1922). Since then the number of publications on the composition of tears has gradually increased.

The composition of tears. The complex composition of tears is a function of the diurnal cycle and also may be influenced by the method of tear collection (Van Haeringen, 1981). Reflex stimulation of basal tear flow manipulation with glass capillaries may result in lower values by a dilution effect, especially if the components under investigation are derived from the cornea or the conjunctiva. Application of filter paper strips or cellulose sponges in order to absorb the tears may augment the concentration of various compounds as a consequence of admixture of tissue fluid that is set free by the microtrauma of manipulation, and of cell content from disrupted epithelial cells.

The components of the tears present as lacrimal fluid on the surface of the eye may be arbitrarily classified as follows: proteins, enzymes, lipids, metabolites and electrolytes (Van Haeringen, 1981). Specific intrinsic lacrimal proteins, present in relative large concentration (1–2 g/l) in human tears, are lysozyme, lipocalin and lactoferrin (Bodelier & Van Haeringen, 1993). Secretory immunoglobulinA (IgA) is the predominant immunoglobulin in tears, as well as in other mucosal surfaces. Relative high concentrations of potassium, up to 42 mmol/l (Van Haeringen, 1981) and of manganese up to 34 μg/l (Frey et al., 1981) are found in tears as compared to blood, where the normal levels are on the order of 4.5 mmol/l and 1 μg/l, respectively.

As mentioned, all terrestrial animals produce tears, but there is an evolutionary divergence in the composition of tears and pronounced species differences have been described in this respect. The tears of humans and higher primates remain unique in their high content of antibacterial lysozyme, lactoferrin and of lipocalin (Bodelier & Van Haeringen, 1993). Lysozyme is absent in the tears of rodents (Van Agtmaal et al., 1985) and ruminants (Padgett & Hirsch, 1967). Rat tears are exceptional in that they contain a very high concentration of the enzyme peroxidase, which also may have antibacterial action (Van

Haeringen et al., 1979). Causes for these differences remain as yet unknown, but an explanation might be that they are attributable to adaptation to the changing environment during the evolution of the various animals. Emotional or psychogenic tears are in fact reflex tears, where the stimulus is emotional rather than irritant-induced.

Frey et al. (1981) investigated the intriguing question whether other not the composition of emotional tears and irritant tears is similar. They found that the protein concentration of emotional tears was 24% greater than that of irritant tears. No differences were found with respect to manganese. It still remains to be established whether there are any differences in the composition of tears of sadness and of joy.

The function of tears. The function of tear fluid is to create a moist environment for the optical quality of the cornea, the lubrication of the surface of the eye during the blinking movement of the eyelids, and above all to defend the eye against microbial infections either by mechanical removal during blinking or by antibacterial and antiviral action. Crying has no direct biological function in the protection of the eye and may serve no physiological purpose whatever. All animal species can survive in their natural environment without the capacity of crying. Darwin (1872/1965) gave the subject of weeping much thought in his masterpiece *The expression of the emotions in man and animals*, but he nowhere ventured a suggestion as to how it has come about in the evolution that man is the only animal that weeps. Montagu (1960) proposed the hypothesis that in man weeping established itself as an adaptive trait in that it served to counteract the effects of more or less prolonged tearless crying upon the nasal mucosa of the infant. Early in the development of man, those individuals who were able to produce an abundant flow of tears would be naturally selected in the struggle for existence, since the tears acted to prevent mucosal dehydration, whereas those who were not so able would be more likely to succumb more frequently at all ages and leave the perpetuation of the species to those who could weep. Frey (see Frey, 1980; Frey et al., 1986) hypothesized that crying served as a means of eliminating some hormones, such as prolactin, ACTH, and leucine-enkephalin, which are known to be released from the pituitary gland in response to stress and enter the blood system.

SEX AND THE LACRIMAL SYSTEM

In humans, but also in rats, mice, guinea pigs and rabbits, significant sexual dimorphism has been demonstrated in the lacrimal glands

(Cornell-Bell et al., 1985). In all these species, the area of the acini in the lacrimal glands of males is larger than that of females. Experiments with rats, in which, in addition to differences in weight, sex differences in morphological appearance and enzyme activity of the exorbital lacrimal glands have been reported, strongly suggest that the influence of androgenic hormones and prolactin may be the underlying cause for these differences (Azzarolo et al., 1995). Sex has not been found a factor in the volume or the rate of flow of tears (Hanson et al., 1975; Henderson & Prough, 1950), except perhaps among young women in the age range of 15–30 years, who demonstrate a greater flow of tears than men, as measured by the Schirmer test (Henderson & Prough, 1950). This finding, however, could not be replicated by Hanson et al. (1975). In a study on ocular variables throughout the menstrual cycle in eleven normal women, no statistically valid relationship could be demonstrated between the different phases of the menstrual cycle and tear production as measured by the Schirmer test (Yolton et al., 1994). Also no significant differences were found in tear production as measured by the Schirmer test between women taking oral contraceptives and women not on the medication (Hanson et al., 1975; Yolton et al., 1994). Wide variations in the test were found in the persons, who were all asymptomatic young women. Contact lens intolerance has been reported in patients on oral contraceptives (Verbeck, 1973), but if it is a true side effect of the birth control pill, its etiology should be sought in factors other than tear production.

Tapaszto (1973) reported that male tears contained 24 times more manganese than female tears, but this could not be confirmed by Frey et al. (1981). Frey et al. (1986) demonstrated the presence of prolactin in the main lacrimal gland and in tears and suggested that this substance may function to stimulate tear production. This might help explain, in part, why male and female children have similar crying behavior (Bell & Ainsworth, 1972; Maccoby & Feldman, 1972), but women cry more often than men once they reach adulthood (see Chapters 4 and 6, this volume). Serum prolactin levels in male and female infants and children are not significantly different; it is only after the age of about 16 that female prolactin levels exceed those of males (Guyda & Friesen, 1973; Jacobs et al., 1972). Prolactin is dramatically increased during pregnancy and in hyperprolactinaemia. However, contrary to expectations, patients with functional hyperprolactinaemia were not more prone to weeping than matched healthy controls (Vingerhoets et al., 1992).

AGING AND THE LACRIMAL SYSTEM

Newborn babies secrete tear fluid already in the first day of their life (Apt & Cullen, 1964), although they do not demonstrate weeping overtly. Premature infants, however, may fail to secrete tears at birth, depending on the degree of prematurity (Patrick, 1974). In most cases, crying with tears starts at about six weeks of age (Spiegler & Mayer, 1993), when the efferent nerve supply to the main lacrimal gland is completely established. Crying thus seems to be both phylogenetically and ontogenetically a late development in the human species. The reflex secretion of tears, as measured by the Schirmer test, decreases significantly with increasing age as was already observed by Schirmer (1903) and by many other investigators thereafter (see Van Haeringen, 1977). The age groups investigated mostly cover the 20–80 years of age range and the reduction in Schirmer values from the youngest to the oldest age group is about 70%. In contrast with these findings, recently others (e.g., Xu & Tsubota, 1995; Nava et al., 1977), investigating also large numbers of persons, failed to demonstrate the decline of the Schirmer value with age. This discrepancy is probably attributable to the acknowledged variability in the performance of the Schirmer test—that is, in the latter studies, the persons were asked to blink normally after placing of the Schirmer strip. It is conceivable that blinking causes an extra irritation, which in elderly people may increase reflex tearing compared with a closed eye condition, such as is employed mostly. The lacrimal drainage capacity also decreases with increasing age.

Normally, the decreased reflex tear secretion capacity in the older eye may be compensated by a reduced lacrimal drainage. The condition of "dry eye" with subjective symptoms including burning, itching, sticking, dryness, mucus discharge, and foreign body sensation, occurring with increasing frequency in older patients may be caused by an imbalance due to abnormal low tear production or a high evaporation rate. The lids may become less taut with age and this interferes with normal blinking function, causing a higher evaporation because the tear film is not properly restored over the ocular surface. In some cases, the dry eye is associated with defective mucus secretion or the presence of particulate matter in the tear film, which usually consists of mucus that is not remaining in solution because of moderately depleted tear flow.

Beyond 60 years of age, women have been shown to have a smaller flow of tears than males (Henderson & Prough, 1950); however, this finding could not be confirmed by other investigators (Hanson et al., 1975). Women are more susceptible than men to Sjøgren's syndrome, an autoimmune disease which is accompanied by markedly impaired lacrimal secretory function resulting in keratoconjunctivitis sicca. Treatment of patients with Sjøgren's syndrome with androgen hormones has been shown to ameliorate the symptoms associated with the keratoconjunctivitis sicca (Prijot et al., 1972).

Pharmacology and Lacrimation

A variety of medications are known to inhibit tear production. The oldest description of this phenomenon is given by the ancient Greek poet Homer (8th century B.C.) in his poem *the Odyssey* (Homer, 8th century B.C.). In the fourth canto of *the Odyssey* it is described that Telemachos landed in Sparta at Menelaos' court in his search for his father Ulysses. As soon as Menelaos discovers that Telemachos is the son of his vanished friend, tears begin rolling down his cheeks. But in order to hide this undesired emotional expression, Helena, Menelaos' wife, resolutely takes action:

> *Then a new thought rose in the mind of Helena, the daughter of Zeus: She put into the wine, which they drank, a drug assuaging grief and allaying wrath,causing forgetfulness of all troubles. Whoever drank this, mixed with the wine, not one tear would he let fall down his cheeks all day long not if his mother or father should die, not if in front of him his brother or beloved son was slain by the bronze and he saw it happen before his eyes. That was one of the potent and beneficial drugs, which the daughter of Zeus possessed, given to her by Polydamna, the wife of Thon, an Egyptian.*

The herb mentioned here probably is henbane (*Hyoscyamus niger*), growing in Egypt and like thorn apple (*Datura stramonium*) and belladonna (*Atropa belladonna*), containing the alkaloid atropine, which as a parasympathicolytic drug is known to cause dry eyes. Alternatively, neuroscientists (see Panksepp, 1998) have suggested that the drug described here was opium or cannabis. Opiates, just like brain opioids, interact with *mu* receptors in the brain, resulting in a strong inhibition of separation distress (at least in animals), which in humans also may strongly reduce the propensity to cry.

Other drugs causing dry eyes by interference with the autonomic nervous system, comprise beta-blockers, ganglion blockers, hypnotics and sedatives such as phenobarbitone and benzodiazepines, neuroleptics such as phenothiazines, tranquillizers such as diazepam and antidepressives such as dibenzazepines and monoamine oxidase inhibitors (Van Haeringen & Bleeker, 1982). Even aspirin exerts an inhibitory effect on tear production (Carreras & Matas, 1973). Therefore, in epidemiological studies it is conceivable that participants considered to be normal or healthy show a diminished tear production, which is inadvertently caused as an unwanted side effect of these drugs. Moreover, if the consumption of these drugs is related to age, then it is possible that in some studies the changes observed in several variables of lacrimal gland function are drug induced rather than age related. Local anesthetics only reduce peripheral sensory reflex secretion through blocking of the afferent arc of the reflex. In the performance of the basic Schirmer test, reflex secretion is deliberately eliminated by topical application of a local anesthetic.

The opposite effect of stimulation of the tear flow, mostly as a side effect of drug therapy, is also possible, although occurring less frequently. Lacrimation is known as a side effect of pilocarpin, a parasympathicomimetic, used in eye drops in the therapy of elevated intraocular pressure (glaucoma). Pilocarpin even has been used in the therapy of dry eyes. However, given by oral route in order to reach effective drug levels in the blood, it exerts unwanted intestinal side effects.

Conclusion

In the present chapter, the anatomy of the lacrimal system and its connections with the central nervous system were described. In addition, there was discussion of the chemical composition of tears and how factors like age, disease and drug use may affect the functions of the lacrimal system. Concerning the differences between emotional tears and irritant tears, it can be concluded that, except for the total protein content, there are no specific proteins that can be associated with emotional tearing. Differences in the concentrations of hormones like prolactin may be related to emotional status. However, this may be more likely a consequence of spilling over from elevated levels in the blood rather than a specific means of elimination from the blood.

References

Andrews, J. (1980). *Peppers: The domesticated capsicums.* University of Texas Press.

Apt, L., & Cullen, B.F. (1964). Newborns do secrete tears. *Journal of the American Medical Association, 189,* 951.

Azzarolo, A.M., Bjerrum, K., Maves, C.A., Becker, L,. Wood, R.L., Mircheff, A.K., & Warren, D.W. (1995). Hypophysectomy-induced regression of the female rat lacrimal glands: Partial restoration and maintenance by dihydrotestosterone and prolactin. *Investigative Ophthalmology and Visual Science, 36*, 216–226.

Bell, S.M., & Ainsworth, M.D.S. (1972). Infant crying and maternal responsiveness. *Child Development, 43,* 1171–1190.

Bloch, E., Penn, R.E., & Revelle, L.K. (1979). Structure and origin of the nion lachrymatory factor. A micowave study. *Journal of the American Chemical Society, 101*, 2200–2201

Bodelier, V.M.W., & Van Haeringen, N.J. (1993). Species differences in tears: Comparative investigation in the chimpanzee (Pan triglodytes). *Primates, 34,* 77–84.

Bogorad F. (1928). The syndrome of the crocodile tears. *Vrachebnoje Djelo (Rus.), 11*, 1328. Quoted by J. Murube in: Dacriologia Basica. Universidad de la Laguna, Las Palmas.

Brodnitz, M.H., & Pascale, J.V. (1971). Thiopropanal S-oxide. A lacrimatory factor in onions. *Journal of Agriculture and Food Chemistry, 19,* 269–272.

Carreras y Matas M. (1973). Mogadon, aspirina y secretion lagrimal. *Review Espagnol de Oto-Neuro-Ofhalmologie, 31,* 245–246.

Cornell-Bell, A.H., Sullivan, D.A., & Allansmith, M.R. (1985). Gender-related differences in the morphology of the lacrimal gland. *Investigative Ophthalmology and Visual Science, 26,* 1170–1175.

Dartt, D.A. (1994). Regulation of tear secretion. *Advances in Experimental Medicine and Biology, 350*, 1–9.

Darwin, C. (1872/1965). *The expression of the emotions in man and animals.* London: Murray.

Dean, J., Aneshansley, D.J., Edgerton, H.E., & Eisner, T. (1996). Defensive spray of the bombardier beetle. A biological pulse jet. *Science, 248,* 1219–1221.

Delp, M.J., & Sackeim, H.A. (1987). Effects of mood on lacrimal flow: Sex differences and asymmetry. *Psychophysiology, 24,* 550–556.

Duke-Elder, S., & Wybar, K.C. (1961). The anatomy of the visual system. In: S. Duke-Elder (Ed.), *System of Ophthalmology Vol. 2* (pp. 559–581). London: Kempton.

Feldman, F., Bain, J., & Matuk, A.R. (1978). Daily assessment of ocular and hormonal variables throughout the menstrual cycle. *Archives of Ophthalmology, 96,* 1835–1838.

Fleming, A., & Allison, V.D. (1922). Observations on a bacteriolytic substance ("lysozyme") found in secretions and tissues. *British Journal of Experimental Pathology, 3,* 252–260.

Fourcroy & Vauquelin (1791). Examen chimique des larmes et de l'humeur des narincs. *Annales de Chimie, 10,* 113–130.
Frankel, S.H., & Ellis, P.P. (1978). Effect of oral contraceptives on tear production. *Annals of Ophthalmology, 10,* 1585–1588.
Frey, W.H. (1980). Not-so-idle tears. *Psychology Today, 13,* 91–92.
Frey, W.H., Desota Johnson, D., Hoffman, C., & McCall, J.T. (1981). Effect of stimulus on the chemical composition of human tears. *American Journal of Ophthalmology, 92,* 559–567.
Frey, W.H., Nilson, J.D., Frich, M.L., & Elde, R.P. (1986). Prolactin immunoreactivity in human tears and lacrimal gland: Possible implications for tear production. In: F.J. Holly (Ed.), *The preocular tear film in health, disease, and contact lens wear* (pp. 798–807). Lubbock, TX: Dry Eye Institute.
Guyda, H.J., & Friesen, H.G. (1973). Serum prolactin levels in humans from birth to adult life. *Pediatric Research, 7,* 534–540.
Habermehl, G.G. (1983). *Venomous animals and their toxins.* Berlin: Springer Verlag.
Hanson, J., Fikentscher, R., & Roseburg, B. (1975). Schirmer test of lacrimation. Its Clinical importance. *Archives of Otolaryngology, 101,* 293–295.
Henderson, J.W., & Prough, W.A. (1950). Influence of age and sex on flow of tears. *Archives of Ophthalmology, 43,* 224–231.
Homer (8th century B.C.) *Odyssey*, cant 4, vers 219–230.
Jacobs, L.S., Mariz, I.K., & Daughaday, W.H. (1972). A mixed heterologous radioimmunoassay for human prolactin. *Journal of Clinical Endocrinology, 34,* 484–490.
Jones, L.T. (1966). The lacrimal secretory system and its treatment. *American Journal of Ophthalmology, 62,* 47–60.
Maccoby, E.E., & Feldman, S.S. (1972). Mother-attachment and stranger-reactions in the third year of life. *Monographs of the Society for Research in Child Development, 37.*
Maugh, T.H. (1974). Skunks: On the scent of a myth. *Science, 185,* 1146.
Mitchell, G.A.G. (1953). Anatomy of the autonomic nervous system. Edinburgh: Livingstone.
Montagu, A. (1960). Natural selection in the origin and evolution of weeping in man. *JAMA, Journal of the American Medical Association, 174,* 392–397.
Müller, L.J., Pels, L., & Vrensen, G.F.J.M. (1996). Ultrastructural organization of human corneal nerves. *Investigative Ophthalmology and Visual Science, 37,* 476–488.
Murphy, C.J., Kern, T.J., & Howland, H.C. (1992). Refractive state, corneal curvature accommodative range and ocular anatomy of the Asian elephant (Elephas maximus). *Vision Research, 32,* 2013–2021.
Nava, A., Barton, K., Monroy, D.C., & Pflugfelder, S.C. (1997). The effects of age, gender, and fluid dynamics on the concentration of tear film epidermal growth factor. *Cornea, 16,* 430–438.

Padgett, G.A., & Hirsch, J.G. (1967). Lysozyme: Its absence in tears and leukocytes of cattle. *Australian Journal of Experimental Biology and Medical Science, 45,* 569–570.

Panksepp. J. (1998). *Affective neuroscience.* New York: Oxford University Press.

Patrick, R.K. (1974). Lacrimal secretion in full-term and premature babies. *Transactions of the Ophthalmolological Society UK., 94,* 283–285.

Pinschmidt, N.W. (1970). Evaluation of the Schirmer tear test. *South Medicine Journal, 63,* 1256

Prijot, E., Barzin, L., & Destexhe, B. (1972). Essai de traitement hormonal de la keratoconjunctivite seche. *Bulletin de Societé Belge d' Ophthalmologíe 162,* 795–800.

Ridley, F. (1928). An antibacterial body present in great concentration in tears and its relation to infection of the human eye. *Proceedings of the Royal Society in Medicine, 21,* 1495.

Ruskell, G.L. (1975). Nerve terminals and epithelial cell variety in the human lacrimal gland. *Cell & Tissue Research, 158,* 121–126.

Sack, R.A., Tan, O.T., & Tan, A. (1992). Diurnal tear cycle: Evidence for a nocturnal inflammatory constitutive tear fluid. *Investigative Ophthalmology and Visual Science, 33,* 626–640.

Schirmer, O. (1903). Studien zur Physiologie und Pathologie der Tränenabsonderung und Tränenabfuhr. *Graefes Archif für Ophthalmologie, 56,* 197–291.

Spanton, S.G., & Prestwich, G.D. (1981). Chemical self-defense by termite workers: prevention of autointoxication in two rhinotermitids. *Science, 214,* 1363–1365.

Spiegler, C., & Mayer, U.M. (1993). Tear production in premature infants, newborn infants and infants. *Klinische Monatsblätter für Augenheilkunde, 202,* 24–26.

Stensen, N. (Steno N) (1992). De glandulis oculorum. In: Observationes anatomicae, Leyden 1662. Quoted by J. Murube. History of the dry eye. In: M.A. Lemp & R. Marquardt (Eds.), *The dry eye.* Berlin/Heidelberg, Springer.

Szolcsanyi, J., & Jancso, G. (1976). Sensory effects of capsaicin congeners. Part II. Importance of chemical structure and pungency in desensitizing activity of capsaicin compounds. *Arzneimittel Forschung (Drug Research), 26,* 33–37.

Tapaszto, I. (1973). Pathophysiology of human tears. *International Ophthalmology Clinics, 13,* 119–147.

Van Agtmaal, E.J., Thörig, L., & Van Haeringen, N.J. (1985). Comparative protein patterns in the tears of several species. *Protides in Biological Fluids, 32,* 395–398.

Van der Werf, F., Baljet, B., Prins, M., & Otto, J. (1996). Innervation of the lacrimal gland in the cynomolgus monkey: A retrograde tracing study. *Journal of Anatomy, 188,* 591–601.

Van Haeringen, N.J. (1981). Clinical biochemistry of tears. *Survey of Ophthalmology, 26,* 84–96.

Van Haeringen, N.J. (1997). Aging and the lacrimal system. *British Journal of Ophthalmology, 81,* 824–826.
Van Haeringen, N.J., & Bleeker, G.M. (1982). Lider und Tränenapparat. In: O. Hockwin & H-R. Koch (Eds.), *Unerwünschte Arzneimittelwirkungen am Auge.* Stuttgart: Gustav Fischer Verlag.
Van Haeringen, N.J., Ensink, F.T.E., & Glasius, E. (1979). The peroxidase-thiocyanate-hydrogen peroxide system in tear fluid and saliva of different species. *Experimental Eye Research, 28,* 243–247.
Verbeck, B. (1973). Augenbefunde und Stoffwechselverhalten bei Einnahme von Ovulationshemmern. *Klinische Monatsblätter für Augenheilkunde, 162,* 612–621.
Vingerhoets, A.J.J.M., Assies, J., & Poppelaars, K. (1992). Prolactin and weeping. *International Journal of Psychosomatics, 39,* 81–82.
Xu, K., & Tsubota, K. (1995). Correlation of tear clearance rate and fluorophotometric assessment of tear turnover. *British Journal of Ophthalmology, 79,* 1042–1045.
Yolton, D.P., Yolton, R.L., Lopez, R., Bogner, B., Stevens, R., & Rao, D. (1994). The effects of gender and birth control pill use on spontaneous blink rates. *Journal of the American Optometric Association, 65,* 763–770.

3 DEVELOPMENTAL ASPECTS OF CRYING: INFANCY, CHILDHOOD, AND BEYOND

Debra M. Zeifman

Why Crying is a Developmental Question

Crying is one of the earliest and most powerful, and occasionally the only, means available to the infant for communicating its needs to a caregiver. This "acoustical umbilical cord," as it has been called (Ostwald, 1972), serves to connect the infant to a caregiver, its source of nurture and protection, and has been posited as an adaptive strategy in the species' struggle for survival over the course of human evolution (Bowlby, 1969). In the first years of life, crying is largely replaced by other means of communication as children learn to express their negative emotions in ways other than crying. Changes in the frequency, eliciting circumstances, and ability to control crying, therefore, provide a vehicle for making inferences about many aspects of early development. For example, when an infant cries because a parent leaves the room, we infer that an emotional bond between that parent and child, or an attachment, has formed. When a two year old injures her finger by catching it in the refrigerator door and starts crying only upon entering the living room where her parents are seated, we understand something about her new-found intellectual ability to anticipate and manipulate the reactions of her audience. And when we see a six year old fighting back tears upon being bullied in the schoolyard, we know he is manifesting a recently acquired ability to inhibit crying and regulate his emotions. Thus, changes in crying across the life span provide a window through which biological, cognitive, and socioemotional development may be viewed.

What makes crying a fascinating developmental puzzle is that it possesses both elements that change drastically and some that remain unchanged across the life span. The most fundamental similarity across ages is that adults, like infants, cry as a result of distressing circumstances such as physical and emotional pain. Full-blown adult crying resembles its infant analog in structural features as well (i.e., crying face and tears, heaving vocalizations, rhythmicity of wails, and so on), suggesting at least some continuity across development. However, while many of the components that constitute crying remain the same, there is an apparent shift in the salience of the various components. Whereas crying is primarily an acoustic signal in infancy, it may be primarily a visual one in adulthood; most adult crying appears to be simply tearing. Because of this, infant crying researchers have emphasized the production of distress vocalizations and adult crying researchers have emphasized the production of tears. A change in emphasis among the components of crying, however, is hardly the sole or primary difference between the infant and adult case. In adulthood, crying is a rare event and the circumstances that might provoke it, such as the wedding or the death of a loved one, are frequently beyond the comprehension of an infant or small child. So, like most developing behaviors, there are both continuities and discontinuities that lend themselves to analysis and demand elucidation.

Perhaps owing to real differences or discontinuities in crying across the life span, or because subfields within psychology tend to be divided by the age of the population being studied rather than the phenomenon, the infant crying and adult weeping literatures have, for the most part, existed as separate entities. Each has independently posed and grappled with the problems of crying's definition, function and meaning, at times arriving at similar conclusions and at others arriving at widely disparate ones. Surprisingly, the developmental literature has little to say about crying after the second year (when its frequency drops off considerably), and the adult literature has little to say about crying before the college years (undergraduates being the most common subjects of social psychological investigation). Neither discipline has systematically considered the continuities and discontinuities in crying—a behavior that heralds the arrival of a healthy human neonate into the extrauterine world and, remarkably, is manifested throughout childhood, adulthood and old age—across the entire life span. In this chapter I will describe crying and its development, with an emphasis on infant crying and early development. I will identify similarities and differences between infant and adult crying wherever possible and argue that much can be learned about adult crying from the study of infant crying or, more broadly, from adopting a developmental approach.

Why do Babies Cry?

EVOLUTIONARY HISTORY AND ULTIMATE CAUSES

The prevalent view among developmental psychologists is that crying enhances infant survival by eliciting care and protection from adult caregivers when the infant is helpless to meet his or her own needs (e.g.,

Bowlby 1969). The human neonate is relatively immature compared to most other species, incapable not only of speaking but also of independent locomotion, feeding or temperature regulation (Oppenheim, 1980). The highly aversive sound of infant crying serves to alert the caregiver that the infant is in need, and motivates him or her to attend to the infant, thereby enhancing the infant's chances of survival. Support for this view of infant crying comes from the fact that across diverse cultures by far the most common response to infant crying is picking up the baby, putting it to the breast, and nursing (Bell & Ainsworth, 1972; Bernal, 1972). In fact, the sound of infant crying itself causes a rise in breast temperature in lactating females (Vuorenkowski et al., 1969) and a milk-let down reflex (Mead & Newton, 1967), presumably mediated by the release of the hormone prolactin (Thoman & Levine, 1970). There is, however, some evidence against the view that infant crying is a strictly adaptive behavior and one that consistently promotes infant survival. For example, crying is the most common complaint brought to pediatricians by parents (Forsyth et al., 1985) and a frequently cited reason for giving up breast feeding (Bernal, 1972). It is also the proximate stimulus reported in many cases of infant abuse and infanticide (Frodi, 1981, 1985; Weston, 1968).

Attempts to resolve the apparent paradox of crying's obvious adaptiveness and seeming maladaptiveness have focused on the fact that modern caregiving practices are dramatically different from those that existed for the majority of human evolutionary history. These differences, in turn, may be causing unprecedented levels of infant crying that are not easily tolerated by caregivers in any human culture, including ours (Barr, 1990a). In hunter-gatherer societies, the environment purported to be characteristic of most of evolutionary history (Lee & Devore, 1968), infants and mothers are rarely physically separated by more than arms length or for extended periods of time (Konner, 1972; 1976). Among one African foraging group, the !Kung, infants are carried by their mothers in a sling (or *kaross*) so that their fussing and whimpering are responded to by offering the breast without delay, and full-blown crying is usually averted (Devore & Konner, 1974; Konner, 1972). In Western societies, by contrast, infants are placed in cribs, playpens, and in separate rooms, and spend most of their day out of the arms, sight, and even earshot of their caregivers (Olmsted, 1979). Even when present, caregivers frequently intentionally ignore crying (Bell & Ainsworth, 1972), a practice historically encouraged by some pediatricians (e.g., Spock, 1968). Fussing or low-intensity crying that is either unheard or ignored is likely to escalate into full-blown, vigorous crying and to become more difficult to console. The resulting durations and intensities of infant crying, up to two and three quarter hours per day for the normal two-month-old infant according to some reports (Brazelton, 1962), may exceed some caregivers' tolerance for the aversive sound and precipitate aggressive rather than sympathetic responses to crying (Murray, 1979).

PROXIMATE OR IMMEDIATE CAUSES OF CRYING

Evolutionary arguments may address the ultimate cause or survival value of infant crying, but what are crying's proximate or more immediate causes? Psychologists and parents alike identify four major cry types—birth, hunger, pain, and attention cries—that correspond to events or conditions known to precipitate crying. Infants reliably cry three to three and one half hours after their last feeding (Stark & Nathanson, 1973), and following painful stimuli such as a vaccination. Cold exposure also leads to crying, which in turn generates heat by increasing motor activity (Lester, 1985). However, by far the most frequent and important triggering stimulus for crying is being alone (Newman, 1985), and most infants will stop crying immediately when picked up or held by their caregiver (Bell & Ainsworth, 1972). For this reason, crying has typically been thought of as an "attachment behavior," of the infant's need to be held and comforted that alerts the caregiver (Bowlby, 1958). Of course, human infants often cry without any known cause, as any parent will gladly tell you.

The crying of nonhuman young in response to pain, hunger, or loneliness usually lasts only as long as the cause for crying persists. In contrast, human infants frequently continue to cry long after the offending stimulus has been removed (Wolff, 1987). Colic, a syndrome characterized by excessive inconsolable crying of unknown etiology, is prevalent at least in western cultures, affecting somewhere between sixteen to forty percent of newborn infants (Hide & Guyer, 1982; Stahlberg, 1984). Once initiated, crying is a self-equilibrating behavioral state; that is, it maintains itself in relative autonomy to the triggering cause (Fentress, 1976). And because of the self-perpetuating nature of infant crying, it can be said that crying itself is a contributing factor to crying. This is analogous to the adult case where a distinction can be drawn between the ability to inhibit crying before it begins and the ability to stop crying that has already begun (see Chapter 7, this volume), the latter being recognized as a far more difficult or perhaps even impossible task. The apparent independence of crying from its precipitating causes in the newborn is perhaps the reason that early infant crying is often thought of as reflexive in nature (e.g., Brazelton, 1962).

The independence of infant crying from external causes has led some writers to infer internal causes for crying which, although not amenable to observation, are widely assumed to contribute to its occurrence. Crying has been purported to release excess energy or tension in the newborn as a means of restoring physiological homeostasis (Brazelton, 1962; 1985). Some have even claimed that crying improves pulmonary capacity in the newborn (Brazelton, 1962). Changes in cry patterns in early infancy (discussed at greater length below) are thought by others to reflect maturational changes in brain structures and shifts in central nervous system (CNS) activity, resulting in periods of unexplained fussiness or crying (Emde et al., 1976). These "internal" explanations appeal to adult notions about the cathartic effects of crying (see Chapter 11, this volume) and provide explanations for this otherwise unexplained phenomenon (not to mention their appeal to exasperated parents). While internal explanations are difficult to confirm or disconfirm directly, there is greater support for the notion of crying as a result of internal causes in the first two to three months of life than afterwards. Nonetheless, infants who are reared in cultures where they are rarely allowed to cry do not appear to suffer ill health or psychological deficits (e.g., Konner, 1977), challenging internal, physiological arguments for the functional significance of infant crying.

ENVIRONMENTAL INFLUENCES ON INFANT CRYING

In addition to considering immediate causes or triggers for crying, one must consider whether other factors in the infant's environment contribute in an indirect way to preventing or facilitating crying. There has long been a debate among pediatricians and psychologists about whether promptly responding to infant crying results in more or less infant crying. Following their landmark study, Bell and Ainsworth (1972) reported that, contrary to the then reigning learning theory, popular opinion, and pediatric advice of the time, promptness of maternal response is associated with a *decline* in the frequency and duration of crying over the course of the first year. In comparing crying over the four quarters of the first year of life, they found that by the end of the first year, individual differences in crying reflected the history of maternal response to crying rather than the infant's constitution or degree of irritability. Mothers who ignored their babies' cries had babies who cried more than those who responded promptly. While some of this study's findings have been criticized on methodological grounds (Gewirtz & Boyd, 1977; Lamb et al., 1985), most modern psychologists believe that fears of "spoiling" a newborn infant are unfounded and that, to the contrary, responding promptly will produce a more independent, efficacious and socially competent youngster.

Other aspects of the broader caregiving environment may facilitate or prevent crying even before it occurs. Increased holding and more frequent feedings are both associated with reduced crying. In a randomized controlled trial, parents assigned to a condition in which they held their infants in a baby carrier for, on average, less than two additional hours per day, had infants who cried substantially less than those who were not assigned to the supplemental holding condition (Barr, 1990a). Another study compared infants of La Leche League members with breast-feeding nonmembers because La Leche League members advocate and practice frequent nursing. Shorter intervals between feedings were associated with less fussing and crying at two months, suggesting that widely spaced feedings may be difficult for infants to digest and cause gastrointestinal distress, a common feature of colic (Barr, 1990a). The investigator reporting these research findings argues that typical caregiving practices in western contexts such as widely spaced feedings and little physical contact may recruit normal physiological functions that potentiate crying (Barr, 1990a). There is certainly clear evidence that practices that increase the contact and closeness of infants and their caregivers reliably reduce infant crying.

The Development of Crying

THE EARLY CRYING CURVE

Despite considerable individual variability, there are predictable changes in the amount and pattern of crying over the course of the first postnatal months that are relatively impervious to caregiving influences. The amount of infant crying gradually increases until it peaks at six weeks, then decreases until four months, at which time it remains fairly stable until the end of the first year (Bell & Ainsworth, 1972; Brazelton, 1962; Hunziker & Barr, 1986; Rebelsky & Black, 1972). Just as adult crying is concentrated in the evening hours (Frey, 1985; Chapter 5, this volume) there is a diurnal pattern to infant crying that may be distinguishable as early as ten days of life (Barr, 1990b). The six-week peak and subsequent decline and the diurnal pattern of early infant crying are both robust across different methods of recording, including diaries, interviews, questionnaires, and electronic recording. Unlike the amount of crying later in the first year and onward, the basic appearance of the early crying curve is apparently unaffected by changes in caretaking style (Barr, 1990b). Because of the robustness of the early postnatal crying pattern, and because this pattern appears to be independent of caretaking style or culture, it has been thought to reflect physiological and maturational transitions during the first months of life.

It has been suggested by some that developmental changes in the pattern of infant crying reflect major changes in behavioral and CNS organization (Emde et al., 1976; Emde & Gaensbauer, 1981). The reduction in crying following its peak at approximately two months coincides with a more general decrease in the lability of infant emotions and may reflect a change from endogenous to exogenous control of crying. That is, reflexive crying that lacks association with external events or caregiver efforts may be replaced by responsive crying, just as reflexive smiling is gradually replaced by responsive or social smiling. The shift away from reflexive crying occurs at a time when other early reflexes drop out and are replaced by voluntary behaviors, perhaps as a result of the development of forebrain inhibitory mechanisms (Peiper, 1963).

CRYING IN LATER INFANCY

Another major behavioral shift occurs at about seven months and corresponds to major cognitive and affective gains that occur at this time. Whereas infants initially cry to promote proximity to their caregivers, towards the end of the first year they are more likely to cry when the caregiver is nearby (Bell & Ainsworth, 1972). This finding is consistent with the popular distinction between early infant crying and the more intentional crying of later infancy and childhood (Gekowski et al., 1983). In addition, the slightly older infant is more apt to direct crying, as well as other bids for attention such as smiling or cooing, toward a particular person, typically the primary caregiver (Bowlby, 1969; Sroufe & Waters, 1976). In fact, Bowlby viewed the infant's selective bids for attention as the hallmark of an emerging attachment relationship. And not only is crying directed toward a target audience, but it is also gradually coordinated with looking, reaching, and other nonverbal and non-vocal orienting behaviors over the second half of the first year (Gustafson & Green, 1991). Quite unlike the red-faced, crying newborn who thrashes about aimlessly in his bassinet, the nine month old might stare at the door where he anticipates his caregiver's arrival, and then decrease the volume of crying and reach in her direction to be picked up when she enters the room. Thus, toward the end of the first year, crying is more narrowly directed, often coupled with visible signs of the infant's expectations about its results, and increasingly coordinated with other communicative acts such as looking or reaching.

Changes in crying that reflect maturation and biobehavioral reorganization also result in changes in the functional significance of crying. The two-month-old crying infant seeks attention from or proximity to the caregiver. But at seven to nine months, wariness or fear of strangers or separation most often prompts crying (Lester, 1985; Ricciuti, 1974). The ability to fear strangers, unfamiliar situations, and abandonment depends on types of cognitive and affective growth that have only recently occurred, such as the ability to compare current and past events and familiar and unfamiliar persons (Kagan, 1982). But this fear of the unknown or unfamiliar is both timely and adaptive, in that it coincides with the emergence of independent locomotion. As the infant, now crawling but on the verge of walking, begins to actively explore his or her environment, the potential for encountering new persons or unfamiliar surroundings is greatly increased. Crying as a result of fear of strangers or strangeness provides a tether of sorts, allowing independent exploration while at the same time limiting it by alerting the caregiver of potential impending danger. To the extent that the older infant can often control his or her own proximity to a caregiver, it is not surprising that proximity seeking recedes somewhat in importance as a motivation for crying. Thus, during early infancy, crying changes not only in terms of its pattern and amount, but the situations that provoke crying and its functional significance change as well.

CRYING IN TODDLERHOOD

In contrast to the extensive literature on infant crying, there is little research on crying in toddlerhood or childhood. After the peak of stranger anxiety at around twenty-four months and concurrent with the rapid development of spoken language, crying appears to drop off rather precipitously. In the second year, crying often makes its appearance in the context of "temper tantrums" or other situations in which children experience frustration, for example, because they do not wish to comply with parental wishes (Gesell & Ilg, 1943). The addition of frustration to the list of causes for crying coincides with the appearance of planful, deliberate action, and the opportunity it presents for goals to be thwarted, towards the end of infancy. Other situations that frequently elicit crying in the few laboratory investigations of early childhood crying include caregiving episodes such as diaper change, physical injuries like a fall, and separation from parents (Kopp, 1992). Furthermore, when older children in a laboratory study cry, for example in defiance of a parental request, crying is usually accompanied by verbal negotiations with the offending parent while tears are streaming down the child's face (Kopp, 1992). Thus, crying does not entirely disappear in early childhood. Rather, it is encompassed within and elaborated by other recently developed communication skills, including, most notably, language. And in addition to pain and separation, frustration becomes a common cause of crying in toddlerhood.

Individual differences in the frequency of crying episodes among preschoolers of similar chronological age provide further insight into the relationship between crying and other means of communication, especially language. In several studies, progress in language development has been inversely related to crying; that is, children with more advanced language skills than their agemates are less likely to resort to crying than children with less developed skills. For example, when incidences of crying during a laboratory study of eighteen month olds were examined, it became evident that children who cried were far more likely than those who did not to be limited to one-word sentences and to have not yet achieved two-word sentences (Kopp, 1992). These data, as well as those from similar studies, strongly suggest that crying is utilized when alternative means of communication don't yet exist or fail to achieve desired goals (Reynolds, 1928).

EMOTION REGULATION IN CHILDHOOD

There is no body of work that focuses exclusively on crying in childhood, perhaps because it is difficult to study unpredictable, rare events and obviously unethical to induce crying in children. However, the vast research concerning emotion regulation in childhood is highly relevant to a discussion of childhood crying. Emotion regulation refers to the ability to express emotions in socially appropriate ways, including and especially negative emotions such as anger, disappointment or distress. Studies suggest that children who are better able to regulate their emotions are more socially competent and better liked than those who are less adept at regulating their emotions (Denham, 1986; Denham et al., 1990; Denham et al., 1997; Eisenberg et al., 1993). While positive affect facilitates social exchanges and the formation of friendships (Sroufe et al., 1984), negative affect can disrupt social interactions (Rubin & Clark, 1983). Further, better regulated children exhibit more prosocial and helping behavior than less regulated children when placed in a situation where a peer is in need of assistance (Fabes et al., 1994), presumably because they are less preoccupied with their own distress (Eisenberg & Fabes, 1992; Eisenberg et al., 1992). The conclusion that emerges from this research is that children who are prone to personal distress are not well regarded by their peers, possibly in contrast to the case of adults (see Chapter 9, this volume).

CRYING IN ADULTHOOD: SIMILARITIES, DEVELOPMENT AND DIFFERENCES

Crying remains a compelling signal of physical or emotional injury in adulthood, as in infancy, and a means of soliciting aid and assistance from others when in need. However, there are several notable changes in crying over the course of development. These changes include a reduction in the frequency of crying episodes, greater cognitive complexity of eliciting circumstances, increased selectivity of target or audience, and inhibition of crying in public places. Changes in crying over the course of infancy and childhood due to maturation foreshadow differences between infant and adult crying and provide insight into their basis. For example, adults are capable of meeting most of their own physical needs independently. Crying is, therefore, relatively rare in adulthood and occurs more often as a result of emotional, rather than physical, deprivation or injury. Interestingly, feelings of aloneness (or loneliness), separation, loss, and helplessness remain the most common reasons given or conjectured for crying at all ages (Vingerhoets et al., in press), despite the dramatically different circumstances leading to these feelings at different ages.

The availability of alternative means of communication in adulthood further reduces the frequency of crying episodes, and possibly contributes to the apparent shift in the salience of the various features of crying as an organized behavior. Crying is metabolically costly to infants (Thureen et al., 1998), and in the environment in which humans evolved may have involved substantial predation risk to both adults and children. None-theless, to be effective in garnering aid from any distance, a pre-mobile infant separated from its caregiver would have had no choice but to utilize an acoustic signal to alert its caregiver. Indeed, there is evidence that deaf parents do not respond appropriately to their crying infants despite the visual cues associated with their crying (Lenneberg et al., 1965). Independent locomotion, however, introduces an alternative strategy for alerting a potential caregiver of one's need: orienting one's body and, more importantly, one's face toward the target of one's distress call and presenting a primarily visual signal to that target. Selection of this less costly alternative may be behind the trend, evident in the first year of life, in which crying becomes directed toward the primary caregiver and is gradually coordinated with attention, looking and reaching. Given the obvious disadvantages of a primarily acoustic signal, a greater dependence on the visual or facial expressive aspect of crying and diminished reliance on vocalization might be expected as the ability to physically traverse the distance between oneself and one's caregiver emerges.

This explanation presumes that adult crying, however muffled, is communicative. But if this is so, why then do adults often seek privacy during a crying episode? One explanation may be the growing awareness with maturity that there are risks to a crier besides the consumption of limited energy and being located by a predator when vulnerable. There is abundant evidence, discussed earlier, that crying provokes abuse, at least in infancy. While motivation to arrest the aversive sound of crying underlies its effectiveness for eliciting care, the same motivation may precipitate abuse (Murray, 1979; 1985). Low birth-weight and premature infants are at increased risk for abuse (Parke & Collmer, 1975) because they are irritable and difficult to care for and, some argue, because their cries are more high-pitched and aversive than those of normal weight and maturity (Frodi, 1985). It is also likely that children who easily come to tears are preferentially targeted for taunting and abuse by their peers. It is not surprising, then, that children learn to direct crying toward individuals who are highly motivated to provide care for them and, in childhood, inhibit crying in the presence of those who may be disinclined to tolerate it. The trend evident across development toward increased selectivity and in choice of target audience and inhibition of public crying extends through adulthood, when crying takes place most often in the privacy of one's own home (Becht & Vingerhoets, 1997; Chapter 5, this volume).

The fact that crying is frequently effective in eliciting care may also account for discrepancies between naturalistic and laboratory studies of the phenomenology of crying and reactions of observers to it. Although adults report feeling better after crying in natural settings, they actually feel worse after crying in laboratory studies where crying is induced. The caregiving that is often forthcoming in the aftermath of a naturally occurring crying bout may, among other things, explain why criers report feeling better after crying at home and not in the controlled environment of a laboratory (Cornelius, 1997). Conversely, those exposed to another person's crying may feel worse in a naturalistic than a laboratory situation because of the demand placed on them to act to alleviate another person's distress. Although adults tend to sympathize with research confederates or videotaped models who cry (Hill & Martin, 1997), the sympathy evoked by one-time exposure to the tears of an actor does not require any action or concession on the part of the observer. In contrast, the tears of a child, spouse, friend, or coworker with whom one has an ongoing relationship necessitate some caregiving intervention and, for that reason, might lead to resentment if perceived as excessive, overused, or inappropriate to the level of intimacy of the relationship. And while child-caregiver relationships are expected to be unbalanced in terms of the provision of care, relationships among peers or adults would be expected to involve reciprocity and mutual emotional support. This fact may explain the low popularity ratings of children who are easily distressed and the irritation of adults who perceive crying as a form of emotional blackmail in relationships (Frijda, 1986) or who view the crying of coworkers as distressing and embarrassing (Plas & Hoover-Dempsey, 1988).

Conclusions and Future Directions for Research

Crying is a compelling social behavior at any age and one that signals a need for assistance when an individual is feeling alone and helpless to meet his or her own needs. Changes in crying over the course of infancy and childhood can be linked to broader biobehavioral shifts in development that often provide insight into the growing capabilities of the child. Adopting a developmental approach to crying is useful not only for understanding the developing child, but also for understanding the case of adult crying. For example, the hypothesis that crying is a form of communication is bolstered by the finding that crying declines precipitously with language development. More studies exploring continuities and discontinuities in crying across the *entire* life span would be helpful toward advancing our understanding of crying at every stage.

In this chapter, I have argued that maturation leads to predictable changes in crying and its function. For example, increased coordination of crying with visual attention and the advent of independent locomotion allow infants to direct crying toward particular persons rather than randomly emit distress signals. One would expect that a signal utilized to alert a caregiver from a distance would differ substantially from one used to communicate distress from nearby. Present studies in our laboratory are exploring whether there are systematic changes in the acoustic and facial variables that comprise crying that correspond meaningfully to crucial transitions in development, such as the transition from crawling to walking. Similarly, the relative importance of various cues associated with crying for communicating distress to observers might be explored empirically. It would be worthwhile to determine if the relative importance of the various crying cues depend on the age of the individual who is crying.

There is a delightful irony in comparing infant and adult crying in this sense: the very behavior that infants use, rather promiscuously, to draw others near is one that adults typically hide from others and seek refuge before engaging in. While culture undoubtedly plays a prominent role in this dramatic change, development, too, has an important and perhaps neglected role in this process. In addition to learning display rules, other aspects of infants' and children's changing capabilities influence the form and function of crying at various stages of development. Drastic differences between infant and adult crying that at first glance suggest little similarity between the two, rapidly fade when the development of crying is considered as a continuous, gradual process. Finally, this is an exciting time to be engaged in psychological research precisely because the barriers between the subfields within psychology are rapidly disappearing and the field is becoming increasingly multidisciplinary. I hope I have convinced at least some adult crying researchers and clinicians to consider developmental aspects of crying in their own work. A developmental approach will lead to a deeper and richer understanding of adult crying.

References

Barr, R.G. (1990a). The early crying paradox: A modest proposal. *Human Nature, 1*, 355–389.

Barr, R.G. (1990b). The normal crying curve: What do we really know? *Developmental Medicine and Child Neurology, 32*, 356–362.

Becht, M., & Vingerhoets, A.J.J.M. (1997, March). *Why we cry and how it affects mood.* Annual Meeting of the American Psychosomatic Society, Sata Fe, NM (Abstracted in *Psychosomatic Medicine, 59*, 92).

Bell, S.M., & Ainsworth, M.D. (1972). Infant crying and maternal responsiveness. *Child Development, 43*, 1171–1190.

Bernal, J. (1972). Crying during the first ten days of life. *Developmental Medicine and Child Neurology, 14*, 362–372.

Bowlby, J. (1958). The nature of the child's tie to his mother. *International Journal of Pediatrics, 39*, 350–373.
Bowlby, J. (1969). *Attachment and loss*, Vol. 1: Attachment. London: Hogarth Press.
Brazelton, T.B. (1962). Crying in infancy. *Pediatrics, 29*, 579–588.
Brazelton, T.B. (1985). Application of crying research to clinical perspectives. In: B. Lester & C.F. Boukydis (Eds.), *Infant crying* (307–323). New York: Plenum Press.
Cornelius, R.R. (1997). Toward a new understanding of weeping and catharsis? In: A.J.J.M.Vingerhoets, F.J. Van Bussel, & A.J.W. Boelhouwer (Eds.), *The (non)expression of emotions in health and disease* (pp. 303–321). Tilburg, The Netherlands: Tilburg University Press.
Denham, S.A. (1986). Social cognition, social behavior, and emotion in preschoolers: Contextual validation. *Child Development, 57*, 194–201.
Denham, S.A., Mitchell-Copeland, J., Strandberg, K., Auerbach, S., & Blair, K. (1997). Parental contributions to preschoolers emotional competence: Direct and indirect effects. *Motivation and Emotion, 21*, 65–86.
Denham, S.A., McKinley, M., Couchoud, E.A., & Holt, R. (1990). Emotional and behavioral predictors of peer status in young preschoolers. *Child Development, 61*, 1145–1152.
Devore, I., & Konner, M. (1974). Infancy in a hunter-gatherer life: An ethological perspective. In: N. White (Ed.), *Ethology and psychiatry* (pp. 113–141). Toronto: University of Toronto Press.
Eisenberg, N., & Fabes, R.A. (1992). Emotion, self-regulation, and social competence. In: M. Clark (Ed.), *Review of Personality and Social Psychology, 14*, 119–150. Newbury Park, CA: Sage.
Eisenberg, N., Fabes, R.A., Bernzweig, J., Karbon, M., Poulin, R., & Hanish, L. (1993). The relation of emotionality and regulation to preschoolers' social skills and sociometric status. *Child Development, 64*, 1418–1438.
Eisenberg, N., Fabes, R.A., Carlo, G., & Karbon, M. (1992). Emotional responsivity to others: Behavioral correlates and socialization antecedents. In: N. Eisenberg & R.A. Fabes (Eds.). *Emotion and its regulation in early development* (pp. 57–73). San Francisco, CA: Jossey-Bass Publishers.
Emde, R.N., & Gaensbauer, T.J. (1981). Some emerging models of emotion in early infancy. In: K. Immelman, G. Barlow, L. Petrinovich, & M. Main (Eds.), *Behavioral development* (pp. 568–588). Cambridge: Cambridge University Press.
Emde, R.N., Gaensbauer, T.J., & Harmon, R.J. (1976). Emotional expression in infancy: A biobehavioral study. *Psychological Issues, 10*, Monograph 37.
Fabes, R.A., Eisenberg, N., Karbon, M., Troyer, D., & Switzer, G. (1994). The relation of children's emotion regulation to their vicarious emotional responses and comforting behaviors. *Child Development, 65*, 1678–1693.
Fentress, J.C. (1976). Dynamic boundaries of patterned behavior: Interaction and self-organization. In: P.P.G. Bateson & R.A. Hinde (Eds.), *Growing points in ethology* (pp. 135–169). Cambridge: Cambridge University Press.

Forsyth, B.W.C., McCarthy, P.L., & Leventhal, J.M. (1985). Problems of early infancy, formula changes, and mothers' beliefs about their infants. *Journal of Pediatrics,* 106, 1012–1017.

Frey, W.H. (1985). *Crying: The mystery of tears.* Minneapolis, MN: Winston Press.

Frijda, N.H. (1986). *The emotions.* New York: Cambridge University Press.

Frodi, A. (1981). Contributions of infant characteristics to child abuse. *American Journal of Mental Deficiency, 85,* 341–349.

Frodi, A. (1985). When empathy fails: Aversive infant crying and child abuse. In: B. Lester & C.F. Boukydis (Eds.), *Infant crying* (pp. 263–278). New York: Plenum Press.

Gekoski, M.J., Rovee-Collier, C.K., & Carulli-Rabinowitz, V. (1983). A longitudinal analysis of inhibition of infant distress: The origins of social expectations? *Infant Behavior and Development, 6,* 339–351.

Gesell, A., & Ilg, F.L. (1943). *Infant and child in the culture of today: The guidance of development in home and nursery school.* New York: Harper Collins.

Gewirtz, J.L., & Boyd, E.F. (1977). Does maternal responding imply reduced infant crying? A critique of the 1972 Bell and Ainsworth report. *Child Development, 48,* 1200–1207.

Gustafson, G.E., & Green, J.A. (1991). Developmental coordination of cry sounds with visual regard and gestures. *Infant Behavior and Development, 14,* 51–57.

Hide, D.W., & Guyer, B.M. (1982). Prevalence of infantile colic. *Archives of Disease in Childhood, 57, 559–560.*

Hill, P., & Martin, R.B. (1997). Empathic weeping, social communication, and cognitive dissonance. *Journal of Social and Clinical Psychology, 16,* 299–322.

Hunziker, U.A., & Barr, R.G. (1986). Increased carrying reduces infant crying: A randomized controlled trial. *Pediatrics, 77,* 641–648.

Kagan, J. (1982). Canalization of early psychological development. *Pediatrics, 70,* 474–483.

Konner, M. (1972). Aspects of a developmental ethology of a foraging people. In: N. Blurton-Jones (Ed.), *Ethological studies of child behaviour* (pp. 285–304). Cambridge: Cambridge University Press.

Konner, M. (1976). Maternal care, infant behavior and development among the Kung. In: R.B. Lee & I. DeVore (Eds.), *Kalahari hunter-gatherers: Studies of the Kung San and their neighbors* (pp. 218–245) Cambridge: Harvard University Press.

Konner, M. (1977). Infancy among the Kalahari Desert San. In: P.H. Leiderman, S.R. Tulkin, & A. Rosenfeld (Eds.), *Culture and infancy: Variations in the human experience* (pp. 287–328). New York: Academic Press.

Kopp, C.B. (1992). Emotional distress and control in young children. In: N. Eisenberg & R.A. Fabes (Eds.). *Emotion and its regulation in early development* (pp. 41–56). San Francisco, CA: Jossey-Bass Publishers.

Lamb, M.E., Thompson, R.A., Gardner, W., & Charnov, E.L. (1985). *The infant-mother attachment: The origins and developmental significance of individual differences in strange situation behavior*. Hillsdale, NJ: Erlbaum.
Lee, R.B., & DeVore, I. (Eds.). (1968). *Man the hunter*. Chicago: Aldine.
Lenneberg, E., Rebelsky, F., & Nichols, I. (1965). The vocalization of infants born to deaf and hearing parents. *Human Development, 8*, 23–37.
Lester, B.M. (1985). There's more to crying than meets the ear. In: B. Lester & C.F. Boukydis (Eds.), *Infant crying* (pp. 1–28). New York: Plenum Press.
Mead, M., & Newton, N. (1967). Cultural patterning of perinatal behavior. In: S.A. Richardson & A.F. Guttmacher (Eds.), *Childbearing: Its social and psychological aspects* (pp. 142–244). Baltimore: Williams & Wilkins.
Murray, A.D. (1979). Infant crying as an elicitor of parental behavior: An examination of two models. *Psychological Bulletin, 86*, 191–215.
Murray, A.D. (1985). Aversiveness is in the mind of the beholder: Perception of infant crying by adults. In: B. Lester & C.F. Boukydis (Eds.), *Infant crying* (pp. 307–323). New York: Plenum Press.
Newman, J.D. (1985). The infant cry of primates. In: B. Lester & C.F. Boukydis (Eds.), *Infant crying* (pp. 307–323). New York: Plenum Press.
Olmsted, R.W. (1979). Infant care: Cache or carry. *Behavioral Pediatrics, 95*, 478–483.
Oppenheim, R.W. (1980). Metamorphosis and adaptation on the behavior of developing organisms. *Developmental Psychobiology, 13*, 353–356.
Ostwald, P. (1972). The sounds of infancy. *Developmental Medicine and Child Neurology, 14*, 350–361.
Parke, R., & Collmer, C. (1975). Child abuse: An interdisciplinary review. In: E.M. Heatherington (Ed.), *Review of child development research* (Vol. 5) (pp. 509–590). Chicago: University of Chicago Press.
Peiper, A. (1963). *Cerebral function in infancy and childhood*. New York: Consultants Bureau.
Plas, J.M., & Hoover-Dempsey, K.V. (1988). *Working up a storm: Anger, anxiety, joy, and tears on the job*. New York: W.W. Norton.
Rebelsky, F., & Black, R. (1972). Crying in infancy. *The Journal of Genetic Psychology, 121*, 49–57.
Reynolds, M.M. (1928). *Negativism of preschool children*. New York: AMS Press.
Ricciuti, H.N. (1974). Fear and the development of social attachments in the first year of life. In: M. Lewis & L.A. Rosenblum (Eds.), *The origins of fear* (pp. 73–106). New York: Wiley.
Rubin, K.H., & Clark, M.L. (1983). Preschool teachers' ratings of behavior problems: Observational, sociometric, and social-cognitive correlates. *Journal of Abnormal Child Psychology, 11*. 273–286.
Spock, B. (1968). *Baby and child care, revised edition*. New York: Pocket books.
Sroufe, L.A., Schorck, E., Motti, F., Lawroski, N., & La Freniere, P. (1984). The role of affect on social competence. In: C.E. Izard, J. Kagan & R.B. Zajonc

(Eds.), *Emotions, cognition and behavior* (pp. 289–319). Cambridge, England: Cambridge University Press.
Sroufe, L.A., & Waters, E. (1976). The ontogenesis of smiling and laughter: A perspective on the organization of development in infancy. *Psychological Review, 83*, 173–189.
Stahlberg, M.R. (1984). Infantile colic: Occurrence and risk factors. *European Journal of Pediatrics, 143*, 108–111.
Stark, R.E., & Nathanson, S.N. (1973). Spontaneous crying in the newborn infant: Sounds and facial gestures. In: J.F. Bosma (Ed.), *Development in the fetus and infant, Fourth Symposium on Oral Sensation and Perception* (pp. 323–352). Bethesda, MD: U.S. Department of Health, Education, and Welfare.
Thoman, E.B., & Levine, S.(1970). Hormonal and behavioral changes in the rat mother as a function of early experience treatment of the offspring. *Physiology & Behavior, 5*, 1417–1421.
Thureen, P.J., Phillips, R.E., Baron, K.A., DeMarie, M.P., & Hay, W.W. Jr. (1998). Direct measurement of the energy expenditure of physical activity in preterm infants. *Journal of Applied Physiology, 85*, 223–230.
Vingerhoets, A.J.J.M., & Becht, M.C. (1997). *International Study on Adult Crying: Some first results*. Annual Meeting of the American Psychosomatic Society, Santa Fe, NM.
Vingerhoets, A.J.J.M., Cornelius, R.R., Van Heck, G.L., & Becht, M.C. (2000). Adult crying: A model and a review of the literature. *Review of General Psychology*, *4*, 354–377.
Vuorenkoski, V., Wasz-Hockert, O., Koivisto, E., & Lind, J. (1969). The effect of cry stimulus on the lactating breast of primipara: A thermographic study. *Experientia, 25*, 1286–1287.
Weston, J.T. (1968). The pathology of child abuse. In: R.E. Helfer & C.H. Kempe (Eds.), *The battered child* (pp. 77–102). Chicago: University of Chicago Press.
Wolff, P. (1969). The natural history of crying and other vocalizations in early infancy. In: B.M. Foss (Ed.), *Determinants of infant behavior* (Vol. 4, pp. 81–109). London: Methuen.
Wolff, P. (1987). *The development of behavioral states and the expression of emotions in early infancy*. Chicago: University of Chicago Press.

4 CRYING FREQUENCY ACROSS THE LIFE SPAN

Janice L. Hastrup, Deborah T. Kraemer, Robert F. Bornstein, and Glenn R. Trezza

Imagine a couple who have sought the help of a clinical psychologist or social worker in resolving some conflicts in their relationship. One partner complains that the other is "too emotional" and cries, on average, once a week; the other says the first is too restrained in showing emotion and rarely cries. Assuming the reports on frequency of crying are accurate, how should they be interpreted?

This couple mirrors the wide range of variation in crying found in the general population: some individuals almost never cry after childhood, while for others it is a daily or almost daily experience. Does it matter what age and gender each partner is? Substantial changes in sources of stress and in coping styles have been associated with both age and gender during the course of adulthood (Lazarus, 1996), at least in Western cultures. Since stressful situations can involve crying, one might wonder whether crying frequency will vary or remain stable over the life span. Lazarus suggests that gender may even interact with age over the course of adulthood, with women and men showing quite different trajectories in coping patterns.

Recent research and theorizing in the field of crying behavior have emphasized crying's social, communicative function (see Chapter 9 of this volume). This interpersonal role is evident even in childhood: children who experience pain cry loudly if a caregiver is available but may only whimper if alone or with peers (O'Hair & Cody, 1994). As roles and settings change over adulthood, one might well expect changes in crying behavior. Ideally, one would collect longitudinal data on crying behavior over the course of individual lives. Such data are not yet available, but one may get a glimpse of the possibilities with cross–sectional research.

Now imagine a clinical psychologist or social worker confronted with parents who complain that their child cries "too much." The parents are asked to keep track of the child's crying; they report the following session that there were four episodes in a week, and that this is typical of their recent experiences with this child.

Is this amount of crying excessive? Or do the parents have a low tolerance for what is really within the normal limits of crying frequency in childhood? These questions are of more than academic interest, since excessive crying is an important source of referrals for psychotherapy (Achenbach & Edelbrock, 1981) and one of the major triggers of child abuse by parents and other caregivers (Krugman, 1983–85; Schmitt, 1985).

To respond to these problems, it would help if the clinician had access to some normative information. What is the range of crying frequencies, on average, during childhood? To answer this question, one would need data on children at various ages, because crying frequency declines markedly between infancy and adulthood, as will be seen.

At what age does it begin to matter which gender the child is? Other researchers have observed no gender difference in crying frequency in infancy, but a marked difference in adulthood. One-year-old boys and girls cry at about the same frequency (Maccoby & Jacklin, 1974). In a sample of 286 women and 45 men between 18 and 75 years of age, Frey (1983) found an average of 5.3 crying episodes per month for women vs. 1.4 for men, a ratio of about 3.8:1.0. When and why does this marked gender difference first appear? And does it persist into older adulthood, say, above age 65? In providing normative information in answer to such questions, this chapter will discuss data we have collected on gender differences in crying frequency across the life span and the implications of these findings for understanding changes in crying.

Crying Frequency of Young, Middle-Aged, and Older Adults

GENDER DIFFERENCES ON THE CRYING FREQUENCY QUESTIONNAIRE

For our discussion of adult crying, we will summarize data collected with an instrument, the Crying Frequency Questionnaire (Kraemer & Hastrup, 1986), that provides annual and previous-month estimates of total crying episodes. One-month retest reliability statistics for the estimate of yearly crying episodes from a young adult sample (Kraemer & Hastrup, 1986) were +.84 for men and +.87 for women. The annual estimate—but not the monthly one—has been validated in two ways. First, the estimate of crying frequency per year predicted the probability of crying in response to a sad film in a laboratory setting (Silverstein et al., 1986), although this effect can be obliterated by specific instructions to cry or to try to inhibit crying (Kraemer & Hastrup, 1988). Second, a sample of college students, stratified to represent low, moderate, and high levels of yearly crying estimates, self-monitored their crying episodes for nine weeks; the data confirmed both the gender differences and the predictive validity of self-reports (details are available in Kraemer & Hastrup, 1986). Bellack and Schwartz (1976) have noted that behaviors that are most likely to be self-monitored accurately are: (a) low in frequency; (b) discrete events; and (c) visible (but see Chapter 1, this volume). Crying episodes fit these criteria well.

Table 1 summarizes annual crying frequency estimates from the CFQ in various samples. These are described in more detail below. For simplicity and also because of limited size samples, the data presented in Tables 2 and 3 do not include standard deviations and ranges. Nevertheless, it is important to realize that the ranges represented by these normative samples are quite large. For example, in one study of college students (Kraemer & Hastrup, 1986), young men reported crying an average of 6.5 times per year, but the individual estimates ranged from 0 to 96 episodes. For women, the average was 47.8 times per year, with a range of 0 to 360. Both distributions show a marked positive skew. Striking gender differences are thus already apparent in young adulthood (university students predominantly between 18 and 22 years old). The differences are also clearly evident in two middle-aged samples drawn from the general public: estimates were obtained from a nonclinical sample of 35- to 45-year-old participants in an "opinions and knowledge" survey (Hastrup et al., 1986a, b), as well as from parents of teenagers with a similar but slightly broader age distribution.

Three samples of psychologists were also surveyed. The first two consisted of school psychologists and psychologists working in university settings (Hastrup et al., 1986a, b). These respondents reported slightly lower crying frequency estimates, but with the same consistent gender ratio. The third sample (Trezza et al., 1988) consisted of clinical psychology practitioners; although the difference is in the expected direction, the ratio of women's to men's mean reported crying episodes is greatly reduced relative to the other adult samples. Inspection of the data suggested that the considerably higher mean rate of crying episodes reported by male practitioners cannot be attributed to a few outliers, but represents a more elevated rate relative to other male samples, although with the substantial increase in

Table 1. Self-reported yearly crying frequency estimates of adult men and women on the Crying Frequency Questionnaire (CFQ) (see text for additional details)

Description of sample	Men	Women	Ratio of women to men	Significant gender differences?	d
College students (Kraemer & Hastrup, 1966): 181 men, 316 women	6.5	47.8	7.36:1	Yes: $t_{(495)} = 9.73$	0.87
General public, ages 35–45 (Hastrup et al., 1986b): 83 men, 81 women	5.8	34.7	5.98:1	Yes: $t_{(162)} = 7.16$	1.13
Parents of teenagers: 75 men, 145 women	5.5	29.3	5.33:1	Yes: $t_{(218)} = 7.82$	1.06
School psychologists (Hastrup et al., 1986b): 33 men, 33 women	4.0	23.4	5.85:1	Yes: $t_{(64)} = 2.88$	0.77
Psychology faculty (Hastrup et al., 1986b): 105 men, 110 women	4.8	19.5	6.60:1	Yes: $t_{(113)} = 4.68$	0.88
Clinical psychology practitioners (Trezza et al., 1988): 99 men, 120 women	18.1	26.0	1.44:1	No: $t_{(117)} = 1.29$	0.24
Older adults, ages 65–71 (Hastrup et al., 1986a): 20 men, 44 women	16.3	31.1	1.91:1	No: $t_{(62)} = 1.04$	0.26

Table 2. Mean frequencies of crying incidents monitored by parents during one week

age	girls	boys	age	girls	boys	age	girls	boys
1	14.6	16.9	5	8.4	4.2	9	4.1	4.1
2	14.9	12.2	6	7.3	7.0	10	2.8	1.2
3	14.2	12.8	7	7.7	7.4	11	3.6	2.5
4	12.9	10.5	8	4.8	3.6	12	1.9	1.8

Table 3. Mean frequencies of crying incidents self-monitored by adolescents during two weeks

age	girls	boys	age	girls	boys	age	girls	boys
10	4.00	1.50	13	2.60	0.82	16	2.08	0.50
11	1.45	1.20	14	2.80	0.88	17	3.33	0.29
12	2.81	1.67	15	1.90	0.08			

variability and the wide range typical of female samples. Other data from this same survey suggest that the clinical psychology practitioners may be quite accepting of patients' and their own crying, and that dealing with human problems may itself trigger additional crying episodes (see also Chapter 10).

Whether these differences persist into older adulthood is more difficult to say; the available data suggest there may be a gender difference, but additional research is needed. In a sample of 20 men and 44 women, all between 65 and 71 years of age, a nonsignificant difference was observed, although in the expected direction (Hastrup, Baker et al., 1986a). Inspection of the data suggested that the women's estimates of yearly crying frequency were similar to those of other groups of women, but the mean for men was substantially elevated by the report of a single respondent who estimated his crying frequency to be 180 times per year. If that respondent is omitted, the estimate would be 6.6 episodes per year—in the same range as all but one (the clinical psychology practitioners) of the adult male groups.

To summarize, women in all these age ranges reported crying more frequently than did men. The differences were significant in young adulthood and in middle-age, although there is some overlap due to large individual differences in crying in every age-gender group.

Discussion of Gender Differences in Adulthood

Because the *a priori* estimates of crying frequency were predictive of subsequent self-monitored crying and of probability of crying in the laboratory, estimates of crying episodes may provide a window into the behavior of middle-aged and older adults for whom self-monitoring data are not currently available. Table 1 includes a column of the ratios of estimated yearly crying frequencies of women to men in the various studies. These are rather consistent for five of the seven samples; the two exceptions will be discussed below.

To assess the overall magnitude of gender differences in crying frequency, a meta-analysis of all seven samples was conducted. The last column of Table 1 shows the d values for each study; d represents the difference in standard deviation units between male and female crying frequencies. Using the procedures described by Rosenthal (1984), a sample-size-weighted average d (each study contributing in proportion to its N) was calculated. On average, women's crying frequency exceeds that of men by about .84 standard deviations. Such a d is considered by Cohen (1977) to be a "moderate" size effect; in meta-analytic terms, this d is actually fairly large. To put it in perspective, a d of .84 is substantially larger than those associated with the magnitude of psychotherapy effects vs. various placebo treatments; the Vietnam combat-PTSD link; or the effects of aspirin on heart attack risk (comparison figures from Meyer & Handler, 1997, Table 7).

A closer examination of the two exceptions to the gender differences in crying frequency (the sample of clinical psychology practitioners and the sample of older adults) may be helpful for an adequate evaluation of these results. It should be remembered as well that these were in the expected direction but not statistically significant. In the first exception, it is possible that self-selection into this kind of emotionally draining work may have increased the number of men who are more emotionally expressive and androgynous; alternatively or in addition, the work itself may increase the likelihood of crying. Longitudinal or gender/personality research (see Chapters 6 and 7 in this volume) may provide a tentative answer to this question. Although the gender difference among older adults was not significant in this study, this may reflect the small size of the sample and the large individual differences in crying. When one of the twenty older adult men who reported the highest frequency of all was excluded, the difference between older adult men and women was almost identical to that of middle-aged participants.

Crying Frequency in Childhood: When Does the Gender Difference First Appear?

BELIEFS ABOUT THE AGE OF APPEARANCE OF GENDER DIFFERENCES IN CRYING: CLINICAL PSYCHOLOGISTS, SOCIAL WORKERS, AND PARENTS OF TEENAGERS

When does the gender difference in crying frequency appear? Perhaps a sample of clinical psychologists and social workers, currently practicing in private, public, or agency settings, could provide some insight. Alternatively, parents of teenagers might contribute information from their own and their siblings' experiences growing up, and from observing their own children.

A sample of 227 practicing clinical psychologists and social workers who participated in a larger survey (Trezza et al., 1988) was asked to estimate the age at which boys and girls cried at different rates. The mean predicted age was 8.4 years, with most estimates during the elementary school years or earlier. This pattern was echoed in data obtained from the parents of teenagers. Thus, in both professional clinician and parent samples, there was a strong expectation that boys would begin to cry less frequently than girls during childhood rather than during adolescence. The popular admonition to children, "Big boys don't cry," echoes these data. But are these estimates correct? Two studies bear directly on the age of first appearance of gender differences in crying frequency. The first involved parent-recorded crying episodes among 1- to 12-year-olds of both genders—we also expected the gender difference to appear in elementary school, if not before. A subsequent study required 10- to 17-year-olds to record crying episodes themselves, without active parental participation.

CHILDREN'S EMOTIONS STUDY: DATA FROM PARENTAL MONITORING

A total of 240 families were recruited by newspaper advertisements offering $10 for participation in a study of children's emotions; children currently under treatment for emotional disorders were excluded from the study. Recruitment procedures provided windows of dates for birthdays, in order to insure that all participants would be within three months of their designated age (e.g., 5-year-olds were 4.75 to 5.25 years of age). Since over half of the sample was too young to self-monitor, it was decided to have parents act as recorders and to maintain this procedure through the entire age range; children who were not with the parents at all times were asked to report on recent crying episodes at mealtimes, but in all cases, parents kept the written records of crying behaviors. Recording periods were designed to distribute children in each age/gender group over the seven possible starting days of the week. All monitoring took place during the summer months when children were not attending school, but family vacations and periods spent out of town were excluded because these might be atypical or present unusual difficulties in monitoring, or both. Insofar as was possible, the recording weeks for each age/gender group were distributed over a period of a month or more in order to reduce the impact of shared environmental events that many children experience (e.g., from news reports or movies).

Mean frequencies of reported crying incidents during a one-week period are shown in Table 2. The number of participants in each gender-age group varies between 8 and 11; the return rate of those initially enrolled was over 90%.

Although there is a substantial decline in reported crying frequency over the course of childhood, these data show no differential decline for girls and boys. An analysis of variance showed no interaction between gender and age and no main effect for gender; the main effect for age alone was significant. An examination of the parents' reports of the reactions of others present showed a decline in the number of encouraging reactions of others present for boys but not for girls between 10 and 12.

These parent-monitored data are actually in accord with those published several decades earlier by a British group (Shepherd et al., 1971); the results of that survey, based on parents' global estimates of crying frequencies, show virtually identical patterns of crying frequency for boys and girls, with no differential decline in crying episodes during elementary school years. Although no statistical tests were reported in the Shepherd et al. study, visual examination of their data suggested that girls and boys begin to diverge in crying frequency around age 13 or 14.

It might be argued that children's reports of crying behavior, when filtered through their parents, are subject to considerable bias, especially with respect to the cultural belief that boys cry less often than girls. If such a bias had influenced data collection in the Children's Emotions

Study, boys and their parents should have minimized their reports of the boys' crying frequency. The absence of either a gender difference or a gender-age interaction in these data argues against the hypothesis that cultural beliefs systematically bias data collected in this way. Overall, these data, and the report by Shepherd et al., suggest that there is a tendency for boys to cry as frequently as girls at least through age 12. Neither the clinicians, nor the parents of teenagers, nor the researchers had predicted this finding.

The research plan was clearly too limited, stopping at age 12. The 12-year-old girls averaged 1.9 episodes per week, according to the parent-monitored data; 12-year-old boys averaged 1.8 episodes. The age at which boys and girls begin to differ on crying frequency must be during the teenage years. The next study used a self-monitoring method instead of parental reports, and included children as young as 10 and as old as 17 to provide three age groups (10-, 11-, and 12-year-olds) which overlapped with the first study.

YOUTH EMOTIONS STUDY: DATA FROM SELF-MONITORING

A total of 157 families were recruited through newspaper announcements, and paid $20 for participating in a two-week study. This study again took place during the summer months, and recruitment procedures ensured that the adolescents would be within 3 months of their designated birthday.

Both parents and the adolescents estimated the adolescent's yearly crying frequency on a survey used in earlier research with college students and adults. The adolescent made this estimate independently, and was then instructed in the two-week self-monitoring procedures for incidents in which she or he cried (defined as "at least watery eyes").

The results of the analysis of variance of self-monitoring data shown in Table 3 reveal a substantial decrease in crying frequency between ages 10 and 17, with boys showing a greater decrease than girls. The sharper decline for boys appears to begin after age 12; tests on differences between boys and girls were significant ($p < .05$) for ages 13, 15, 16, and 17, with a marginal difference ($p < .06$) at age 14 as well. Acknowledging the admittedly small samples involved (between 7 and 12 per age-gender group), it appears that there are large gender differences in self-monitored crying frequency. The frequencies reported are difficult to compare with those of Shepherd et al. (1971) because of the use of ordinal category descriptions in that study (e.g., "about once every two weeks"), but the age of greater decrease for boys than for girls is similar, occurring around age 13.

Unlike earlier research with college students (Kraemer & Hastrup, 1986), it appears that 10- to 17-year-olds are not good estimators of their frequency of crying. For girls, estimates of yearly crying frequency correlated only +.06 with the number of self-monitored incidents; for boys, the correlation was +.08 (both correlations are nonsignificant). The parents' estimates, however, were somewhat more predictive. Mothers' estimates of the children's yearly crying frequency correlated +.33 ($p < .01$) with their sons' self-monitored crying frequencies and +.18 ($p = .10$) with their daughters' totals. Fathers' estimates correlated +.28 ($p < .05$) with their daughters' self-monitored crying frequencies and +.42 ($p < .01$) with their sons' totals. This may mean that crying frequency is relatively stable, but younger children might not have the cognitive skills to estimate it accurately.

Summary of Age-Related Patterns in Gender Differences in Crying Frequency

It appears that the gender difference in crying behavior observed in college students and other adult samples appears around age 13. Both parental monitoring and self-monitoring by adolescents support this view. In the parent-monitored data, there were no significant differences between boys and girls, and the crying frequencies for one week averaged 1.9 for girls and 1.8 for boys at age 12. In the self-monitored data for two-week recording periods, there was a minor, nonsignificant difference at age 12 (2.81 incidents for girls vs. 1.67 incidents for boys). Beginning at age 13, however, girls reported a substantially greater number of incidents (between 1.90 and 3.33) than did boys (between .08 and .88). The total of crying incidents reported for boys up through age 12 in the Children's Emotions Study was fully 85% of that total for girls; in contrast, all the adult samples show reports of women crying 1.4 to 7 times as often as men. Whether the gender difference persists into older adulthood is not clear, although the limited available data are suggestive. There was no evidence of a substantial decline among older adults, as had been suggested by several writers (Borgquist, 1906; Heilbrunn, 1955; Löfgren, 1966). Although sources of stress and coping strategies may vary over the adult years (Lazarus, 1996), it may well turn out that individual crying frequencies are fairly stable.

Need for additional research. The data described in this chapter provide cross-sectional windows on crying frequency across the life span. They suggest that at least the gender difference that appears during adolescence is fairly consistent through much of adulthood in the U.S. general population samples so far obtained.

A recent report from the Netherlands (Van Tilburg et al., 1999) suggests a gender difference appearing at a somewhat earlier age. Their data show significant gender differences as early as age 11, the youngest age tested. There are a number of differences between that study and the present one, however. In addition to the possibility of cultural influences, there are some methodological differences. Their study is stronger in having a larger sample size with which to detect variation by gender. It is quite possible that there is a small difference by age 10 or 11 (perhaps reflecting those with earlier onset of adolescence) that becomes more distinctive during the early teens. On the other hand, their data consist of self-reported estimates of crying frequency, which in the adolescent study reported in this chapter were not significantly related to self-monitored crying frequencies. A laboratory-based evaluation of crying proneness has been used with college-aged subjects (Silverstein et al., 1986), and may prove useful in verifying the age of gender divergence on this behavior.

We need more information, however, on the stability of individual differences within gender groups. For example, are boys and girls who are in the upper deciles of crying frequency likely to report greater crying frequencies in adulthood? Does the increased mean frequency of crying episodes of male clinicians reflect some pre-existing individual difference and self-selection into the profession, or a response to their work, or both? Clearly, longitudinal research could provide some answers to these tantalizing questions raised by existing cross-sectional data.

It has already been observed that the cultural belief about boys beginning to cry less frequently than girls during elementary school years did not influence parents' reports substantially, since the patterns of parent-monitored crying frequency for boys and girls are virtually identical up through age 12. Although there is also evidence (Silverstein et al., 1986) that college students' crying frequency predicts likelihood of crying in response to a sad film in a laboratory setting, further validation studies are needed to show that reports from other ages (childhood, older adulthood) are also accurate. An alternative to the use of laboratory settings might be unobtrusive recording; however, the low overall frequency of crying behavior and its often private nature may make observational studies of adults virtually impossible.

Caution is needed in generalizing the present results to other cultures. The studies described in this chapter all involved samples based in the United States. Infants in China have been reported to cry less often than European American infants, suggesting the possibility of cultural and/or temperament differences (Camras et al., 1998). Likewise, we cannot be sure that substantial cohort effects will not appear in some other era with different attitudes toward emotional expression. And, lastly, we should note that all of the child and adolescent participants, and nearly all of the college-age participants, were living in upstate New York communities at the time; there may be subcultural and regional variations within the United States that have not yet been addressed, including perhaps ethnic differences as well.

The major, unanswered question is: why do gender differences appear during adolescence and persist into adulthood? Why do boys' crying episodes decline sharply in frequency around age 13, while girls' episodes stabilize? Depression is an unlikely explanation, since depressive symptoms and crying frequency are only weakly correlated (Hastrup, Baker et al., 1986a), and mean age of onset of Major Depressive Disorder does not occur until approximately 15 years of age (Lewinsohn et al., 1994). If these differences in the U.S. samples resulted from social pressures and expectations alone, they probably would have appeared before adolescence. Responses of others present in the Youth Emotions Study were clearly less encouraging of crying for boys than for girls in the 10- to 12-year-old age range, but were not accompanied by a differential decline in crying frequency until age 13 and beyond. There are a number of other behavioral events during early adolescence (e.g., negative interpersonal events and stressors involving family, peer and intimate relationships; Compas & Phares, 1991), however, which remain to be explored as possible explanations. One must be cautious in interpreting group data, however; secular trends toward earlier onset of puberty have been suggested, at least for girls; racial and ethnic differences in age at puberty may also need to be considered (Herman-Giddens et al., 1997).

If not social pressures or other psychosocial events, then perhaps there is something biological that reduces the frequency of, but does not eliminate, crying episodes in men. One obvious answer is some component of puberty, since the sudden sharp decline in crying episodes at age 13 coincides with the average age of pubertal onset for males (Tanner, 1975 see also Chapter 6, this volume).

Two available indices of biological puberty might be employed with minimal risk to participants. Panksepp (1998) found an age-related decline in distress vocalization in two species of small mammals, and observed that testosterone injections themselves could inhibit such vocalizations. Total testosterone measurement in humans requires only 3–5 ml of venous blood, and provides an estimate of the functioning of the endocrine variables underlying somatic puberty in males (Tanner, 1975; Voorhess, 1978). Ratings of physical pubertal development using Tanner's stages are an alternative; this standard clinical assessment procedure used by endocrinologists to assess puberty could be obtained independently of crying frequency, which could then be self-monitored to assess both cross-age and within-age-category relationships. If biological maturation contributes substantially to the decline in crying frequency for boys, one might expect that boys with a later onset of puberty will have a delayed decline in crying frequency relative to boys with an early onset. Coupling such a study with the laboratory assessment of crying proneness and self-monitoring of crying suggested above would provide more definitive information on the development of gender differences. One might also expect wealthier nations with earlier onset of puberty in their populations to show gender differences at an earlier age than would be observed in poorer nations.

Concluding Comments

What do we know about crying frequency across the life span? First, we know that crying persists through adulthood for most individuals despite ample development of verbal ways of communicating emotion. Second, there are huge individual differences in this behavior; some people rarely cry while for others, it is a daily occurrence. Third, there may be some predictable variations in crying frequency; for example, clinical psychology practitioners demonstrated a relatively small and nonsignificant gender difference compared with most other adult samples described in this survey. Fourth, there are clearly gender differences which arise during adolescence and persist through much or all of adulthood.

The final answer to the question of the origins of individual differences in crying frequency will likely be a complex one, perhaps reflecting variations in temperament (e.g., Scarpa et al., 1995), as well as gender, culture, social environment, and biological processes. Crying is an integrative topic for psychologists, with relevance to biological, developmental, clinical, social, and cognitive psychology. That means that crying is also a complicated behavior for investigation, with a need for expertise in many different areas. The difficulties do not appear insurmountable;

the consistency of age-related patterns of crying in samples of fewer than a dozen suggests parent- and self-monitoring can provide reasonable estimates of frequency for clinical purposes. The social benefits of enhancing our knowledge about crying include better assessment of the role of children's behavior in situations of abuse, and better understanding of the function of crying throughout the life span.

References

Achenbach, T.M., & Edelbrock, C.S. (1981). Behavioral problems and competencies reported by parents of normal and disturbed children aged four through sixteen. *Monographs of the Society for Research in Child Development, 46* (1, Serial No. 188).

Bellack, A.S., & Schwartz, J.S. (1976). Assessment for self-control programs. In: M. Hersen & A.S. Bellack (Eds.), *Behavioral assessment: A practical handbook*. (pp. 111–142) New York: Pergamon.

Borgquist, A. (1906). Crying. *American Journal of Psychology, 17,* 149–205.

Camras, L.A., Oster, H., Campos, J., Campos, R., Ujiie, T., Miyake, K., Wang, L., & Meng, Z. (1998). Production of emotional facial expressions in European American, Japanese, and Chinese infants. *Developmental Psychology, 34,* 616–28.

Cohen, J. (1977). *Statistical power analysis for the behavioral sciences*. New York: McGraw-Hill.

Compas, B.E., & Phares, V. (1991). Stress during childhood and adolescence: Sources of risk and vulnerability. In: E.M. Cummings, A.L. Greene, & K.H. Karraker (Eds.), *Life-span developmental psychology: Perspectives on stress and coping* (pp. 111–129). Hillsdale, NJ: Erlbaum.

Frey, W.H., Hoffman-Ahern, C., Johnson, R.A., Lykken, D.T., & Tuason, V.B. (1983). Crying behavior in the human adult. *Integrative Psychiatry, 1,* 94–98.

Hastrup, J.L., Baker, J.G., Kraemer, D.L., & Bornstein, R.F. (1986a). Crying and depression among older adults. *The Gerontologist, 26,* 91–96.

Hastrup, J.L., Phillips, S.M., Scheiner, J., McAfee, M.P., & Kraemer, D.L. (1986b). *Individual differences in natural crying frequency among psychologists and the general public.* Fifty-seventh Annual Meeting of the Eastern Psychological Association, New York, NY.

Heilbrunn, G. (1955). On weeping. *Psychoanalytic Quarterly, 24,* 245–255.

Herman-Giddens, M.E., Slora, E.J., Wasserman, R.C., Bourdony, C.J., Bhapkar, M.V., Koch, G.G., & Hasemeier, C.M. (1997). Secondary sexual characteristics and menses in young girls seen in office practice: A study from the Pediatric Research in Office Settings Network. *Pediatrics, 99, 505–512.*

Kraemer, D.L., & Hastrup, J.L. (1986). Crying in natural settings: Global estimates, self-monitored frequencies, depression and sex differences in an undergraduate population. *Behaviour Research and Therapy, 24,* 371–373.

Kraemer, D.L., & Hastrup, J.L. (1988). Crying in adults: Self-control and autonomic correlates. *Journal of Social and Clinical Psychology, 6,* 53–68.

Krugman, R.D. (1983–85). Fatal child abuse: Analysis of 24 cases. *Pediatrician, 12,* 68–72.

Lazarus, R.S. (1996). The role of coping in the emotions and how coping changes over the life course. In: C. Magai & S.H. McFadden (Eds.), *Handbook of emotion, adult development, and aging* (pp. 289–306). New York: Academic Press.

Lewinsohn, P.M., Clarke, G.N., Seeley, J.R., & Rohde, P. (1994). Major depression in community adolescents: Age at onset, episode duration, and time to recurrence. *Journal of the American Academy of Child & Adolescent Psychiatry, 33,* 809–818.

Löfgren, L.B. (1966). On weeping. *International Journal of Psychoanalysis, 47,* 375–381.

Maccoby, E.E., & Jacklin, C.N. (1974). *The psychology of sex differences.* Stanford, CA: Stanford University Press.

Meyer, G.J., & Handler, L. (1997). The ability of the Rorschach to predict subsequent outcome: A meta-analysis of the Rorschach Prognostic Rating Scale. *Journal of Personality Assessment, 69,* 1–38.

O'Hair, H.D., & Cody, M.J. (1994). Deception. In: W.R. Cupach & B.H. Spitzberg (Eds.), *The dark side of interpersonal communication* (pp. 181–213). Hillsdale, NJ: Erlbaum.

Panksepp, J. (1998). *Affective neuroscience.* New York: Oxford University Press.

Rosenthal, R. (1984). *Meta-analytic procedures for social research.* Beverly Hills, CA: Sage.

Scarpa, A., Raine, A., Venables, P.H., & Mednick, S.A. (1995). The stability of inhibited/uninhibited temperament from ages 3 to 11 years in Mauritian children. *Journal of Abnormal Child Psychology, 23,* 607–618.

Schmitt, B.D. (1985). Colic: Excessive crying in newborns. *Clinics in Perinatology, 12,* 441–51.

Shepherd, M., Oppenheim, B., & Mitchell, S. (1971). *Childhood behaviour and mental health.* London: University of London Press.

Silverstein, S.M., Hastrup, J.L., & Kraemer, D.L. (1986). *Individual differences in crying in a laboratory setting.* Fifty-seventh Annual Meeting of the Eastern Psychological Association, New York, NY.

Tanner, J.M. (1975). Growth and endocrinology of the adolescent. In: L.I. Gardner (Ed.), *Endocrine and genetic diseases of childhood and adolescence* (2nd ed.) (pp. 14–63). Toronto: W.B. Saunders Co.

Trezza, G.R., Hastrup, J.L., & Kim, S.E. (1988). *Clinicians' attitudes and beliefs about crying behavior.* Fifty-ninth Annual Meeting of the Eastern Psychological Association, Buffalo, NY.

Van Tilburg, M.A.L., Unterberg, M.L., & Vingerhoets, A.J.J.M. (1999). *Crying during adolescence: The role of gender, menstruation and empathy.* Second International Conference on The (Non)expression of Emotions in Health and Disease. Tilburg, The Netherlands.

Voorhess, M.L. (1978). Normal and abnormal sexual development. In: R. Kaye, F.A. Oski, & L.A. Barness (Eds.), *Core textbook of pediatrics* (pp. 224–279). Philadelphia: J.B. Lippincott Co.

5 THE SITUATIONAL AND EMOTIONAL CONTEXT OF ADULT CRYING

Ad J.J.M. Vingerhoets, A. Jan W. Boelhouwer,
Miranda A.L. Van Tilburg, and Guus L. Van Heck

Why and when crying occurs in adults is still a mystery in many respects. The close association of crying with sadness is well known. However, crying is also common in situations where sadness is not the predominant emotion. For example, people cry at positive events like weddings, sports games and situations in which they are not threatened themselves in any way but feel happy or empathize with others who are in distress. Moreover, persons can employ crying in order to manipulate others. In addition, several personal and situational characteristics can influence the probability of shedding tears. For example, fatigue, the presence of others, and so-called display rules all may inhibit or facilitate crying.

In babies and young children, crying is generally considered as a sign of distress or pain in response to pre- or perinatal trauma, unfulfilled basic needs, overstimulation, developmental frustrations, physical pain, or frightening experiences (e.g., Lester, 1984; Solter, 1995; see also Chapter 3, this volume for the developmental aspects). The situation for adult crying, however, appears to be far more complex. As a matter of fact, it is still largely unknown why adults cry and what the function of their crying is.

In adults, both contextual and intrinsic factors can trigger a crying response and lower or increase the threshold for shedding tears. Emotional and situational contexts are very powerful in this respect. Particular situations or events will induce a specific emotional state if they have been appraised in a certain way. In turn, particular emotional states can elicit a crying response. However, several personal, social and cultural factors may stimulate or inhibit tears. It is our conviction that a thorough investigation of the context of crying may help to broaden our insights into the crying phenomenon. Therefore, this chapter focuses on the antecedents and context of adult crying, specifically, the emotional and situational context. For personal, social and cultural aspects see Chapters 7, 8, and 9 in this volume.

First, this chapter explores what makes people cry. Theories on crying will be discussed briefly, as they are relevant for the present issue. Then, the literature on situations and emotions which stimulate a flow of tears will be reviewed and summarized. In the last part of this chapter, attention is directed to the broader context of crying. Major questions are: *where* and *when* do people cry, and *who is present* when they do? Attention will be paid to the circumstances in which people cry and to temporal aspects such as the most likely time of the day for crying episodes. So, the aim is to summarize what is known about the conditions that trigger and moderate a crying response.

Theorizing About the Causes of Crying

Scientists with different backgrounds, such as ethologists, anthropologists, psychologists, psychoanalysts, and biochemists, have speculated on essential factors making people cry. Some of them have emphasized biochemical processes, others have focused on cognitive or communicative aspects (see, e.g., Kottler, 1996). Here, the primary interest is in what theorists say on the general nature of the internal and external factors that induce crying.

Psychoanalysts have developed an overflow theory of crying in which crying is seen as a kind of hydraulic/ overflow process, that is, as a safety valve (e.g., Breuer & Freud, 1895/1955; Koestler, 1964; Sadoff, 1966). Tears are thought to represent the overflow of emotions that have passed a certain critical level. In this way, an excessive build-up of emotions is avoided. There occurs a draining off of energy that has been mobilized during the period of distress. This perspective might have fueled the common logic that crying in some sense is good for one's health (cf. Cornelius, 1981). Thus, a crucial aspect in this conception is the *high intensity* of emotions.

Crile (1915) argued that crying results when we have developed tension anticipating some physical action which we then do not carry out. Crying is then considered as helpful in the release of built up nervous energy, resulting in a quick return to our normal state. In a similar vein, Bindra (1972) postulates that tears reflect the emotions and feelings that cannot be worked off in action, but can be consummated only in biological processes which result in an overflow of tears. In this view, tears help to discharge tension in situations in which an individual is unable to cope effectively. Frijda (1986) agrees with philosophers like Thomas Hobbes (1658) and Helmuth Plessner (1970) that crying occurs in situations in which people feel overwhelmed and experience a loss of control. So, crying is conceived of as a sign of helplessness and powerlessness.

Alternatively, the two-factor theory of Efran and Spangler (1979) proposes that crying is the result of a *reappraisal of the factors* that induce arousal, which leads to a resolution of an emotional conflict. Arousal is assumed to be the consequence of an attempt to assimilate events that are incongruent with or interrupt schema-based expectations. Recovery is facilitated when a psychologically relevant event leads to the *giving up of the original schema,* thus making further assimilation efforts unnecessary. This theory is cognitive by nature because the induction of arousal and the recovery are defined in cognitive terms. Tears are considered to be a sign of tension release. They indicate the shift from arousal to recovery. At the moment that the individual feels that the worst is over, crying is assumed to be associated with increased parasympathetic activation after having been associated with increased sympathetic arousal in an earlier phase.

Labott and Martin (1988) have argued that both the two-factor theory and the overflow theory have serious limitations, in particular because both fail to predict when other responses, such as laughing, will occur. Therefore, they have proposed a combination of these two theories, stating that *incongruity* and *schema-change associated with high arousal*, caused by some specific emotions, most adequately predict emotional tears.

With respect to situations, the overflow model implies that, in fact, any episode can elicit crying. The particular context that actually elicits crying is like the drop that causes the bucket's running over. Also, the Efran and Spangler model (1979) and the extended framework proposed by Labott and Martin (1988) do not refer to a specific group of situations or a particular type of environment.

Authors like Kottler (1996) have emphasized the communicative aspects of crying. Here, crying is considered to help in facilitating attachment and strengthening the mutual bonds between people. Stated otherwise, crying may induce sympathy, empathy, and comfort. In this view, adults cry for similar reasons as babies and young children, namely, to signal distress and consequently to induce a helping response from others. It has been suggested by Kottler that humans are the only animal species that cries because they are helpless and dependent on parents and/or caretakers for a relatively long time after birth. This dependence explains the need for a powerful and effective behavior pattern elicit help from others. Thus, one can speculate that crying occurs in particular in situations in which individuals feel alone and in which it is important for them to receive emotional support and sympathy. Crying may be useful to mobilize help from others in emergency situations (e.g., Kottler, 1996;

Roes, 1990), to inhibit aggressive impulses of potential aggressors (Roes, 1990), or to induce sympathy, pity, and comfort from others (Mélinand, 1902; Borquist, 1906; Cornelius, 1981, 1997). This suggests that crying is under some control of the will. Indeed, crying can be used more or less consciously as a tactical manoeuvre (cf. Buss, 1992), similar to, for example, pain behavior which is also subject to operant conditioning processes.

According to Murube et al. (1999), emotional tearing is a facial manifestation of communication. Originally, tears served as a symbol for suffering, with the intention to request help in order to overcome the problem (pain, loneliness, fear, etc.). Later in phylogenetic development, emotional tears (without the auditory accompaniments) also became associated with offering help, when feeling sympathy, admiration, or helpfulness. The authors emphasize that their distinction does not overlap with the simple negative emotion—positive emotion classification. Although requesting help always results from negative feelings, offering help may be accompanied by positive or negative emotions. For example, crying when expressing condolence and when witnessing an athlete winning an important game are both considered as examples of offering help tears.

Although some theories, like the one put forth by Murube et al. (1999), might be partly applicable to joyful situations, like weddings, reunions, the happy end of a movie or passing an exam, most theories of crying strongly emphasize negative or distressing situations as causes of crying. As a consequence, the relevance of current models of coping is rather limited with respect to crying associated with positive events and experiences. In any case, it does not produce a clear picture. For example, applying the psychoanalytical overflow theory to crying at a wedding results in the hypothesis that the happy emotions during a wedding might be so overwhelming that they build up and reach a certain critical level just like sad or distressing events would do. Tears might function as an overflow process to drain off energy in both cases. However, psychoanalysts (see Wood & Wood, 1984) have also suggested that happy events might induce crying because of memories of unhappy experiences or unfulfilled wishes. A third view, explaining that crying occurs when the behavioral expression of positive emotions is considered as less appropriate in the given situation, is more conceptually linked to helplessness and the inability to work out emotions behaviorally. For example, one can think of the winner of a gold medal at the Olympic Games, who thinking of the efforts, hardships and tough times during the training, has to stand still on the rostrum to listen respectfully to the national anthem with no opportunity for expressing emotions.

To summarize, it is clear that there is a wide variety of opinion on the background of crying. Some interpretations are basically untestable, but other perspectives may stimulate research which may yield answers to some basic questions in this area. Unfortunately, until now, most theories have not been tested empirically and research has generally not been theory-driven. Thus, it is hard to judge the value of these theories in explaining why and in what situations people cry. Therefore, in the following the focus will be on empirical data. Several mainly explorative and descriptive studies have been directed at the antecedents of crying spells, especially their situational and emotional context. The results of these studies will be discussed next.

Research on the Causes of Crying

Only in pathological conditions linked to neurological processes (see Chapter 15, this volume), do clear triggers inducing crying seem to be absent. In normal conditions, crying can be elicited by external (situations, events) or internal (thoughts, memories) stimuli. In the literature, several different approaches for studying the question of *why* people cry can be identified. For instance, the focus of attention can be on the emotions and feelings elicit crying, or on social episodes and other external situations. However, the precise status of emotions and feelings is not clear. Should they be considered as the real triggers of crying or just as mediators? Or perhaps even as the consequence of crying? Should one focus on the external events or internal stimuli that evoke those feelings and emotions? Or is it important to find out why sadness is often associated with crying, while in a substantial number of other cases this emotional state does not elicit tears? We will start our efforts to address these critical problems by providing a chronological summary of the results of studies focusing on the emotions and antecedent situations which are associated with crying. We consider this as an important first step to come to a better understanding and a more parsimonious explanation of crying in adults.

SITUATIONAL AND EMOTIONAL ANTECEDENTS OF CRYING

Without going into details about the specific causes, Darwin (1872/1965) has written about crying associated with both positive and negative moods. For example, he speaks of reading a pathetic story, but also of suffering bodily pain and mental distress. Borgquist (1906) was probably the first investigator who studied aspects of crying more systematically using questionnaire data as well as information from ethnologists and missionaries to take into account cultural background. He concluded that crying for grief, anger and joy was frequently mentioned in historical and legendary accounts in many cultures. Based on a thorough analysis of introspective descriptions of the crying state, Borgquist identified three types of crying situations: (i) grief or sadness, (ii) anger, and (iii) joy. However, he also pointed to sympathy and fear as important accompanying feelings. In addition, he mentioned physical conditions like nervousness, fatigue, and pain. It was concluded that crying occurs predominantly in conditions containing elements of sadness, helplessness, or hopelessness. According to Borgquist, the crying response is "the physical accompaniment of a mental state which is a recognition of an inability to remove certain painful or oppressive conditions; the cry appears when the feeling has reached a certain intensity" (p. 165).

Lund (1930) based his report on observations at funerals, weddings, theatrical and musical performances, and other events in which the expression of emotions normally occurs. In addition, he asked psychology students to describe in detail cases of crying observed during the past trimester. These observations led to the following labeling of crying causes: (i) laughter; (ii) relief from tension; (iii) loss; (iv) joy; (v) sympathy and self-pity; (vi) dramatic events; and, finally, (vii) aesthetic experiences.

Young (1937) collected more than 1000 reports of crying and laughing episodes in a large sample of college students. He classified the reasons for crying as follows: (i) disappointment or discouragement, (ii) lowered self-esteem and a sense of personal inadequacy, (iii) unhappy mood, (iv) organic state, (v) special events, and (vi) laughter to the point of tears. This investigation further emphasized that, whereas some of the conditions were organic in nature (e.g., fatigue, nervousness, headaches, illness, bodily injuries), the environmental causes of crying were almost uniquely social. In particular, actions, words, or attitudes of others were identified as the most important triggers.

Koestler (1964) listed the following situations which may cause crying: (i) raptness, (ii) mourning, (iii) relief, (iv) sympathy, and (v) self-pity. He further addressed crying in pain and when hungry, both in particular pertaining to babies and children. It is not clear upon what kind of data he based this classification.

Löfgren (1966) also did not indicate on which kind of data he based his conclusions. He proposed the following classification: (i) frustrating encounters with persons or things; (ii) bodily injury and pain; (iii) object loss; (iv) shame and humiliation; (v) pity (and self-pity); (vi) 'just moods', 'happy endings', weddings, joy, rage, etc; (vii) danger of various kinds, accompanied by a subjective experience of fear; and (viii) 'pathological weeping'.

Based on student data, Bindra (1972) derived information on instigating situations as well as initial emotional states. Concerning the latter, he made the following distinction: (i) elation—typically associated with situations like reunion, reciprocation of love, or music, (ii) dejection—often linked with separation and/or loss, and (iii) anguish—frequently associated with conflict, humiliation, or failure. In addition, he noted that in the majority of the cases the triggering situation was one in which the participant had actively initiated the event that induced the emotional state, whereas in approximately 20% of the cases the reporter was actually not participating, but an observer, reader, or listener. Furthermore, it was found that the majority of these latter situations were concrete and real (e.g., novel, film, song) and only 10% were imagined situations including recall or extrapolations of past experiences.

Another investigation using recent actual crying episodes is the 1-month diary study by Frey et al. (1983). These investigators applied a rather global classification of crying-inducing events and observed that 40% of the events were associated with interpersonal issues, 32% with media matters (books, movies, TV), and 7% with sad thoughts. Later on, Frey (1985) specified the interpersonal situations into two broad categories: positive (e.g., reunion, weddings) and negative situations (e.g., conflicts), without providing information concerning their relative importance.

More recently, Kottler (1996) presented the following taxonomy: (i) physiological response (to irritants, but also to exhaustion or even orgasm); (ii) reminiscence; (iii) redemption and release; (iv) 'in connection to others'; (v) grief and loss; (vi) despair and depression; (vii) joyful and aesthetic transcendence; (viii) vicarious experience; (ix) anger and frustration; and (x) manipulative tears.

Williams and Morris (1996) observed gender differences in their sample of English and Israeli university students and faculty members. Women cried more often in conflict situations and in situations inducing anger. In contrast, men rarely cried for these reasons. For men, tender situations stimulated tears more easily. Recent research in Malaysia (Joseph, 1996) yielded rather similar findings.

Wagner et al. (1997) investigated crying behavior and attitudes towards crying in medical doctors, nurses, and medical students, applying self-report questionnaires. Their results showed that the main reasons for crying were identification and bonding with suffering and dying patients and their families. Other relatively frequently mentioned causes were humiliation, being criticized or in other ways not treated well by supervisors, frustration, and being overtired or overworked and exhausted.

Vingerhoets et al. (1997), examining a sample of 250 persons, mainly women, came to the following conclusions. Adults cry in response to discrete emotional events, but also without any clear external trigger when reflecting on their lives or situations. Stimuli that are essentially weak and neutral may evoke strong memories to traumatic or very emotional events, in that way resulting in crying. Second, the inducing situations are often conflicts, feelings of personal inadequacy, and/or loss events. In this study, the authors came across several problems regarding the identification and classification of the various triggers of crying.

A first problem concerned the difficulty in distinguishing between objective situations and closely related feelings and emotions. For example, people described as a major inducing situation that they *felt* lonesome, rejected, homesick, or humiliated. A second problem was that people appear to be able to suppress crying and suspend it to a later and more appropriate moment, for instance, after having isolated oneself from a group or an opponent in a conflict situation. Or people first start crying when lying in bed reflecting on what has happened at work or in response to a message that they received earlier that day. Since books, films and TV-reports or official memorial ceremonies may induce memories of tragic events in some, but not in others, it is difficult to obtain a good understanding of what actually makes people cry in such situations.

Vingerhoets and Becht (1997), within the context of the International Study on Adult Crying (ISAC), found substantial agreement in crying inducing situations between males and females. Conflicts, loss experiences, and witnessing suffering were reported most often by all participants. In addition, as already reported by William and Morris (1996), men appeared to let their tears flow more easily when experiencing positive events as compared to women.

Unterberg (1998) asked adolescents (11–16 years) to report their most recent crying episode. Interestingly, her results indicated that girls cry frequently for empathic reasons, whereas boys mentioned this rather seldom. In addition, it appeared that boys cry more often because of physical pain than girls. The interpretation of these findings requires caution. For example, we do not know to what extent boys and girls actually differ in terms of number of pain experiences. One might speculate, for instance, that these differences are mainly due to a differential pattern of participation in competitive sports and game activities that for boys are generally somewhat rougher than for girls.

Finally, Damen (1999) conducted a semi-structured interview among 20 students, collecting information about the antecedents of being touched to tears. In a second study, 103 students completed a 33-item questionnaire, developed on the basis of the results of the interviews. The interviews revealed that people often become tearful by the beauty of nature, music, film, others they love, or an out of the ordinary, and hence, significant gesture or communication. Factor analysis yielded the following three factors: (i) aesthetics related items (e.g., beautiful art, nature, poems, songs, etc.); (ii) film and sentimental issue related items (e.g., puppy dogs, etc.); and (iii) social event related items (e.g., weddings, reunions, making love, etc).

Whereas until now the focus was mainly on crying eliciting situations, Vingerhoets et al. (1997) and Vingerhoets and Becht (1997) additionally assessed accompanying emotions. It appeared that often there is a blend of emotional feelings. Only exceptionally respondents report just one emotion. Sadness, powerlessness, anger, anxiety, and frustration prevail. There is also a not infrequent combination with powerlessness. One experiences anger and powerlessness or sadness and powerlessness. Therefore, in spite of the rich data that have been reported, it is not easy to summarize the most important situational and emotional antecedents of crying identified in empirical research and to draw any firm conclusions. On the other hand, helplessness and powerlessness are recurrent themes at least since Thomas Hobbes (1658).

Problems in Current Crying Research

It is still largely unknown why people experiencing the same emotions respond to them with different experiences of crying. Is this just a matter of intensity or do other factors play a substantial role? Results vary to a large degree and authors seldom differentiate between emotions and crying-inducing situations. In addition, all authors use their own categorization systems, which prevents the possibility of a more systematic comparison of the empirical data. One can only say that recurring themes are grief/loss, interpersonal conflicts, anger/frustration, joyful experiences, and powerlessness.

Another major problem is that many investigators fail to recognize the difference between, on the one hand, *crying proneness* or *crying propensity* and, on the other hand, *actual crying behavior*. This distinction is important because there are substantial differences between the answers to questions regarding the type of situations that most likely elicit tears and the kinds of situations in which people *most often* cry. The discrepancy between these two questions may be best illustrated by referring to the death of loved ones, which ranks very high among the situations that will make people cry. However, fortunately, this is such a rare event that it does not often show up in the answers to requests to describe the situation that has made one cry most recently. In other words, people cry most often for reasons that do not rank very high among the situations that are very likely to trigger the crying response. And vice versa, people generally do not often cry for those events that are rated as very likely making them cry. Only some exceptional situations rank relatively high on both lists. For male students, the loss of intimate relationships is such an item. In the ISAC study (Vingerhoets & Becht, 1997), this item was rated among the most situation to induce tears and as the most frequently reported actual trigger. For female students, watching sad movies is important in both respects. Thus, a different picture emerges when subjects are requested to report on the crying inducing potential of situations (e.g., the death of a loved one, broken relationships, weddings, etc.) versus when asked to provide information about their last actual crying episode. In the latter case conflicts, being rejected, and personal inadequacy appear to be more important. Remarkably, physical pain is hardly ever mentioned by adults.

Whereas measurements of actual crying frequency rely on the recall of recent episodes, crying proneness is usually measured by describing hypothetical situations to respondents and asking them to indicate how likely it is that they will shed tears in these particular situations. Examples are the Crying "Frequency" Scale (CFS; cf. Kraemer & Hastrup, 1986), a 20-item checklist, covering "a wide variety of emotions and events", developed by Lombardo et al. (1983), the Weeping "Frequency" Scale (Labott & Martin, 1987), the Crying Questionnaire (Williams, 1982), and a 30-item list constructed by Williams and Morris (1996). Studies using these scales show a strong correspondence in findings: the death of intimates, broken love relationships, and sad movies or television programs rank highest. Among the positive situations, weddings, music and reunions occupy top positions. Probably the most extensive list is the one developed by Vingerhoets (1996; see also appendix) within the context of the ISAC-project, which contains a total of 55 negative and positive situations and emotions that are more or less likely to elicit tears in several cultures. Situations included in this list that rank highest for their crying inducing potential were 'tragic' events, funerals, loss of relationship, sad movies and television programs, and the state of despair.

In a recent laboratory study, we adopted an alternative approach in order to learn more about the type of situations that are likely to evoke crying. One hundred female students volunteered watching the dramatic movie "Once Were Warriors" (Tamahori, 1995). This movie depicts the family life of a dysfunctional Maori family in New Zealand. The father has lost his job and spends most of the time drinking with his friends. He mistreats his wife and does not pay any attention to his children. After the film, participants were asked to report the scenes during which they had cried (ranging from just wet eyes to full sobbing). The results indicate that the most potent crying-inducing event was the scene where the mother finds her daughter who has committed suicide (approximately one-third of the subjects reported shedding tears). Subjects also rated the scenes of the daughter's funeral, a rape scene, and the assault and battery of the mother as highly emotional. A major advantage of this methodology is that it allows for an accurate comparison of actual crying behavior in different groups (men vs. women, young vs. old, different personality types, etc.) or under different circumstances (e.g., alone vs. accompanied by friends or strangers, at different times during the day, during different phases of the menstrual cycle, etc.). We feel that such an approach may be very helpful in learning more about inhibiting and facilitating factors of the crying response, since it gives the researcher at least some control and possibilities for manipulation.

In conclusion, a number of authors have more or less systematically collected data on situations and emotions that induce crying. In addition to problems in making adequate categorizations of the qualitatively different conceptualizations of situations, we feel that another serious problem is the fact that emotions are seldom experienced in a pure form. Very often it is a rather complex blend of feelings with helplessness or powerlessness as crucial elements that elicit the crying response. This all implies that it is not easy to come to an adequate classification of the experiences that trigger the crying response.

Vingerhoets et al. (1997) made clear that there is a whole complex of factors that determine whether or not tears will be shed (see Figure 1). More precisely, in addition to the exposure to potentially crying inducing situations and a specific appraisal of those situations, there are moderating factors that play a major role as determinants of whether or not crying will be inhibited or facilitated. Within these moderators, a

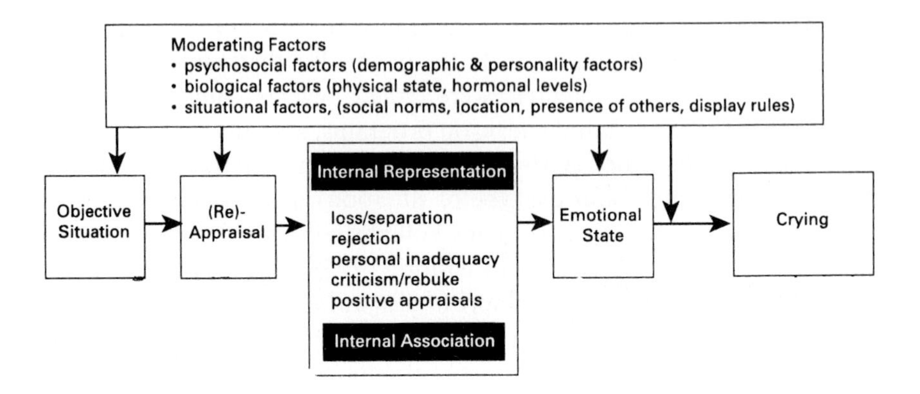

Figure 1. Preliminary model of adult crying, focusing on the antecedents and moderating factors.

differentiation can be made between person factors (both physical and psychological state and trait variables), social factors (e.g. the absence or presence of (specific) others), and cultural factors (in particular, so-called cultural "display rules"). Being exposed to an emotional situation or a strong emotional memory is, although necessary, not a sufficient condition to evoke crying. For a more accurate prediction of whether crying will occur, additional information is needed concerning the moderating variables.

Even if the situation seems at first sight likely to induce crying, a crying response may not necessarily occur. As already indicated, Vingerhoets et al. (1997) found that people may delay their crying reaction until a later, more appropriate moment. It is not entirely clear why people do this, since this issue has not been dealt with specifically in systematic research. Vingerhoets et al. suggested the following explanations for postponing crying: waiting until a specific person is absent or present, exposure to cues that reactivate memories, or a discussion resulting in a reappraisal of the situation. In the ISAC study (Vingerhoets & Becht, 1997), participants were asked to estimate the time interval between the actual event and the crying response. It appeared that in nearly 75% of the cases the response is rather direct, i.e., the interval is less than 15 minutes. In about 10% of the cases, the crying-eliciting event happened more than one day ago. This once more illustrates that whether or not a person will cry in a certain situation not only depends on the exposure to an emotional situation, but also on moderating factors that make a crying response more or less likely to occur. Below we will summarize recent data that further enhance our understanding of the *'where and when'* of crying.

The Situational Context of Crying

Concerning the timing of crying, Frey (1985) reported a dramatic increase in female crying frequency between 7 P.M. and 10 P.M., whereas no significant variations were detected between 9 A.M. and 7 P.M. Vingerhoets et al. (1997) and Becht and Vingerhoets (1997) substantiated these findings and demonstrated further that the propensity to cry shows a gradual increase between 4.00 AM and 11.00 PM (see Figure 2). Van Tilburg and Vingerhoets (2000) found a similar gradual increase when collecting diary data on mood and crying during two full menstrual cycles among 82 female students.

There are several potential reasons why crying occurs predominantly in the evening. First, there is an increased opportunity for conflicts due to the fact that this is the time when partners and children are together. Second, it is also the time to watch television, which is a powerful stimulus for (especially female) tears. Third, one may feel tired after a full day of working, which may lower the crying threshold. Fourth, feeling safe, alone or only with intimates might affect the proneness to cry. A final theoretical explanation is that crying also shows a circadian rhythm,

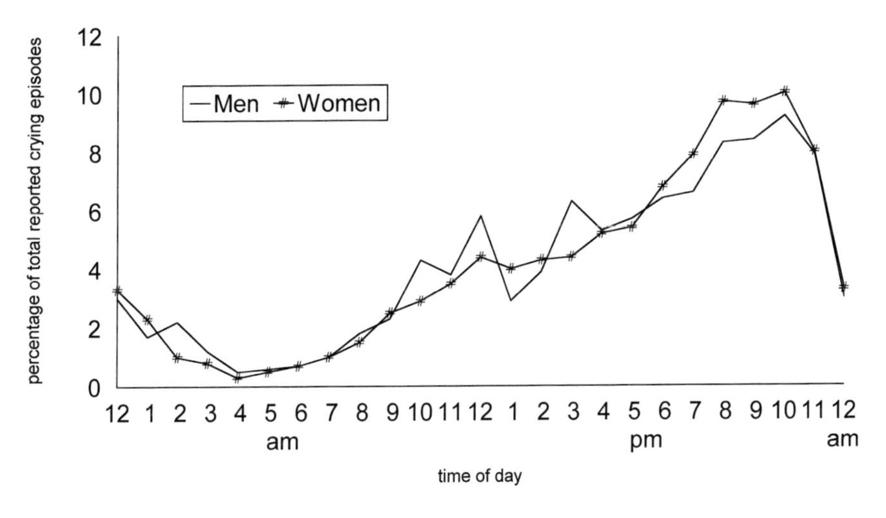

Figure 2. Crying by time of the day.

just like many other behaviors and physiological and psychological processes. Indeed, there appears to be some correspondence in findings obtained in babies and newborns and the present adult data (e.g., St. James-Roberts et al., 1994).

Research has further yielded evidence that the likelihood that someone will cry is also dependent on the presence of others in the situation. Vingerhoets et al. (1997) and Becht and Vingerhoets (1997) established that in the majority of the cases (75%), people cry at home, with no one (37%) or just one person (29%) present. The effects of the presence of others on crying are summarized by Cornelius and Labott (Chapter 9, this volume). Generally, it has been found that people report a greater likelihood of crying when alone than when with others. It is often feared that crying may be perceived as a sign of self-indulgence or weakness and, as such, crying is an act that should be controlled or inhibited. For example, Truijers (1999) found that particularly among adolescent males, crying was among the most important acts that induced shame. In addition, people may attempt to control crying because they do not want to upset others or make other people lose control (Littlewood, 1992). For example, parents may not want their small children to see them crying, because it may cause severe distress in them.

The opposite situation, when one cries when in the company of others, should not be ignored. On some occasions people may cry just because others are present. For example, people can use tears to signal or to manipulate others (e.g., Buss, 1992; Lackie, 1977). We probably all know from our own experience that children having hurt themselves often do not start crying until they see one of their parents. Lackie (1977) further gives an interesting description of a widow of an alcoholic and one of her daughters who, according to the author, used crying to control others. Furthermore, crying can be contagious. Seeing other people cry may make the beholder cry too, as may happen during funerals or other public situations, like memorial meetings, etc. It has been found that in particular women are sensitive to this form of emotional contagion (Doherty et al., 1995; Hatfield et al., 1994).

To summarize, the presence of others may either inhibit or stimulate crying, depending on the relationship with them, their display of emotion, etc. Progress in understanding the effects of a social context on crying may be hindered by a lack of consensus on whether or not crying must be considered as a communicative display (cf. Cornelius, 1997; Fridlund, 1994). There is evidence that adult crying has a communicative function, similar to infant crying which attracts attention, help, solace and instrumental support from nearby persons. However, adults can use many other verbal and nonverbal behaviors to communicate feelings of distress. Moreover, as shown above, empirical data indicate that adults often cry when alone. Why would people display behavior with an essentially communicative function when alone? According to Fridlund (1994), private displays of emotions might be considered social because we may treat ourselves as an audience or as another interacting person, we may act as if others were present or imagine they are, we may forecast or rehearse interactions with others, and we may treat animals or inanimate objects as participants in the interaction. Following this reasoning, crying alone means that the communicative function is maintained, while the negative reactions of the social environment and of the individual him/herself (shame) are, at the same time, avoided.

Vingerhoets et al. (1997) and Vingerhoets and Becht (1997) focused on some additional aspects of the crying-inducing situation. In terms of the responsibility for the situation, it was found that most frequently persons see themselves as responsible, but often also the partner, family or relatives are "blamed". Furthermore, relatively frequently, the events and the crying response were anticipated. In only 11% of the cases did tears come as a complete surprise. In addition, most respondents felt powerless and unable to cope with the situation and in the majority of cases one did not try to hide one's tears. These findings seem to indicate that crying implies an inability to cope in any other way with the situation.

Summary and Conclusion

When studying the context of crying, one should be aware of the fact that all the studies discussed so far have been conducted in (modern) Western cultures, more specifically, the United States and Northern European countries. Therefore, it is not clear to what extent one may generalize the findings presented here to other cultures (see Chapter 8, this volume). As one example of the strong cultural influences on crying, we would like to refer to the work by Wellenkamp (1988, 1992) on crying and its meaning in the Toraja tribe in Indonesia. According to the traditional beliefs of the Toraja, it is taboo for adults to cry (audibly), except in two very well described situations: (i) after a death and during the funeral or a secondary burial which is a kind of ritual, and (ii) when women are unable to become pregnant. In the latter case, women who are unable to conceive a child are expected to cry with other women at a rock said to be inhabited by a spirit as a kind of remedy for infertility. The prohibition on crying is as important as the ban on adultery and cursing someone. This is not to say that crying does not occur in a wide variety of situations like marital quarrels or departures. However, in those cases one has to make a sacrificial offering to atone for violating the prohibition. In addition, Lutz (1999), who vividly describes crying in a historic perspective, makes clear how crying in former times was related to religious experiences and heroism. These examples clearly illustrate the importance of social and cultural rules and norms when investigating the context of crying. In addition to personal and sociodemographic factors, culture also may be expected to influence emotion regulation, resulting in either increased or decreased emotional expression accompanied by tears.

It appears that people report an increased likelihood of shedding tears in uncommon situations like the death of intimates, broken love relationships, weddings, etc. In contrast, conflicts, being rejected, and the experience of personal inadequacy appear to be important in eliciting actual crying episodes. Grief, sadness, joy, anger, frustration, self-pity, helplessness and powerlessness are the most frequently reported emotions associated with adult crying. It is not clear yet whether these feelings are part of the appraisal processes that cause crying behavior, or are simply feelings that accompany the shedding of tears. As for the temporal context, the tendency to cry increases during the day and peaks in the evening. Furthermore, crying can be postponed if the situation does not allow one to cry at the moment itself.

Little is known about the social context of crying, in spite of the obvious power of others to influence emotional experience and expression. So, it is not clear how crying is shaped by the social context, especially by the reactions of others. However, data suggests that the presence of others may either stimulate or inhibit crying. It has been shown that gender plays a role, although the influence of gender on crying may be changing somewhat over time. In addition, cultural factors play a crucial role. What is clear from this chapter is the important distinction between potential triggers, on the one hand, and moderating variables, on the other hand. We are convinced that this model may help to design future studies to examine the complex interplay of the different relevant factors to crying behavior. Experiencing an emotion is one thing, but expressing that emotion via tears is quite a different thing, with its own rules and influenced by other factors. We are just beginning to understand the different influences on these two different processes, but we hope to have given an important impetus with the present chapter.

References

Becht, M., & Vingerhoets, A.J.J.M. (1997). *Why we cry and how it affects mood.* Paper presented at the Annual Meeting of the American Psychosomatic Society, Santa Fe, NM (Abstracted in *Psychosomatic Medicine, 59*, 92).

Bindra, D. (1972). Weeping, a problem of many facets. *Bulletin of the British Psychological Society, 25*, 281–284.

Borgquist, A. (1906). Crying. *American Journal of Psychology, 17*, 149–205.

Breuer, J., & Freud, S. (1955). *Studies on hysteria* (J. Strachey, Trans.). London: Hogarth Press (original work published 1895).

Buss, D.M. (1992). Manipulation in close relationships: Five personality factors in interactional context. *Journal of Personality, 60*, 477–499.

Cornelius, R.R. (1981). *Toward an ecological theory of crying and weeping.* Nineth Annual Meeting of the Eastern Psychological Association, Buffalo, NY.

Cornelius, R.R. (1997). Toward a new understanding of weeping and catharsis? In: A.J.J.M. Vingerhoets, F.J. Van Bussel, & A.J.W. Boelhouwer (Eds.), *The non-expression of emotions in health and disease* (pp. 303–321). Tilburg, The Netherlands: Tilburg University Press.

Crile, G.W. (1915). *The origin and nature of the emotions.* Philadelphia, PN: Saunders.

Damen, F. (1999). *Ontroering [Being touched].* Unpublished Master's thesis, Department of Social Psychology, University of Utrecht.

Darwin, C. (1872). *The expression of emotions in man and animals.* London: John Murray (1965, Chicago: University of Chicago Press).

Doherty, R.W., Orimoto, L., Singelis, T.M., Hatfield, E., & Hebb, J. (1995). Emotional contagion: Gender and occupational differences. *Psychology of Women Quarterly, 19*, 355–371.

Efran, J.S., & Spangler, T.J. (1979). Why grown-ups cry; A two-factor theory and evidence from The Miracle Worker. *Motivation and Emotion, 3*, 63-72.

Frey, W.H. (1985). *Crying: The mystery of tears.* Minneapolis, MN: Winston Press.

Frey, W.H., Hoffman-Ahern, C., Johnson, R.A., Lykken, D.T., & Tuason, V.B. (1983). Crying behavior in the human adult. *Integrative Psychiatry, 1*, 94–100.

Fridlund, A.J. (1994). *Human facial expression: An evolutionary view.* San Diego, CA: Academic Press.

Frijda, N.H. (1986). *The emotions.* New York: Cambridge University Press.

Hatfield, E., Cacioppo, J.T., & Rapson, R.L. (1994). *Emotional contagion.* New York: Cambridge University Press.

Hobbes, T. (1658|1966). De Homine. In: W. Molesworth (Ed.), *The English works of Thomas Hobbes of Malmesbury.* Aalen, DE: Scientia.

Joseph, C. (1996, August). *The antecedents and context of adult crying: Findings in a Malaysian sample.* Paper presented at the International Conference on the (Non)expression of Emotions in Health and Disease, Tilburg, The Netherlands.

Koestler, A. (1964). *The act of creation.* London: Hutchinson.
Kottler, J.A. (1996). *The language of tears.* San Francisco, CA: Jossey-Bass.
Kraemer, D.L., & Hastrup, J.L. (1986). Crying in natural settings. *Behaviour Research and Therapy, 24*, 371–373.
Labott, S.M., & Martin, R.B. (1987). The stress-moderating effects of weeping and humor. *Journal of Human Stress, 13*, 159–164.
Labott, S.M., & Martin, R.B. (1988). Weeping, evidence for a cognitive theory. *Motivation and Emotion, 12*, 205-216.
Labott, S.M., Martin, R.B., Eason, P.S., & Berkey, E.Y. (1991). Social reactions to the expression of emotion. *Cognition and Emotion, 5*, 397–417.
Lackie, B. (1977). Nonverbal communication in clinical social work practice. *Clinical Social Work Journal, 5*, 43–52.
Lester, B.M. (1984). A biosocial model of infant crying. In: L. Lipsitt, & C. Rovee-Collier (Eds.), *Advances in infancy research* (pp. 167–212). Norwood, NJ: Ablex.
Littlewood, J. (1992). *Aspects of grief: Bereavement in adult life.* London: Tavistock/Routledge.
Löfgren, L.B. (1966). On weeping. *International Journal of Psychoanalysis, 47*, 375–381.
Lombardo, W.K., Cretser, G.A., Lombardo, B., & Mathis, S.L. (1983). Fer cryin' out loud—There is a sex difference. *Sex Roles, 9*, 987–995.
Lund, F.H. (1930). Why do we weep? *Journal of Social Psychology, 1*, 136–151.
Lutz, T. (1999). *Crying. The natural and cultural history of tears.* New York: Norton.
Mélinand, C. (1902, June). Why do we cry? The psychology of tears. *Current Literature, 32*, 696–699.
Murube, J., Murube, L., & Murube, A. (1999). Origin and types of emotional tearing. *European Journal of Ophthalmology, 9*, 77–84.
Plessner, H. (1970). *Laughing and crying: A study of the limits of human behavior*. Evanston, IL: Northwestern University.
Roes, F. (1990). Waarom huilen mensen? [Why do people cry?]. *Psychologie, 10*, 44–45.
Sadoff, R.L. (1966). On the nature of crying and weeping. *Psychiatric Quarterly, 40*, 490–503.
Solter, A. (1995). Why do babies cry? *Pre- and Perinatal Psychology Journal, 10*, 21–43.
St. James-Roberts, I., Bowyer, J., Varghese, S., & Sawdon, J. (1994). Infant crying patterns in Manali and London. *Child: Care, Health, and Development, 20*, 323–337.
Tamahori, L. (Director) (1995). *Once were warriors* [Film]. (Distributed by Polygram).
Truijers, A., & Vingerhoets, A.J.J.M. (1999). *Shame, embarassment, personality and well-being.* Second International conference on The (Non)Expressions of Emotions in Health and Disease. Tilburg, the Netherlands.

Unterberg, M.L. (1998). *Huilen en sekseverschillen bij adolescenten: Leeftijdtrends en de rol van menstruatie [Gender differences in crying in adolescents: Age trends and the role of menstruation]*. Master thesis. Tilburg: Department of Psychology, Tilburg University.

Van Tilburg, M.A.L., & Vingerhoets, A.J.J.M. (2000). Menstrual cycle, mood, and crying. *Psychosomatic Medicine, 62*, 146.

Vingerhoets, A.J.J.M. (1995). Huilvragenlijst voor volwassenen [Adult Crying Inventory]. Department of Psychology, Tilburg University, Tilburg, the Netherlands.

Vingerhoets, A.J.J.M., & Becht, M.C. (1997). *International Study on Adult Crying: Some first results*. Poster presented at the Annual Meeting of the American Psychosomatic Society, Santa Fe, NM (Abstracted in *Psychosomatic Medicine, 59*, 85–86).

Vingerhoets, A.J.J.M., Van Geleuken, A.J.M.L., Van Tilburg, M.A.L., & Van Heck, G.L. (1997). The psychological context of crying episodes: Toward a model of adult crying. In: A.J.J.M. Vingerhoets, F.J. Van Bussel, & A.J.W. Boelhouwer (Eds.), *The (non)expression of emotions in health and disease* (pp. 323–336). Tilburg, The Netherlands: Tilburg University Press.

Wagner, R.E., Hexel, M., Bauer, W.W., & Kropiunigg, U. (1997). Crying in hospitals: A survey of doctors', nurses' and medical students' experiences and attitudes. *Medical Journal of Australia, 166*, 13–16.

Wellenkamp, J.C. (1988). Notions of grief and catharsis among the Toraja. *American Ethnologist, 15*, 486–500.

Wellenkamp, J. C. (1992). Variation in the social and cultural organization of emotions: The meaning of crying and the importance of compassion in Toraja, Indonesia. In: D.D. Frank & V. Gecas (Eds.), *Social perspectives on emotion* (Vol. 1, pp. 189–216). Greenwich, CT: JAI Press.

Williams, D.G. (1982). Weeping by adults: Personality correlates and sex differences. *Journal of Psychology, 110*, 217–226.

Williams, D.G., & Morris, G.H. (1996). Crying, weeping or tearfulness in British and Israeli adults. *British Journal of Psychology, 87*, 479–505.

Wood, E.C., & Wood, C.D. (1984). Tearfulness: A psychoanalytic interpretation. *Journal of the American Psychoanalytic Association, 32*, 117–136.

Young, P.T. (1937). Laughing and weeping, cheerfulness and depression: A study of moods among college students. *Journal of Social Psychology, 8*, 311–334.

6 MALE AND FEMALE TEARS: SWALLOWING VERSUS SHEDDING? THE RELATIONSHIP BETWEEN CRYING, BIOLOGICAL SEX AND GENDER

Marrie H.J. Bekker and Ad J.J.M. Vingerhoets

One of the most pervasive stereotypes of sex differences in our culture is that of the emotional, labile women versus the rational, strong man. A more behavior-oriented version of this stereotype is the crying woman alongside the man who, under all circumstances, knows how to control his feelings and to withhold his tears. To what extent does this stereotype reflect reality? Do women really cry more often? And, if so, what explanations can be offered for this difference?

It is the aim of this chapter to review the relationship between crying and gender. First, the relation between crying behavior and biological sex is discussed. The major issue here is the extent to which it is valid to consider crying a predominantly female reaction. In the remainder of the chapter, various explanations for this relationship are critically reviewed. After exploring the possible role of biological factors, we focus on the reasons why men and women cry and on the types of situations in which they cry. Finally, considering crying as a coping mechanism, we investigate how it is used by men and women, respectively.

Defining Crying

Current cultural stereotypes would have us believe that tears are most commonly shed by infants and women. This picture indeed seems to be supported by the scientific literature. Vingerhoets and Scheirs (2000) (see Table 1) identified 14 studies that have yielded evidence that women cry more frequently and more intensely than men. Data collected in the context of the International Study on Adult Crying (see Chapter 8, this volume) once more corroborate this view. Figure 1 presents the distributions of the male and female four-week crying frequencies reported in this project.

Table 1. Summary of studies on gender differences in crying

Article	N	Method/period covered	Frequency	Intensity	Duration	Proneness[1]	Reasons for actual crying[2]	Effects on mood
Young 1937	48 men and 8 women	questionnaire/past 24 hrs	women more frequently	—	—	—	no data on gender differences reported	—
Bindra 1972	25 men and 25 women	questionnaire/ description of recent crying episode	—	women more intensely	women longer	—	women more due to "anguish"; men more due to "elation" and "dejection"	—
Williams 1982	70 men and 70 women	questionnaire/last year	—	women more intensely	—	women always more prone. No gender differences found with regard to evoking situations[3]	—	—
Frey et al. 1983	45 men and 286 women	record keeping for 30 days	women more frequently	women more intensely	no gender difference found	—	data only presented for women	men: 73% felt better; women: 58% felt better afterwards
Lombardo et al. 1983	285 men and 307 women	questionnaire/no reference to specific period ("How often do you cry?")	women more frequently	women more intensely	—	women always more prone. No gender differences found with regard to evoking situations	—	no gender difference in relative importance of feelings, but all feelings stronger in women
Ross and Mirowsky 1984	680 husbands and 680 wives	questionnaire/last week	women more frequently	—	—	—	—	—

Table 1. (*Continued*)

Hastrup et al. 1986	77 husbands and 145 wives wives in younger group. 20 men and 44 women in older group	questionnaire/last year	women more frequently; difference not significant in oldest subjects	—	—	—	“no gender differences for specific causes of crying”. No exact data reported.	—
Kraemer and Hastrup 1986	23 men and 33 women	questionnaire followed by record keeping for 9 weeks	women more frequently	—	—	—	no gender differences found	—
Choti et al. 1987	58 men and 56 women	questionnaire after watching films	women more frequently	—	–	—	—	—
Delp and Sackeim 1987	37 men and 43 women	direct observation: measuring the wetting of filter paper after experimental manipulation of mood	—	post manipalation values only were higher for women. Men and women react differently to mood manipulation	—	—	—	—
Williams and Morris 1996	224 men and 224 women in all. There were equal sized subgroups from two countries	questionnaire/one year in general	women more frequently	women more intensely	women longer	women always more prone. Differences between sexes were smallest for “death of someone close” and for several positive emotions	—	—

Table 1. (*Continued*)

De Fruyt 1997	25 men and 79 women	questionnaire/no reference to specific period	—	—	—	women more prone in general. No data on specific situations reported[3)]	no data reported	NO gender differences found for negative and positive emotions following crying
Wagner et al. 1997	83 men and 169 women (health prof-essionals)	questionnaire/no reference to specific period ("Did you ever cry in the hosp-ital workplace?")	more women had cried than men	—	—	—	—	—
Vingerhoets and Becht[4)]	1687 men and 2280 women (30 countries)	questionnaire/last four weeks	women more frequently	women more intensely	women longer	women always more prone. Differences between sexes were smallest for positive emotions	women more due to "con-flict"; men more due to "loss" and "positive events"	improvement of mood in both men and women; effect somewhat larger in women

from: Vingerhoets and Scheirs (2000)

— aspect of crying not investigated

1) the label "proneness" refers to the power of different situations or emotions to elicit crying. Subjects were asked to indicate how likely it was that they would cry in certain situations.

2) subjects were asked to describe the precipitating factors of the crying episode that had occurred on a recent occasion and that was still vivid in their memories.

3) proneness to cry was erroneously called "weeping frequency" in this study.

4) the data of this large cross-cultural study were first presented at "The international conference on the (non)expression of emotions in health and disease", which was held at Tilburg University (The Netherlands) in August 1996. The data are graphically presented in Figure 1.

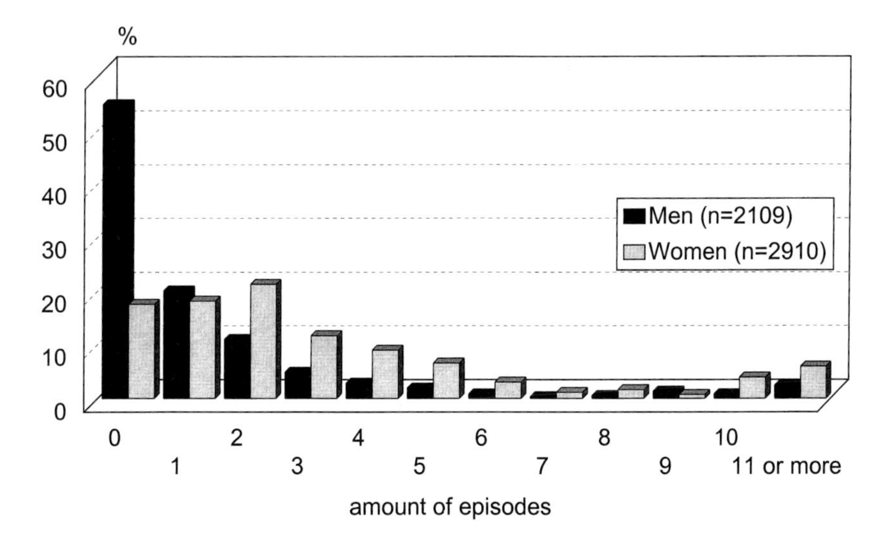

Figure 1. Distributions of the self-reported four week crying frequencies of male and female respondents in the ISAC project.

With respect to *crying proneness*, the (self-reported) likelihood that one will cry when being exposed to a certain situation or experiencing a certain emotion, women usually also show higher scores than men (also see Chapter 8, this volume). According to Vingerhoets and Scheirs, further research is needed to determine whether women also cry for longer durations and whether they report more problems with it than men do (cf. Illovsky, 1991).

Nonetheless, one may wonder whether the available data justify the conclusion that women are substantially more often moved to tears than men. When trying to answer this question, it is necessary to agree upon an exact definition of crying and to have a closer look at the ways in which crying is usually measured. To this end, we make use of a model that is based on Gross and Muñoz (1995), describing two forms of emotion regulation. In this model, emotional cues elicit emotional response tendencies which manifest themselves at the behavioral, the experiential and the physiological level. These response tendencies may or may not be followed by a (manifest, observable) emotional response. The fundamental definitive question then is whether crying is restricted to the actual shedding of observable tears or should also include the (antecedent) feeling of being moved to tears (the emotional response

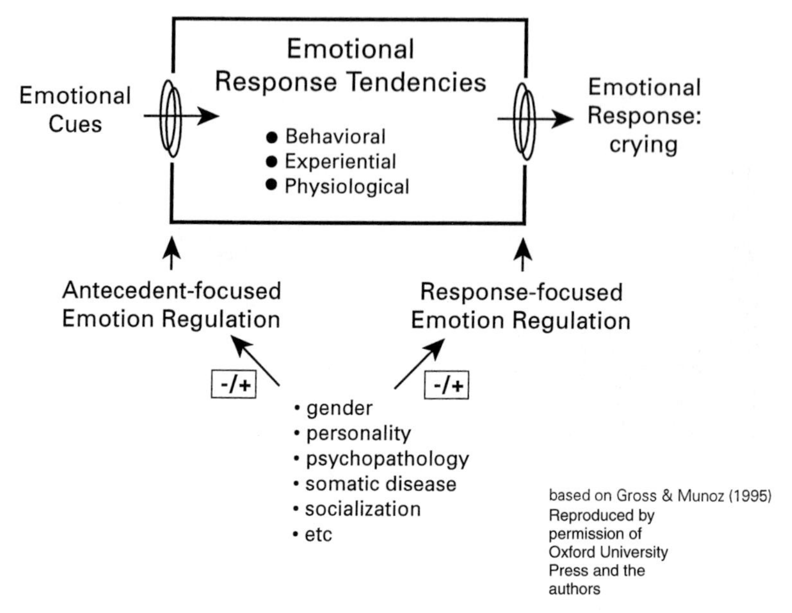

Figure 2. Emotion regulation model after Gross and Muñoz (1995), adapted for crying. Both forms of emotion-regulation may be independently linked to biological, psychological, and cultural factors.

tendency). In the case of feeling moved to tears, the actual shedding of tears might or might not be realized; the tears can be "swallowed." This is what Gross and Muñoz (1995) label as "response-focused emotion regulation"; that is, the emotion program has already been activated, but the individual modulates the response tendencies that have been generated.

What implications may this broader definition of crying have for sex differences in crying? If a narrow definition of crying, just shedding tears, is used, then there is no doubt that women are more frequent criers than men. Although many researchers seem to be aware of the possibility of distinguishing between actual shedding tears and feeling moved to tears, the latter has rarely been attended to. Therefore, it is not known to what degree the sexes differ in feeling moved to tears.

Men generally tend to avoid the expression of "weak" emotions. For instance, there are indications that men are reluctant to admit to (phobic) anxiety. Such masculine avoidance of coming forward with phobic complaints or their differential expression of depression (e.g., Oliver & Toner, 1990; Vredenburg et al., 1986) can result in artificially low prevalences of these disorders in men (Bekker, 1996). Analogously, men may more actively avoid showing their tears (Kottler, 1996) because of current cultural stereotypes of masculinity, or, as Darwin (1872) put it: "[because of] its being thought weak and unmanly by men (....) to exhibit bodily pain by any outward sign" (p. 153). Thus, it is possible that men and women feel moved to tears equally often, but that men prevent themselves from actually shedding tears. On the other hand, women generally experience more negative affectivity (e.g., Gijsbers van Wijk, 1995), and thus seem to have more to cry about. Although the evidence with respect to frequency of sadness in both sexes seems to be inconclusive, women usually report more intense sadness (as well as other emotions) and they more often indicate that they feel helpless and powerless (Fischer, 1993).

In what follows, we would like to further expand on what Gross and Muñoz (1995) refer to as antecedent-focused emotion regulation strategies, "things that we do before an emotion starts, that affect whether a given emotion occurs ... modifying the inputs to the emotional system" (p. 153). Male preventive behavior regarding crying might thus not only involve the swallowing of tears in situations that evoke a crying response, but also the avoidance of situations that men know are likely to make them cry. It could thus be expected that crying frequency and crying proneness in men, as well as their willingness to report these variables in research, would be higher if men felt less inhibited by current social norms concerning masculinity to give in to their tears and to share this with others (see also Lutz, 1999). At least partial support for this view is provided in the study by Ross and Mirowsky (1984), in which it was shown that there were large differences in (self-reported) crying frequency between traditional and non-traditional men. Moreover, Vingerhoets et al. (1992) showed the importance of male self-esteem for (admitting to) crying. In addition, it has been found that male therapists also report crying much more often than men in other professions (see Chapter 4, this volume). Additionally, Kottler (1996) suggests that women at higher levels of responsibility in the workplace cry less often than the average woman, which also seems to imply that the professional context strongly influences if not the experience, then at least the expression of emotions (or, of course, the other way around, how one deals with emotions may influence one's ambitions and may contribute to one's professional success).

Another way of looking at sex differences in crying is to assume that they are an artefact of sex differences in personality. Women generally obtain higher scores on measures of empathy, neuroticism, depression, and distress (e.g., Feingold, 1994; Heller, 1993). One could argue that sex differences in crying would disappear after controlling for such personality features. Peter et al. (2000) explicitly tested this hypothesis in an adult sample, controlling for the Big Five personality dimensions; emotional stability, extraversion, autonomy, conscientiousness, and agreeableness. After correction for these variables, sex differences in crying were maintained. In contrast, Van Tilburg et al. (1999), studying sex differences in crying proneness in a sample of adolescents, reported that boys and girls no longer differed significantly with respect to crying proneness, when controlling for empathy. These data suggest that such sex-bound differences in personality should be taken into account in future research on sex differences in emotional responding.

In sum, it seems plausible to conclude that, in general, women cry more frequently and more intensely, or at least show more *manifest* crying behavior than do men. However, more research is needed to determine whether the sexes differ in crying *potential* or, in terms of Gross and Muñoz' model (1995), the crying response tendency, after having controlled for social desirability and other social factors that may interfere with the receipt of an honest answer to questions about crying. Current cultural norms concerning masculinity might influence not only the actual tendency to shed tears, shorten the duration of crying, and inhibit intense or frequent crying in specific situations, they may also strongly influence the self-reports of men concerning this issue. Therefore, there is a strong need for observational studies, preferably in standardized conditions, to obtain a better insight into the nature of the sex differences.

Developmental Issues

Interestingly, sex differences in crying do not exist in newborns or children up to two years old (Feldman et al., 1980; St. James-Roberts & Halil, 1991). Studies by Kohnstamm (1989), Moss (1967), and Philips et al. (1978) even suggest that male toddlers cry more often than girls. However, note that the absence of a sex difference in newborns and very young children may not necessarily be considered a strong indication for a predominantly socio-cultural determination of later sex differences. Since boys and girls develop their most important physical and hormonal distinctive characteristics in early adolescence (see also Chapter 10, this volume), these same biological factors may influence the sex differences in crying. It is therefore important to obtain a better understanding of the development of these sex differences in children, with adequate attention paid to parental rearing style and other relevant social environmental factors. For example, the higher crying frequency in very young boys can be the result of sociocultural factors already active in this early stage of human development, such as sex-specific interactions between the child and its caregivers and their peers (e.g., Langlois & Downs, 1980; Philips et al., 1978). In addition, one may wonder to what extent the higher prevalence of certain childhood diseases in boys and the higher risk of hurting themselves due to their more rough play activities is responsible for the higher crying frequency in boys at young ages. Future studies should therefore not only focus on mere crying frequency, but also on the kind of situations that make boys and girls cry. Observational studies in kindergarten and classrooms would yield very valuable data in this respect.

Since research on emotional expressions including crying has largely been confined to very young children and adults, there are hardly any published studies that have addressed crying behavior in school children or adolescents (see Chapter 4, this volume). The suggestion that the higher crying frequency of women compared to men becomes first manifest about age thirteen and then further develops (Hastrup, cited in Frey, 1985) has not been substantiated with empirical findings. On the contrary, data collected by Van Tilburg et al. (1999) show that the sex difference is already manifest at the age of eleven and starts to increase from that age, in particular, as a consequence of a decrease in boys' crying frequencies. How can this be explained? At least two alternative interpretations are possible. First, the development of the sex difference in adult crying coincides with the sex typing processes starting to shape girls' and boys' behavior in very specific domains, namely, dating and struggles with parents involving conflicting views of autonomy and independence (Unger & Crawford, 1996). Crying, being part of a more feminine sex role pattern, might thus from that age on become more and more integrated into the girls' and more and more excluded from the boys' behavioral repertory. Moreover, teenage girls may have more reasons to cry than boys of the same age, because of culture's ambivalence about the mature female body and about the role of sexuality in women's lives, which makes puberty more conflictual for girls than for boys (Unger & Crawford, 1996). Second, biological developmental aspects, in particular, hormonal changes, might be an important factor in early adolescence (see Miller Buchanan et al., 1992). However, this view was not supported for girls in the just mentioned study of Van Tilburg et al. (1999), because they failed to find a higher crying proneness and crying frequency in menstruating than in non-menstruating 12- to 14-year old girls. Furthermore, the possible inhibiting role of male sex hormones should also not be excluded beforehand, as will be shown below.

Biological Factors

Which hormonal factors play a role as co-determinants of the sex difference in adult crying? The predominantly female hormone prolactin has been subjected to investigation because of its presumed threshold lowering effects on shedding tears (Frey, 1985; Vingerhoets & Scheirs, 2000; Chapter 10, this volume). Neither the possibility that other female hormones play a role nor that certain male hormones have a threshold-heightening effect on shedding tears has been considered yet, although mood changes during the post-partum period, the premenstrual phase and during the menopause have been associated with (changes in) other female sex hormones, including estrogen and progesteron. In addition, animal data suggest a role for male hormones; studies on the isolation cries of chickens and guinea pigs show that testosterone injections reduced the crying in young animals, whereas the removal of the male sexual glands affected the age-related decline in crying (see Panksepp, 1998). In addition, there is some evidence that empathy, which is presumably related to crying, at least in women (see Van Tilburg et al., 1999), is negatively related to androgen levels (Moir & Jessel, 1995). Until now, there has been a lack of data concerning this issue, which should be addressed more directly with adequate research methods in future studies. Moreover, few surveys have been done with respect to sex differences in crying in the elderly (see Chapter 4, for an exception). This is particularly relevant because of the major changes in blood concentrations of several female sex hormones in post-menopausal women. Other interesting data could be obtained from transsexuals undergoing hormonal treatment and men with prostrate cancer, whose testosterone levels are chemically lowered (cf. Slabbekoorn et al., submitted).

One may wonder why the biological system should support differences in crying of men and women. In this respect, Murube's (1997) views deserve attention. This author comes up with an intriguing but basically untestable hypothesis about the nature of sex differences in crying, emphasizing the role of evolution. He postulates that, in the course of history, men took up their role as defenders of the tribe and group against aggressors. Because tears would interfere too much with their fighting and protecting capacity, giving women and children a feeling of not being safe, men in the course of evolution showed a decrease in crying, which made them better fit for their environment and more attractive as a sexual partner. In this way, the ability to withhold tears contributed to the survival and selection of men.

The need for further research on the role of biological factors in crying is also clear from some findings that seem to challenge the role of biological factors. For example, Horsten et al. (1997) described impressive cross-cultural differences in the percentages of women, mainly arts or social sciences students, reporting an association between crying proneness and phase of the menstrual cycle. Of the total sample of 2018 participants, 44.9% answered positively to the question "Is your crying tendency dependent on the phase of your menstrual cycle?" However, the percentages ranged between as low as 15.4% and 18.9% in countries like China and Ghana, to as high as 69.2% and 68.9% in Australia and Turkey. These large variations suggest that biological (i.e. hormonal) influences are of marginal importance in female crying at best. On the other hand, the role of education and the media in emphasizing the role of hormonal factors may be important in sensitizing women to experience and report an association between phase of the menstrual cycle and mood changes, including crying (Gurevich, 1995). Other evidence limiting the importance of hormonal factors in sex differences concerns the impressive inter-individual and inter-group differences. We already mentioned Kottler (1996), who conjectures, unfortunately without presenting any relevant research findings, that women who are employed in tertiary professions, are less prone to crying than other women. Williams and Morris (1996) suggest that the relatively low crying frequencies of Israeli men and women (as compared to British data) may be a consequence of having been in the army, where one learns to inhibit emotional expression. Male therapists appear to cry more often than men in other professions (see Chapter 4, this volume). There is no reason to assume that these groups differ with respect to levels of hormones that might be relevant in the context of crying, although Frankenhaeuser (1991) provided evidence for a more male-like psychobiological response of women in "male" professions than of women in more traditional female professions.

Given the enormous variation in male and female sex stereotypes among different countries or cultures, further support for a substantial cultural influence on sex differences in crying might be found in the huge variation in the size of the sex-differences in crying frequencies among countries as reported by Vingerhoets and Becht (1997) (see also Chapter 8).

In summary, biological factors might play a role in sex differences in crying, but further research to determine their precise role is needed. The possible relationship between male sex hormones and crying in men and the influence of female sex hormones, in particular, prolactin, in women need further exploration. Of course, the importance of biological factors does not rule out the role of sociocultural factors. We consider a complex interaction between social and biological factors as most promising for investigating the nature of sex differences in shedding tears. Therefore, below we will focus further on relevant sociocultural factors.

Sex-Specific Daily Life and Crying

Scrutinizing possibly relevant psychological or sociocultural factors in determining sex differences in crying, one of the first questions that comes to mind is why women would cry more often than men? Apart from the differential reactions of the social environment and social learning processes, which we will discuss briefly later on, we think that this question should be considered at least at the following three levels: (1) a possible differential exposure to emotional events; (2) a differential appraisal of similar events; and (3) a more dramatic impact of events on physical and/or psychological state variables that may moderate (i.e. facilitate or inhibit) crying, e.g., by affecting the crying threshold.

Concerning the first point, are there any indications that women are exposed more often to negative events? At the level of input or emotional cues, it is difficult to determine whether men and women quantitatively differ in the experience of daily life situations and life-events with a crying-evoking potential. This is partly due to the fact that an event or situation is difficult to distinguish from a subject's appraisal of that event or situation or from its meaning (Gore & Colten, 1991) which, in turn, is narrowly connected with one's preferred coping strategies. And, similar to daily life circumstances, appraisal and coping preferences are also highly sex-specific processes (Nolen-Hoeksema & Girgus, 1994; Ptacek et al., 1994; Vingerhoets & Van Heck, 1990), which makes it difficult to determine whether women have more to cry for than men. A relevant concept in this respect is gender role stress (Eisler & Blalock, 1991; Gillespie & Eisler, 1992). Gender role stress refers not only to the existence of sex-specific stressors, such as being in unemotional relationships or being victimized for women, and, for men, being subordinated to a female superior or failing in competition. The concept also includes a specific rigid way of coping with these stressors. Until now, unfortunately, it is not clear whether the feminine gender role stressors have higher tear-provoking potential than the masculine ones. Empirical evidence has shown that women are more sensitive to others, in particular, to their wishes and problems (Bekker, 1993) and that they define themselves more in relation to others and show more empathy (Chodorow, 1989). Women may also be more affected by the sadness of others. If we consider these "sad" moods and cognitions as concomitants or antecedents of crying, then we might conclude that women indeed are more often exposed to crying-inducing situations. Moreover, in literature, there is some evidence of more frequent exposure to stressful or traumatic events with high tear-provoking potential of women than of men (see Vingerhoets & Scheirs, 2000). This differential exposure might be related to differences in professions, leisure activities, the nature of relationships with significant others, and, more general, differences in life style and interests.

Possible examples of situations that might affect the crying tendency of women are the following: sexual abuse, pregnancy, and widowhood. Rape and sexual abuse are more often experienced by women than by men. They have a major impact upon their health and well-being (Finkelhor & Browne, 1988; Hanson, 1990). However, the literature suggests that such experiences may often result in an inhibition, rather than a facilitation of crying (e.g., Hanson, 1990). Indeed, an inability to experience emotions and to cry has been described repeatedly in victims of sexual abuse and other traumatic experiences, due to a mechanism that would protect them against overwhelming emotions.

Another example of a situation that—particularly older—women are more often than men confronted with is the loss of a beloved person, such as one's partner which is due to the fact that women live longer than men (Verbrugge, 1989). However, here again, the question is whether the experience of such an event has a long lasting impact on one's crying propensity. Moreover, this factor of course cannot explain why young women also cry more often than men. Thus, while both of the above mentioned factors may have their influence on female crying, as determinants of sex differences, they must be dismissed.

Laan (2000, personal communication) conducted a study on crying and the expression of emotions among 207 male and female nurses and police officers. Thirty-nine per cent indicated that their crying proneness had changed considerably and more permanently after having experienced a life event. Frequently reported events were the birth of a child (24.7%), the death of a intimate or family member (23.5%), disease or major injury of the person him/herself (16.0%), and divorce or break of a romantic relationship (13.6%).

Fischer et al. (in press) show that women cry more often in conflict situations than men, and that men cry relatively more often for a loss and for positive reasons. The fact that women cry more often in conflict situations is further substantiated by the finding that among the items that show the largest sex differences on the Adult Crying Inventory is one that reads "I cry when I am involved in quarrels or conflicts". These data are also in agreement with the findings of Rubin (1983) indicating that women show more emotionality in intimate, heterosexual conflicts, and with those of Komter (1985) suggesting that women experience or at least perceive more powerlessness in marital relationships. Note that it is not clear whether women are also exposed more often to conflicts and men more frequently to positive situations or that these kinds of situations simply more likely elicit tears in women and men respectively. In terms of emotions, women's crying episodes are more often accompanied by anger and powerlessness, whereas male tears accompany positive feelings relatively frequently.

As examples of sex differences in leisure activities, one may think of reading specific kinds of books and watching sad movies or television programs (Van der Bolt & Tellegen, 1995–96). These are indications that women actively seek specific stimulation that makes them feel sad, because of a certain pleasure in the experience of feeling moved to tears. In Western culture, a predominantly female practice is watching soap operas and movies with a high tear eliciting capacity. Ang (1985) called the realism of the soap opera Dallas *emotional* realism: "What is recognized as real is not knowledge of the world, but a subjective experience of the world: a 'structure of feeling'" (p. 45). Meier and Frissen (1988) reported that their female respondents mainly watched some specific television series because they consciously sought its emotion-inducing power. One of them reported: "(When I am watching) I am no longer in my own surroundings; I am completely swooning then. I am whining wonderfully. When I am in the right mood, I perfectly like such a swoon movie... handkerchief at hand ... wonderful!" The authors noticed that being alone is probably a sine qua non for these women's enjoyment. A remarkable characteristic of this behavior is that getting emotional and producing tears is actively sought, making use of tear-jerkers or tear-jerking practices.

The professional context may not only determine the exposure to emotional situations, it may also facilitate or inhibit crying (Kottler, 1996). To mention just one example, women are overrepresented in the lower functions of healthcare work, such as nursing, in which the responsibility for other persons who are sick and in other ways dependent evokes a lot of emotion (OECD, 1993). There is indeed some empirical support for the notion that female nurses cry relatively often in their professional context (e.g., Wagner et al, 1997), whereas men in technical professions may be confronted with far fewer emotional situations. It would be interesting to learn more about crying in men and women in similar professions, like police officers, in particular how they respond to emotional events like suicide and traffic accidents with children involved.

We are not aware of studies describing situations in which men's crying frequency exceeds women's, although sports settings, e.g., victory and defeat, elicit high levels of emotional reactivity including shedding tears in men (cf. Kottler, 1996). It is tempting to speculate about explanations for this rather isolated phenomenon; why are men so selective in their "choice" of situations in which shedding tears and emotionality is openly exhibited and shared with other men? What makes sports so special for men? Kottler (1996) speculated that men cry almost uniquely in response to feelings that are part of their core identity, which in his opinion is, in most cultures, framed in their roles of provider, protector, warrior, athlete, husband, father, and team player. Therefore, male tears would be more likely to express pride, bravery, loyalty, victory, and defeat. However, until now empirical evidence for these hypotheses has been lacking.

In short, there is evidence that women and men both willingly and unwillingly differ in their exposure to emotional events. Probably, because women have fewer problems with crying, they are concerned with avoiding emotional situations (both in professions and in their leisure) than men do.

Concerning the second point, the differences in appraisal, Vingerhoets and Scheirs (2000) point to the identification of sex-specific stressors that are perceived to be more stressful by either men or women. In addition, women tend to perceive themselves more often than men as powerless. Finally, there is some evidence (e.g., Jorgensen & Johnson, 1990) that women appraise life stressors as having greater impact on their lives and they indicate that they need more time to recover from them.

Some examples of situations that occasionally may have a longer lasting influence on one's crying threshold are the following. One of the authors (AV) some time ago was approached by two women who reported that their tendency to cry had increased significantly and permanently (at least for two years) after the birth of their first child. Reading newspapers and watching news reports on television displaying sad news facts made their tears run day after day, while there were no signs of mental or physical problems. On the other hand, this author also came across the case of a women who reports never having shed any tear in the past 25 years after having experienced a stillbirth. The final example concerns a man who was very easily moved to tears after having experienced a myocardial infarction. Even the delicious taste of food could stimulate his tears. These examples illustrate that one's psychological and/or physical state can be affected so strongly by certain events, resulting in extreme, either low or high, thresholds for shedding tears. In such cases the nature and intensity of the emotional(?) stimuli seem to play only a marginal role as determinants of crying.

While we are aware of the anecdotal nature of these accounts, they nevertheless invite more research to learn more about these phenomena and the underlying mechanisms. These data further suggest that some situations may not only be regarded as tear provoking events, but possibly also as moderating factors with a more or less permanent influence on crying proneness. However, one should realize that most women before a certain age (e.g., students participating in crying research) only exceptionally have experienced any of these events and still report crying more often than men. The power of these factors to explain sex differences in crying is thus rather limited.

In sum, at this moment it is not yet fully clear how women's and men's daily life experiences influence their crying behavior, although there are data strongly suggesting a relationship between crying and one's educational and professional status, as well as one's leisure activities. Although women definitely seem to have more to cry for, that is, more exposure to situations that, in our culture, are associated with crying, future research should further investigate the role of each of these situational factors. Figure 3 schematically represents the factors that we hypothesize to be relevant as determinants of sex differences in crying.

Crying as a Sex-Specific Coping Behavior

Focusing on crying as a coping behavior, we follow Vingerhoets and Scheirs (2000) in recognizing its emotion-focused (tension relief,

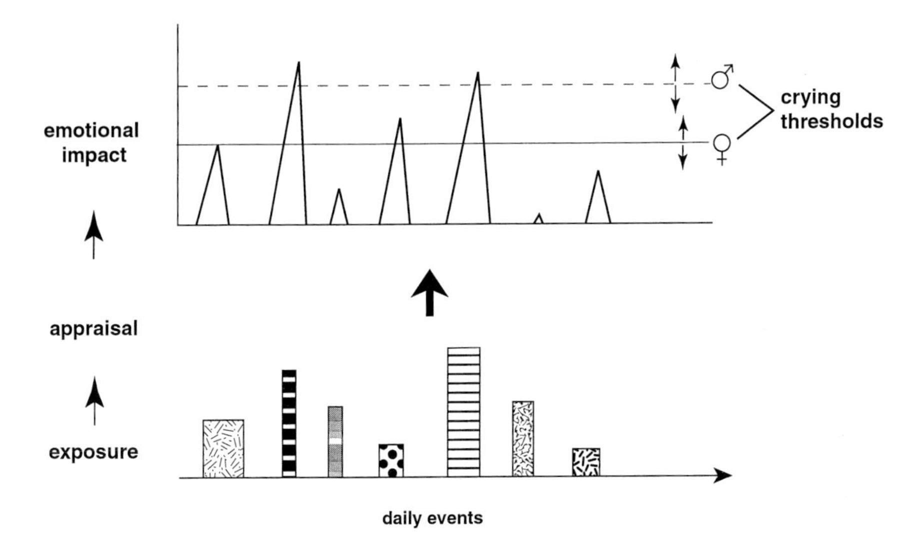

Figure 3. Schematic representation how different daily events may have more or less emotional impact, which may or may not exceed a crying threshold. It is assumed that men generally have a higher threshold than women.

catharsis) and emotional support seeking functions (comfort, compare Lutz, 1999), as well as its problem-focused functions ("manipulating people;" Kottler, 1996). However, we wonder whether these latter are just derivative and secondary functions. We consider the signal to the attachment figure that care is needed as the primary function of crying. In early infancy, crying is part of the dependency system, facilitating attachment behavior. Crying is a clear and effective signal to arouse attention and to mobilize help from the attachment figure (Bowlby, 1973). Perhaps, when people have grown up, crying also gains some functions which seem to become independent of any attachment figure, e.g., when crying alone, although one may wonder whether attachment figures play a role imaginarily, as suggested by Fridlund (1994). Although we are aware of the developmental aspects (see Chapter 3, this volume), we nevertheless wonder if there is any good reason to assume that adult crying has lost its primary function: its function as a signal to (potential) attachment figures, expressing the need for care, e.g., for comfort and to be taken seriously like during conflicts. From this perspective, adult crying can be regarded as a "tend and befriend" response to stress (Taylor et al., 2000). Concerning the assumed problem-focused coping aspects of crying, consider the situation in which a woman cries during a conflict with her husband. Her crying will not solve the conflict, but it may influence the husband's willingness to take her more seriously, to comfort her, which in turn may contribute positively to a problem-solving conversation. Although crying in itself does not solve the conflict and hardly can be seen as problem-focused coping, it nevertheless paves the way and sets the conditions for a more likely solution of the problem.

The Role of the Social Environment

Crying, like any other behavior, is subject to operant learning processes. Reinforcement and punishment of crying behavior, even in the most subtle forms of reward or negative reinforcement such as (lack of) attention, will affect the likelihood of future manifestations. Elaborating from this perspective, existing sex differences in crying may be better understood: current stereotypes of masculinity are not compatible with behaving dependent, infant-like and "weak", which will receive little positive reaction. Regarding women, there is more tolerance regarding emotionality, at least, within "feminine"-labeled situations (Fischer, 1998). Women in general receive more consolation, support, help, comfort etc. when they show their powerlessness by crying than men (see also Fischer, 1993). Given the changing positions of women and men in society and the increasing number of women in traditionally "male" professions, future research should be aimed at the reactions to sex-specific crying in specific professional settings.

In addition, it would be interesting to examine how men and women appraise their own tears in different situations. Kottler (1996) claims that men feel regret and resolve to show more self-control in the future after crying. He further argues that, while women feel sympathetic and accepting toward men who cry, men see women's crying as neutral or positive, but view other men doing so as inappropriate and as a clear sign of weakness. On the other hand, Frijda (1997) suggests that, in particular in conflicts, men may interpret women's crying as a form of blackmail. Moreover, according to Kottler (1996), men show less tolerance toward crying children and also feel more internal disruption when they hear children crying. It is not clear to what degree these descriptions are empirically supported.

There is evidence that women generally feel more confident in expressing emotions, including crying, while especially traditional men (have learned to?) dislike crying (Fischer, 1993; Ross & Mirowsky,

1984). Men apparently can be distracted more easily from their sadness and indulge in escape behaviors, whereas women ruminate more on their sadness, which includes crying, talking to others about their feelings, and identifying causes for them (Nolen-Hoeksema & Girgus, 1994). Nolen-Hoeksema's (1987) assumption that these response styles develop through the socialization of sex-appropriate behavior was confirmed in a study by Conway et al. (1990) showing that higher femininity was associated with more rumination and higher masculinity with more distraction.

Conclusions

Reviewing the literature on sex differences in crying allows for the following conclusions. First, given the existence of a strong sex difference stereotype in terms of crying—men don't, women do—it is remarkable how many aspects of the relationships between sex and crying have rarely been investigated. Nevertheless, there is little doubt that women indeed cry more often than men do. On the other hand, applying the emotion regulation model of Gross and Muñoz (1995), we wonder whether the *tendency to cry* and the first physical manifestations of crying are also in agreement with the sex difference stereotype. Second, cross-cultural differences in crying frequency, certain variations within the female population, in particular those linked to one's profession, and the absence of sex differences at very young ages, suggest that sociocultural factors play a dominant role in sex differences in manifest crying, probably in interaction with hormonal influences starting in puberty. Third, when trying to identify relevant sociocultural factors, two lines for future research might be fruitful. The first emphasizes situational influences. Women experience emotions such as sadness, depression and powerlessness, which are associated with crying, more often. It would be worthwhile to obtain more insight into the role of specific, gender-related situational and interactional influences in generating these emotions (or transforming other emotions such as anger into these ones) and the tendency to cry. For example, what are the functions and effects of crying in conflict situations within intimate relationships or in work situations, at what specific moments is the crying signal given? What is the relationship between couples' and workers' gender role orientation and men's and women's crying behavior? A second line of possible future research focuses on crying as part of coping strategies. The present review showed that women use crying in other and more ways than men, and also seem to elicit other, generally more positive, reactions from the social environment. More research is needed to determine how crying as well as crying inhibition are evaluated by women and men themselves, to what degree their evaluation varies among situations and how their crying behavior as well as its evaluation is related to their general well-being.

Acknowledgement

Parts of this chapter are based on Bekker, M.H.J., & Vingerhoets, A.J.J.M. (1999). Adam's tears: The relationship between crying, biological sex, and gender. *Journal of Psychology, Evolution, and Gender*, *1*, 11–31, Copyright 1999, with permission from Taylor & Francis

References

Ang, I. (1985). *Watching Dallas: Soap opera and the melodramatic imagination.* London: Methuen.

Bekker, M.H.J. (1993). The development of a new Autonomy-scale based on recent insights into gender identity. *European Journal of Personality* *7*, 177–194.

Bekker, M.H.J. (1996). Agoraphobia and gender, a review. *Clinical Psychology Review* *16*, 129–146.

Bekker, M.H.J., & Vingerhoets, A.J.J.M. (1999). Adam's tears: The relationship between crying, biological sex, and gender. *Journal of Psychology, Evolution, and Gender*, *1*, 11–31.

Bowlby, J. (1973). *Attachment and loss. Volume 2: Separation.* New York: Basic Books.

Chodorow, N. (1989). *Feminism and psychoanalytic theory*. Cambridge: Polity Press.

Conway, M., Giannopoulos, C., & Stiefenhofer, K. (1990). *Sex Roles, 22*, 579–587.

Darwin, Ch. (1872/1965). *The expression of emotions in man and animals.* London: John Murray (1965. Chicago: University of Chicago Press).

Eisler, R.M., & Blalock, J.A. (1991). Masculine gender role stress: Implications for the assessment of men. *Clinical Psychology Review, 11*, 45–60.

Feldman, J.F., Brody, N., & Miller, S.A. (1980). Sex differences in non-elicited neonatal behaviors. *Merrill-Palmer Quarterly, 26*, 63–73.

Feingold, A. (1994). Gender differences in personality: A meta-analysis. *Psychological Bulletin, 116*, 429–456.

Finkelhor, D., & Browne, A. (1988). Child sexual abuse: A review and conceptualization. In: G.T. Hotaling, D. Finkelhor, J.T. Kirkpatrick, &

M.A. Straus (Eds.), *Family abuse and its consequences: New directions in research* (pp. 270–284). Newbury Park, CA: Sage.

Fischer, A.H. (1993). Sex differences in emotionality: Fact or stereotype? *Feminism and Psychology, 3*, 303–318.

Fischer, A.H. (1998). *De top m/v: De paradox van emoties [The top m/f: The paradox of emotions]*. Inaugural lecture. University of Amsterdam.

Fischer, A.H., Bekker, M.H.J., Vingerhoets, A.J.J.M., & Becht, M.C. (In press). Femininity, masculinity, and the riddle of crying. In: I. Nyclicek, L. Temoshok,, & A.J.J.M. Vingerhoets (Eds.), *Emotional expression, well-being, and health.* Reading UK: Harwood.

Frankenhaeuser, M. (1991). The psychophysiology of sex differences as related to occupational status. In: M. Frankenhaeuser, U. Lundberg, & M. Chesney (Eds.). *Women, work, and health: Stress and opportunities.* (pp. 39–61). New York: Plenum.

Frey, W.H. (1985). *Crying: The mystery of tears.* Minneapolis, MN: Winston Press.

Fridlund, A.J. (1994). *Human facial expression: An evolutionary view.* San Diego, CA: Academic Press.

Frijda, N.H. (1997). On the functions of emotional expression. In: A.J.J.M. Vingerhoets, F.J. van Bussel, & A.J.W. Boelhouwer (Eds.), *The (non)expression of emotions in health and disease* (pp. 1–14). Tilburg: Tilburg University Press.

Gillespie, B.L., & Eisler, R.M. (1992). Development of the feminine gender role stress scale: A cognitive-behavioral measure of stress, appraisal, and coping for women. *Behavior Modification, 16*, 426–438.

Gore, S., & Colten, M.E. (1991). Gender, stress, and distress: Social-relational influences. In: J. Eckenrode (Ed.) *The social context of coping* (pp. 139–163). New York/London: Plenum Press.

Gross, J.J., & Muñoz, R.F. (1995). Emotion regulation and mental health. *Clinical Psychology: Science and Practice, 2*, 151–164.

Gurevich, M. (1995). Rethinking the label: Who benefits from the PMS construct? *Women & Health 23*, 67–98.

Gijsbers van Wijk, C.M.T. (1995). *Sex differences in symptom perception; A cognitive-psychological approach to health differences between men and women.* Unpublished PhD thesis. University of Amsterdam.

Hanson, R.K. (1990). The psychological impact of sexual assault on women and children: A review. *Annals of Sex Research, 3*, 187–232.

Heller, W. (1993). Gender differences in depression: Perspectives from neuropsychology. *Journal of Affective Disorders, 29*, 129–143.

Horsten, M., Becht, M., & Vingerhoets, A.J.J.M. (1997). *Crying and the menstrual cycle.* Poster presented at the Annual Meeting of the American Psychosomatic Society, Santa Fe, NM (Abstracted in *Psychosomatic Medicine, 59*, 102–103).

Illovsky, M.E. (1991). A comparison of the physical and mental health of doctoral level women scientists. *Journal of College Student Psychotherapy, 5*, 99–110.

Jorgensen, R.S., & Johnson, J.H. (1990). Contributors to the appraisal of major life changes: Gender, perceived controllability, sensation seeking, strain, and social support. *Journal of Applied Social Psychology, 20*, 1123–1138.

Kohnstamm, G.A. (1989). Temperament in childhood: Cross-cultural and sex differences. In: G.A. Kohnstamm, J.E. Bates, & M.K. Rothbart (Eds.), *Temperament in childhood* (pp. 483–508). Chicester: John Wiley & Sons.

Komter, A. (1985). *De macht der vanzelfsprekendheid [The power of naturalness]*. Den Haag: Vuga.

Kottler, J.A. (1996). *The language of tears*. San Francisco, CA: Jossey-Bass.

Langlois, J.H., & Downs, A.C. (1980). Mothers, fathers, and peers as socialization agents of sex-typed play behaviors in young children. *Child Development, 50*, 1237–1247.

Lutz, T. (1999). *Crying. The natural and cultural history of tears*. New York: Norton.

Meier, U., & Frissen, V. (1988) Zwijmelen tussen de schuifdeuren: Televisie kijken. In: L. van Zoonen (Ed.), *Tussen plezier en politiek: Feminisme en media* [Between pleasure and politics: Feminism and the media]. Amsterdam: SUA.

Miller Buchanan, C., Eccles, J.S., & Becker, J.B. (1992). Are adolescents the victims of raging hormones: Evidence for activational effects of hormones and behavior at adolescence. *Psychological Bulletin, 111*, 62–107.

Moir, A., & Jessel, D. (1995). *A mind to crime*. London: Michael Joseph.

Moss, H.A. (1967). Sex, age, and state as determinants of mother-infant interaction. *Merrill-Palmer Quarterly, 13*, 19–36.

Murube, J. (1997). Emotional tearing: A new classification. *Rizal Journal of Ophthalmology, 3*, 27–35.

Nolen-Hoeksema, S. (1987). Sex differences in unipolar depression: Evidence and theory. *Psychological Bulletin, 101*, 259–282.

Nolen-Hoeksema, S., & Girgus, J.S. (1994). The emergence of gender differences in depression during adolescence. *Psychological Bulletin, 115*, 424–443.

Oliver, S.J., & Toner, B.T. (1990). The influence of gender role typing on the expression of depressive symptoms. *Sex Roles, 22*, 775–790.

Organisation for Economic Co-operation and Development (OECD) (1993). *Women, Work and Health*. Synthesis report of a panel of experts. Paris (General Distribution).

Panksepp, J. (1998). *Affective neuroscience. The foundations of human and animal emotions*. New York: Oxford University Press.

Philips, S., King, S., & DuBois, L. (1978). Spontaneous activities of female versus male newborns. *Child Development, 49*, 590–597.

Ptacek, J.T., Smith, R.E., & Dodge, K.L. (1994). Gender differences in coping with stress: When stressors and appraisals do not differ. *Personality and Social Psychology Bulletin, 20*, 421–430.

Ross, C., & Mirowsky, J. (1984). Men who cry. *Social Psychology Quarterly, 47*, 38–146.

Rubin, L.B. (1983). *Intimate strangers*. New York: Harper & Row.

Slabbekoorn, D., Van Goozen, S.H.M., Gooren, L.J.G., & Cohen-Kettenis, P.T. (Submitted). Effects of cross-sex hormone treatment on emotionality in transsexuals.

St. James-Roberts, I., & Halil, T. (1991). Infant crying patterns in the first year: Normal community and clinical findings. *Journal of Child Psychology and Psychiatry, 32*, 951–968.

Taylor, S.E., Klein, L.C., Lewis, B.P., Grunewald, T.L., Gurung, A.R., & Updegraff, J.A. (2000). Biobehavioral responses to stress in females: Tend-and-Befriend, not Fight-or-Flight. *Psychological Review, 107*, 411–429.

Unger, R.K., & Crawford, M. (1996). *Women and gender: A feminist psychology*. New York: McGraw-Hill.

Van der Bolt, L., & Tellegen, S. (1995–96). Sex differences in intrinsic reading motivation and emotional reading experience. *Imagination, Cognition and Personality, 15*, 337–349.

Van Tilburg, M.A.L., Unterberg, M., & Vingerhoets, A.J.J.M. (1999). *Crying in adolescence: The role of gender, menstruation and empathy*. Poster presented at the Second International Conference on the (Non)Expression of Emotions in Health and Disease, Tilburg, The Netherlands.

Verbrugge, L.M. (1989). The twain meet: Empirical explanations of sex differences in health and mortality. *Journal of Health and Social Behavior, 30*, 282–304.

Vingerhoets, A.J.J.M., & Van Heck, G.L. (1990). Gender, coping and psychosomatic symptoms. *Psychological Medicine, 20*, 125–135.

Vingerhoets, A.J.J.M., & Becht, M.C. (1997). *International study on adult crying: Some first results*. Poster presented at the Annual Meeting of the American Psychosomatic Society, Santa Fe, NM (Abstract in *Psychosomatic Medicine, 59*, 85–86).

Vingerhoets, A.J.J.M., & Scheirs, J. (2000). Gender differences in crying: Empirical findings and possible explanations. In: A. Fischer (Ed.), *Gender and emotion. Social psychological perspectives* (pp. 143–165). Cambridge: Cambridge University Press.

Vingerhoets, A.J.J.M., Van den Berg, M., Kortekaas, R.Th., Van Heck, G.L., & Croon, M. (1992). Weeping: Associations with personality, coping, and subjective health status. *Personality and Individual Differences, 14*, 185–190.

Vredenburg, K., Krames, L., & Flett, G.L. (1986). Sex differences in the clinical expression of depression. *Sex Roles, 14*, 37–49.

Wagner, R.E., Hexel, M., Bauer, W.W., & Kropiunigg, U. (1997). Crying in hospitals: A survey of doctors', nurses', and medical students' experience and attitudes. *Medical Journal of Australia, 166*, 13–16.

Williams, D.G., & Morris, G.H. (1996). Crying, weeping or tearfulness in British and Israeli adults. *British Journal of Psychology, 87*, 479–505.

7 PERSONALITY AND CRYING

Ad J.J.M. Vingerhoets, Miranda A.L. Van Tilburg,
A. Jan W. Boelhouwer, and Guus L. Van Heck

There is ample evidence showing wide inter-individual differences in crying frequency and crying proneness. For example, Frey et al. (1983) reported substantial differences in crying frequency, in particular, among women. Other studies, such as those by Williams and Morris (1996) and Vingerhoets and Becht (1997; see also Chapter 6, this volume), also revealed sizeable differences in crying frequency among individuals. Moreover, exposing individuals to an emotional movie shows in rather large inter-individual differences in the kind of scenes that evoke crying as well as in its amount and intensity. Even the most powerful scenes evoked a crying response in no more than approximately one-third of the participating women (see Chapter 5, this volume).

Kottler (1996) described individual as well as group differences in crying when discussing crying in a professional context. According to Kottler, social class, education and occupation are better predictors of crying than cultural and religious background. He asserts that higher educational levels, a flexible definition of gender roles, and having a people-oriented job, increase the likelihood of crying in response to a variety of situations. With respect to the cultural context of work settings, Kottler writes: "Therapists cry. A lot. Engineers don't. Stockbrokers don't, although they often feel like it. Truck drivers don't cry (except in country-and-western songs). Soldiers don't generally cry unless they reach a place of prominence in which they are permitted to do so on behalf of all the others who would like to weep. Nurses cry. Nurses *have* to cry in order to deal with the pain they get so close to. Doctors, however, rarely cry. They insulate themselves from pain—their own as well as that of their patients" (p. 119). Kottler further shows how persons, in particular women, have to play two different roles. At home in the family setting, they are more prone to cry, whereas in the work setting they don't cry because, according to Kottler, they have to take the role of 'the macho warrior' in a setting that can best be characterized as bureaucratic, impersonal and authoritarian, where aggression, power, ambition, and productivity are highly valued.

Differences in crying and crying propensity may be related to *stable* person characteristics reflecting sex, socio-economic status, social learning practices, and education, as well as *state* features like physical condition, fatigue and, for women, phase of the menstrual cycle and pregnancy. In addition, contextual variables like time of the day, cultural expectations, and specific situational context variables are important as facilitating or inhibiting factors (see also Chapter 5, this volume). Moreover, one should be aware of the role of age. We also know some examples showing that the experience of major life events (in these particular cases, becoming a mother and being struck by myocardial infarction; see chapter 6 for more information) may have an enduring effect on emotionality and the propensity to shed tears.

To what extent are personality traits important in explaining these inter-individual differences in crying in addition to or in close association with the just mentioned person and context characteristics? Since personality reflects, among others things, specific tendencies to experience and express certain emotions, one can expect that personality may at least partly explain individual differences in crying behavior. Keltner (1996) describes three ways in which specific emotion tendencies can be related to complex personality features. First, specific emotion tendencies are likely to be associated with habitual beliefs and percepts that are part of personality; second, emotions prepare individuals for actions and consequently these action tendencies characterize the more habitual behavioral repertoire; third, expressive style reflecting personality may mediate the influence of personality on the social environment. Moreover, given the close association between emotional expression and particular coping strategies (cf. Vingerhoets et al., 1993), a further argument in support of the role of personality is provided by the description of coping as "personality at work during stressful conditions."

In this chapter, we focus on the role of personality as a determining factor of crying. First, studies on personality and crying will be reviewed. Second, we will critically scrutinize these studies and put forth suggestions for a more appropriate, theoretically based, approach. Third, a number of personality features that theoretically might be relevant for future research are put forward. We will also briefly discuss the relationship between coping and crying, because of the intertwining of personality with coping and the suggestion that crying can be regarded as a coping mechanism.

Before we continue, however, we want to emphasize the difference between *crying proneness* and actual *crying frequency*. As outlined before (Vingerhoets & Scheirs, 2000), until now the literature has failed to make a distinction between these two different concepts. We suggest reserving the term 'crying frequency' for estimates of actual crying episodes within a given period either measured retrospectively or in a concurrent design. In contrast, 'crying proneness' refers to the general reported propensity to cry in a certain situation. This distinction is important, because people might actively try to avoid or seek situations that will be likely to make them cry, which may explain possible discrepant findings. It seems reasonable to assume that crying proneness reflects a more stable personality characteristic, whereas crying frequency is more dependent on specific environmental conditions. Indeed, Lensvelt and Vingerhoets (unpublished data) obtained data in groups of monozygotic and dizygotic female twins that strongly suggest a genetic basis for crying proneness, but not for crying frequency, which appeared to be more environmentally determined.

In conclusion, we feel that the best way to study the relationship between personality and crying is either to focus on crying proneness, a rather stable characteristic, or to concentrate on actual crying in specific situations. In that case, the nature of the situation is controlled for, allowing a much better estimate of the importance of personality. Following the reasoning by Keltner (1996), we assume that in emotionally strong situations (like the loss of an intimate) personality is particularly important as a determinant of withholding tears, whereas in relatively weak emotional situations, personality may play a major role in triggering the crying response.

Crying and Personality: A Review of the Literature

Williams (1982) was among the first to explore systematically the role of personality in crying. He collected data using a questionnaire that measured crying proneness (rather than crying frequency, as he labeled it), and personality, assessing sex-role, empathy, neuroticism, psychoticism, and extraversion. Significant positive correlations were found between crying proneness and femininity, empathy, and in the case of male respondents, neuroticism. For men, a negative correlation with masculinity was also observed.

Frey (1985) reported an investigation of the relation between crying frequency and personality which failed to yield any systematic associations. In this study, well-being, social potency, achievement, social closeness, stress, alienation, aggression, impulsiveness, danger-seeking, authoritarianism, and absorption (in fantasy) were measured in female participants. For none of the variables, was a significant relationship with crying found.

Choti et al. (1987) did not rely on self-report measures of general crying proneness. Instead they examined personality correlates of crying in a laboratory setting. In pairs, students were exposed to films depicting sad material. Some of the personality characteristics that were assessed were similar to those studied by Williams (1982), i.e., empathy, introversion, extraversion, and masculinity-feminity. In addition, social desirability, ego strength, stress, and moods were assessed. Empathy and extraversion were positively associated with self-reported crying during the films. Male participants showed a strong negative association between ego strength and crying ($r = -.42$), while positive correlations were observed between reports of film-induced crying and pre-film levels of tension and depression. Furthermore, the findings of Schlosser (cited in Goldberg, 1987) revealed a negative relationship between hardiness and crying proneness in women. This is not an unexpected result given the increasing awareness among critics that hardiness can be defined best as the mere absence of neuroticism or negative affectivity (e.g., Funk, 1992; Williams et al., 1992). Hardy women further reported that, when confronted with stressful encounters, crying made them feel even more depressed.

Buss (1992) focused on the purposeful use of crying in order to manipulate others. Kottler (1996) and Frijda (1997) also addressed this issue explicitly. Buss identified 12 tactics that partners use in close relationships and explored the links between applying these tactics and personality. In this context, the manipulation tactic referred to as regression is of particular relevance. High loadings on this factor were obtained by the following items: 'Pout until (s)he does it'; 'Sulk until (s)he does it'; and 'Whine until (s)he does it'. It was found that this tactic was applied more frequently by women (see also chapter 6, this volume). Substantial associations were found between emotional stability and this specific tactic: neurotics tend to use these behaviors to influence their partners to a greater extent than emotionally stable individuals.

Vingerhoets et al. (1993) examined the relationship between crying and both personality attributes and temperament variables. Their first study with a sample of 131 women revealed negative associations between crying proneness (in this study also referred to as crying frequency!), on the one hand, and alexithymia and the coping strategy 'distancing', on the other hand. In addition, positive associations were found between crying proneness and the coping factors, self-blame, day-dreams and fantasies, and expression of emotions/seeking social support. In a regression analysis, 22 per cent of the variance could be explained, with distancing and alexithymia as significant predictors. In a second study, Vingerhoets et al. (1993) reported, for men as well as for women, significant associations between crying proneness and alexithymia (negative) and neuroticism (positive). On the one hand, one would expect a negative relationship with alexithymia, because alexithymic people lack the ability to experience emotions. However, on the other hand, alexithymic individuals are known for their experience of diffuse, unpleasant, and undifferentiated arousal accompanied by crying spells (Taylor et al., 1997). Thus, the negative association between crying proneness and alexithymia needs more attention in future research. In addition, in female subjects, significant negative correlations were observed between crying propensity and the temperament factors strength of excitation and strength of inhibition. Of further interest was the emergence of self-esteem as a significant positive predictor of crying proneness. The higher the individuals' self-esteem, the greater the crying propensity. Of course, it remains to be established whether individuals with high self-esteem actually cry more often or consider admitting that they cry sometimes less threatening.

Recently, De Fruyt (1997) confirmed the association between neuroticism and crying proneness. In addition, he could demonstrate a positive link between extraversion and mood change after crying; extraverts generally felt better after a crying episode than introverts.

In a study by Peter et al. (in press), 48 men and 56 women provided Five-Factor personality data (extraversion, agreeableness, conscientiousness, emotional stability, and autonomy), and ratings of crying frequency and crying proneness (separately for positive and negative reasons). In addition, general mood changes after crying were assessed. As expected, female subjects reported a higher crying frequency over the past 4 weeks (2.6 for women vs. 1.0 for men) and more proneness to cry for both positive and negative reasons. No gender differences were found for mood changes. For women, negative associations were found between emotional stability and all crying indices except mood changes after crying. Gender differences were substantial, even after controlling for differences in personality. Finally, no clear relations were found between the basic personality traits of the five-factor model and self-reported mood changes after crying.

It is interesting to compare these findings with those reported by Keltner (1996) on the facial expression of emotions during a 6-min fragment of a bereavement interview regarding the participant's deceased spouse. First, contrary to expectations, a positive correlation was found between extraversion and the expression of sadness. This finding resulted in an adaptation of the original hypothesis, which stated that extraversion was related to the experience of only positive emotions. Keltner's study suggests that extraversion not only encourages the expression of positive emotions but also of those negative emotions which may facilitate positive social contact. Furthermore, neuroticism was found to be connected with increased facial expression of negative emotion.

Vingerhoets et al. (1998) conducted a study with the aim of obtaining better insight into the relationship between indices of crying and repression. Since previous studies have demonstrated large gender differences in crying behavior, these relationships were studied separately for men and women. One-hundred-and-eight women and 77 men provided data on crying, social desirability, and manifest anxiety. Based on their scores on social desirability and anxiety, the respondents were classified as 'low-anxious', 'high-anxious', 'repressors', or 'defensive high-anxious'. No significant differences were found between the four male groups. In contrast, women showed a strong main effect, with repressive women scoring significantly lower on crying proneness than women in the other three groups. This result could not be due to the repressive women's high social desirability scores alone, since defensive high-anxious women scored significantly higher on the crying proneness scores than repressive women. Similarly, repressors' low crying proneness scores could not be explained solely by their low anxiety-score, as repressors scored significantly lower on crying proneness than low-anxious women. These results suggest that repressive women try to repress crying, attempt to maintain their self-image of emotional control, and to keep the expression of negative affect to a minimum. However, it still has to be established whether the failure to find such an association in men should be explained by their low baseline values of crying proneness or by a genuine gender difference in the appreciation of crying.

In addition to an adolescent version of a crying measure, Unterberg (1998) measured empathy in 217 male and 264 female adolescents. Correlations between crying proneness and empathy were .13 and .39, and .38 and .49 for crying frequency and crying proneness for boys and girls, respectively. Thus, stronger associations were found in female adolescents. In this respect, it is interesting to note that there are indications of a negative link between plasma androgen levels and empathy, which may also be considered as a partial biological explanation for the gender difference in crying (see chapter 6, this volume).

Of further relevance may be the positive association between neuroticism and the occurrence of the maternity blues in post-partum women reported by Kendell et al. (1984).

In conclusion, it is not easy to summarize the rather diverse findings on personality and crying. The reviewed studies used a wide range of different samples. Furthermore, a variety of personality measures have been employed. However, in spite of this, some results appear to be consistent across studies. Repeatedly, substantial positive associations have been found between crying proneness, on the one hand, and neuroticism, extraversion, and empathy, on the other hand. Research outcomes further suggest that the role of personality in crying may differ substantially for men and women. However, due to a lack of a firm theoretical framework, these studies have contributed little to a better understanding of the role of personality as a determinant or moderator of crying. Therefore, in the next section, we will put forth some suggestions for starting more theory-driven research, with adequate attention for the different possible mechanisms behind (not) shedding tears.

Theoretical Models for the Study of Personality in Crying

The studies reviewed above have shown that there are indeed substantial associations between personality and crying. However, most of these studies suffer from serious drawbacks. Most importantly, the reasons for focusing on certain personality features and not on others have not always been made clear. Often, it was decided to use a widely used personality measure without having a clear rationale behind it. In future studies investigators should be more critical and more theory-driven when selecting personality measures. The following two theoretical models of crying might be of help in doing so.

Relevant for the issue at hand is the emotion regulation model of Gross and Muñoz (1995; see Chapter 6, this volume). In this model, a distinction is made between *antecedent-focused* and *response-focused* emotion regulation strategies. Antecedent emotion regulation strategies include the avoidance of situations like funerals or sad movies that are likely to elicit emotional feelings and/or crying. In addition, it also includes cognitive strategies to filter out certain environmental information or to use selective attention. In contrast, response-focused emotion regulation implies the conscious inhibition and suppression of emotional response tendencies such as not letting one's tears go, keeping up a stiff upper lip, which are related, among others, to impression management.

In addition, Kennedy-Moore and Watson (1999) have provided a model that provides a systematic way of thinking about the process of emotional expression. This model can be applied to either a general disposition (not) to express or to specific instances of emotional (non)expression, and to either positive or negative emotions. According to these authors, emotional (non)expression can be thought of as a trait in that there is considerable consistency across time and across situations in the degree to which individuals express their emotions. Following this model, crying can result directly from initial prereflective reactions, bypassing cognitive-evaluative steps, or it can reflect an expressive style that manifests itself at different points in the process of emotional expression, *viz.* conscious perception of the affective reaction (e.g., crying), the labeling and interpretation of this affective response, the evaluation of crying as acceptable or not in terms of beliefs and goals, and finally, at the level of perceiving the social context for expression.

Also, the more elaborated crying model of Vingerhoets et al. (1997) is helpful in illustrating the many ways in which crying and personality may be linked. This model (see also chapter 5, this volume) is based mainly on emotion and stress models. It distinguishes between the following components: (i) objective situations; (ii) (re-)appraisal, resulting in a subjective internal representation of the situations, such as loss, personal inadequacy, conflict, etc; and (iii) an emotional response. Together with (iv) moderating variables (both person factors and situational context variables), these factors determine whether or not a crying response will occur. It is assumed that both inter- and intra-individual differences in crying behavior can—at least partly—be explained by differences in any of these components of the model. The reasons why people do or do not express their emotions may vary widely. Some don't express their emotions because they are not aware of their feelings, others do not shed tears because they do not want others to see them doing so, and a third group may rarely express emotions because they avoid or filter out emotional stimulation. Both emotional stability and psychopathology are expected to correlate negatively with crying, but for quite different reasons. Psychopaths lack empathy which prevents the awareness of the pain and suffering of others. Emotionally stable people do recognize suffering in others but deal with it in a specific way. In a broader context, these differences may be important when exploring the relationship between crying and health.

These models suggest that for a good understanding of the relationship between crying and personality, the following questions should be addressed: (i) what is the relationship between personality and exposure to crying-inducing situations?; (ii) what is the relationship between personality and appraisal?; and (iii) what is the link between personality and the inhibition/facilitation of specific emotional responses, in particular, crying? Finally, in a somewhat broader sense, one may also wonder to what extent persons will adhere to cultural rules and norms and will behave according to the current display rules. Scrutinizing the literature on personality and, respectively, stressor exposure, appraisal, coping, and emotional expressiveness points out that there are many studies showing associations among these variables. Personality features reflecting stress resistance or stress vulnerability like optimism, hardiness, neuroticism or negative affect, depression, self-esteem and internal locus of control have been found to be associated with more than one of the above aspects mentioned (Van Gennip & Vingerhoets, 1998). For instance, neuroticism has been defined in terms of the tendency to *experience* and *express* distress. More precisely, this personality feature can lead to distress through exposing people to a greater number of stressful events, through increasing their reactivity to those events, or through mechanisms unrelated to environmental events (e.g., Bolger & Schilling, 1991). Regarding depression, one can say that depression-prone individuals may be more inclined to avoid people or just to seek the company of peers, watch only specific TV reports, or read more about problematic life situations. In addition, they may be more prone to appraise situations as more threatening and difficult to cope with. Finally, they may have a low threshold for showing typical depressive reactions, like social withdrawal or crying. This example shows that people in general are quite able to manipulate their feelings and mood by transforming their environment and the way they perceive situations. However, it should not be overlooked that crying, conceived of as coping behavior with both emotion- and problem-focused aspects, can only be understood fully when personal intentions, related to goals, belief systems and appraisals, are taken into account.

Personality and Exposure to and Appraisal of Crying-Inducing Situations

According to Mischel (1977, p. 248), "some of the most striking differences between persons may be found not by studying their responses to the same situation but by analyzing their selection and construction of stimulus conditions." So, it is quite plausible that exposure to crying-inducing situations is determined by a wide variety of factors including gender, profession, and personality. To mention some examples, it has been shown that women, more than men, apply mood management strategies by selecting specific films or books (Zillmann, 1988a, b). In addition, work-related situations might be more or less emotion-inducing. The likelihood of being confronted with sad events is rather high in, for example, firemen, policemen, caretakers and health care professionals. Personality factors also play a major role in career selection. Furthermore, there is evidence that the nature of female friendships is different from male friendships and that women's empathic skills might be held responsible for a more extensive confrontation with emotional events, not only in family contexts but also in wider social networks.

Apart from seeking or avoiding emotional situations, other relevant factors imply the appraisal of these situations and the ability to apply certain coping techniques successfully in order to lessen the threat of a situation. For example, applying imagery may be very helpful in minimizing acute pain (and consequent crying) in a medical setting. Other well-known examples are denial and monitoring/blunting. *Denial* is not a personality feature in a strict sense but refers to a specific way of dealing with serious and dramatic events. It is considered to be a coping mechanism, applied when people are confronted with situations that are too painful to face. In particular, when patients are confronted with the diagnosis of a serious chronic condition or a life-threatening disease, denial takes a central role in the early phase of the confrontation with the stressor. Since this defense mechanism is important when striving to put aside emotions temporarily, people applying this defense mechanism will generally cry less when exposed to stressful encounters. This particular defense mechanism is thus a clear example of antecedent-focused emotion regulation. *Monitoring/blunting* relates to the processing and filtering of possible threatening information and as such is also an illustration of antecedent-focused emotion regulation. Equivalents of this dimension are the coping strategies (i) approach-confrontation versus avoidance and (ii) vigilance-sensitization-cognitive confrontation versus repression-cognitive avoidance. One would expect that blunters and/or avoiders cry less often because they tend to protect themselves against the most painful aspects of the situations with which they are confronted.

Empathy refers to the capacity of a person to be sensitive to the needs and suffering of others and to be inclined to share positive and negative emotions with others. The concept has some relationship with susceptibility to emotional contagion. Positive associations between empathy and crying have been found (e.g., Unterberg, 1998; Williams, 1982).

Furthermore, personality features like optimism, internal control, self-esteem, and negative affectivity (including neuroticism, depression, trait anxiety or, their antipode, hardiness) show associations with primary as well as secondary appraisal (e.g., Chang, 1998; Hemenover & Dienstbier, 1996; Peacock & Wong, 1996). Recently, Vingerhoets (unpublished data) designed a study aimed at exploring the links between personality, on the one hand, and experience and appraisal of crying inducing situations and emotions, on the other hand. Forty men (mean age 47.6 years) and 48 women (mean age 44.3 years) indicated (i) whether they had experienced particular situations or feelings in the past four weeks; (ii) to what extent they were moved or affected by these situations/feelings; (iii) to what extent the situations/feelings were evaluated as pleasant; and (iv) to what extent they felt in control over the situations/emotions and knew what to do. In addition, personality measures were administered including a Big-Five inventory, Eysenck's personality questionnaire, an instrument for assessing trait anxiety, and a social desirability scale.

We will summarize some of the findings as far as they are relevant for the current issue, i.e., the relationship between personality, on the one hand, and experience and appraisal indices of the ten highest ranking items that potentially induce crying, on the other hand. A regression analysis with the number of high crying events/feelings as the dependent variable and gender and personality variables as predictors, revealed that women reported more events/emotions. In addition, extraversion and emotional stability emerged as negative predictors, and conscientiousness and autonomy as positive ones. The extent to which one felt touched by the event or feeling could be predicted on the basis of agreeableness (positive) and emotional stability (negative). For controllability, emotional stability was the only significant positive predictor. Pleasantness of the situation or of the feeling could not be predicted on the basis of these personality factors. Thus, the results of this exploratory study suggest that gender and personality, in particular, emotional stability, play an important role as determinants of both the experience of potentially crying inducing situations/feelings and the appraisal of them.

Personality and Response-Focused Emotion Regulation: Speculation and Expectations

Regulation of the emotion response can also be expected to be associated with personality characteristics. Until now there have been only a very limited number of studies addressing the relationship between crying and personality in which the expression of emotions was a crucial element. Below we will discuss some variables and concepts that may yield promising results. We do not pretend to provide an exhaustive enumeration of relevant concepts. We only want to emphasize that it is important to examine the relationship between emotional expressivity in general and crying in particular. A clear relationship should not be taken for granted, because it can be argued that the expression of some emotions can be employed to prevent oneself from getting tears in one's eyes. In addition, one of the most intriguing questions still to be answered is why people express some emotions more than others and, especially with respect to crying, what makes the difference between just experiencing a certain emotion and expressing it with tears.

Emotion self disclosure (Snell et al., 1988) refers to the willingness to display one's feelings and emotions to others. We are not aware of any data on the relationship between this concept and crying proneness or crying frequency.

Emotional expressiveness (King & Emmons, 1990; see also King et al., 1994) focuses on the general tendency to express emotions. Until now, no data have been available, but it would be worthwhile to obtain insight into the relation between more general expressiveness of emotions and crying in particular. Another relevant concept reflecting individual differences in the strength of affective reactions is *affect intensity* (Larsen & Diener, 1987).

Gross and John (1995) developed the Berkeley Expressivity Questionnaire to measure emotional expressivity. Although this instrument was designed to assess the general strength of emotion, emotional response tendencies and the degree to which such tendencies were expressed as behavior, their results showed that it makes sense to differentiate at least between the expressivity of positive and the expressivity of negative emotions. They are now working on a measure that further differentiates the three subscales into specific emotion categories such as joy, amusement, anger and sadness. One may wonder whether crying needs a separate measure or subscale to obtain a reliable and valid indication of crying proneness.

King and Emmons (1990) also identify *the ambivalence over emotional expression* identity concept, which involves reconsiderations with respect to the expression of emotion in terms of what it implies for the image of the individual and how others will appreciate it. No data are available that link this concept to aspects of crying.

Crying can also follow a failure to achieve the goal of emotional control. This goal is held by people who value appearing calm and unemotional at all times and especially in interactions with others (Kennedy-Moore & Watson, 1999). *Emotional control* can be assessed by several instruments.

The Emotion Control Questionnaire (Roger & Najarian, 1989) contains four subscales labeled Emotional Inhibition, Aggression Control, Benign Control, and Rehearsal. One should be aware that these four subscales are not expected to measure a unidimensional concept of 'emotion control'. For crying research in particular, the first subscale may have relevance. A related trait reflecting an extreme form of the goal of emotional self-control is *rationality-antiemotionality* (e.g., Grossarth-Maticek et al., 1985).

Emotional intelligence and *psychological mindedness* both refer to the cognitive skills to monitor and self-regulate emotions effectively (Mayer & Salovey, 1995). This means that emotionally intelligent persons can accurately appraise their emotions and use them in adaptive ways. They are able to comprehend the feelings of others and to make empathic responses. Given the preliminary evidence showing strong and impressive positive associations with several outcomes, it would, of course, be of utmost interest to obtain more insight into how these people deal with crying. Unfortunately, until now no data have been available for examining links with crying.

Crying has also rarely been investigated in relation to defensiveness and repression, except for the above mentioned study of Vingerhoets et al. (1998) which showed a strong negative association between repression and crying proneness for women, but not for men. From a theoretical point of view, these concepts may be attractive, in particular because of their hypothesized relation with psychobiological processes and health.

Most characteristic of the *type C cancer prone personality* as described by Temoshok and Dreher (1992) is the willingness to try to rise above pain and despair, and to spend much effort to keep a happy face for the world. Type C persons are characterized as generally not allowing themselves to indulge in tears. According to these authors, many of these people lost the capacity for crying in childhood, when parents either prohibited, ignored, punished and/or disapproved this behavior.

The *internalizer/externalizer* and *repression* concepts refer directly to the expression of feelings. Repressors and internalizers show little if any overt response, whereas externalizers are more likely to express their feelings (Turvey & Salovey, 1993/94). It has been suggested (cf. Berry & Pennebaker, 1993) that physical and/or sexual abuse often result in the chronic inhibition and repression of emotions. A possible explanation of this is that the display of emotions (be it negative or positive) may trigger an aggressive reaction of a violent father or caretaker. However, the role of temperament factors also cannot be excluded. We are not aware of any studies focusing on this variable in relation to crying. So, all these speculations are based on generalizations from data on facial expression and other forms of emotional display. It is not clear to what extent outcomes obtained in studies on expressiveness also holds for crying.

Personality features like *social desirability* and *impression management* can also be expected to influence crying or withholding tears. However, this holds in particular for specific situations rather than predicting how often a person generally cries. For example, individuals high in social desirability may be expected to comply better with culturally determined display rules (Hochschild, 1979), which may imply more crying in some situations and less crying in other circumstances. A final concept relevant in this respect is *self-monitoring* (Snyder, 1974). Mattutini and Cornelius (1997), in a study involving a sample of college students, found that high self-monitors, who are generally thought of as being highly sensitive to situational cues, indicated that they had significantly more control over whether or not they began crying in a particular situation than did low self-monitors. High and low self-monitors did not differ in whether or not they thought they could stop themselves once they had started crying.

A further concept that we would like to pay attention to is *alexithymia*. This construct has its roots in the work of psychosomatically oriented clinicians (see Taylor et al., 1997). As it is presently defined, the construct is composed of the following characteristics: (i) difficulty in identifying feelings and in distinguishing between feelings and bodily sensations of emotional arousal; (ii) difficulty in describing feelings to other people; (iii) limited imaginary processes and fantasy; and (iv) an externally oriented cognitive style. Some investigators emphasize other features as well, such as infrequent recollection of dreams, paucity of facial emotional expressions, a stiff, wooden posture, and reduplication, i.e., that they see others as duplicates of themselves. In the literature, a distinction has been made between primary and secondary alexithymia. Primary alexithymia refers to a more or less stable trait, whereas secondary alexithymia is considered to result from the exposure to dramatic stressors like rape, sexual and/or physical abuse, and other traumatizing events. *Emotional numbness* is another concept that has been described in this respect. One should be aware that this description of alexithymia does not imply that alexithymic individuals never express emotions. Sometimes they manifest outbursts of weeping, anger, or rage, but they lack the ability to relate them to memories or specific events. As mentioned earlier, Vingerhoets et al. (1993) examined the relationship between alexithymia and crying proneness and found negative associations indicating that alexithymic individuals report a reduced tendency to cry.

To summarize, there are a number of personality concepts in which the (non)expression of emotions is a crucial feature. However, once more we want to emphasize that care should be taken not to extrapolate too easily from the expression of positive to negative emotions and from the more general concept of emotional expressivity to crying in particular. It may be argued that the association between these concepts is weak at best. The expression of emotions has a strong interpersonal effect and may regulate how individuals interact. In this way, the escalation of situations and conflicts may be prevented, even resulting in a possible decrease of exposure to crying inducing situations. Future research has to focus on these specific associations in order to broaden our understanding of the link between emotions, the expression of them, and crying. It would be of particular interest to study crying in response to a standard stimulus (such as a sad movie) as a function of the personality measures described here.

Crying and Coping

What is the precise relationship between crying and coping? There are good reasons to suppose that crying may be regarded as a coping mechanism (see Figure 1), whereas, on the other hand, it can also be considered as an outcome of coping. It has indeed been suggested that crying is some sort of last resort which occurs when all coping efforts have failed and were in vain (Frijda, 1986). On the other hand, it is clear from other chapters (e.g., Chapters 9, 11 and 13, this volume) that crying may serve at least two important coping functions. First, there is much

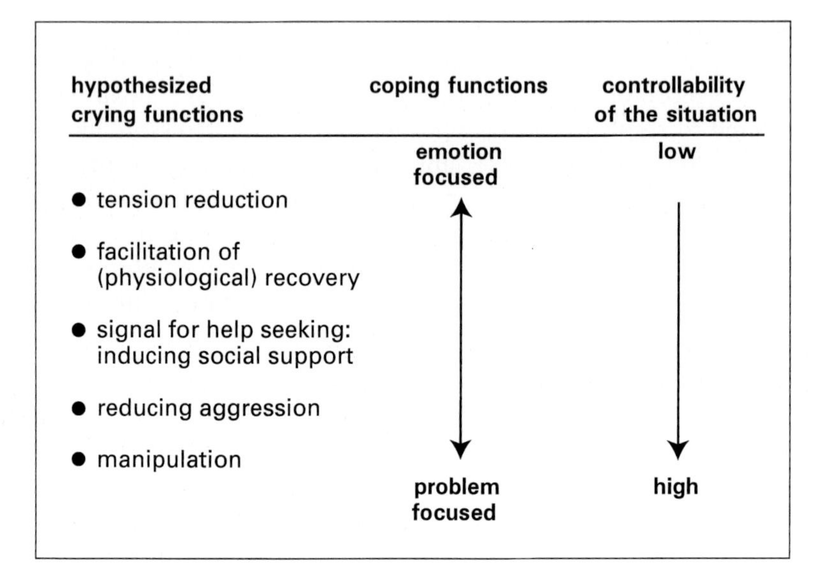

Figure 1. Schematic representation of the postulated functions of crying in relation to the controllability of the situation.

speculation that crying reduces tension and brings relief, which could be considered as a kind of emotion-focused coping, as has been discussed in chapter 13. On the other hand, crying is also a powerful stimulus, which has a large impact on the social environment. It makes people behave in ways they would otherwise not do. Crying generally induces sympathy, comfort and emotional support. It has further been posed that crying *inhibits* aggression, although in some conditions, especially if it does not stop, it can also *elicit* aggression, suggesting a possible role as problem-focused coping. Given this strong impact on the behavior of others, it is tempting to postulate that crying is sensitive to conditioning processes, because it is rewarded by the social environment. An interesting parallel can be made with chronic pain behavior and other somatization complaints, which have also been thought to result from operant conditioning processes. In the clinical literature (e.g., Vingerhoets et al., 1999; see also Chapter 12) there are some illustrations of how excessive crying can be successfully treated applying behavioral therapeutic approaches.

To summarize, crying can be conceived of as a behavior that may serve both an emotion-focused and a problem-focused function. In addition, it may be considered as an ultimate attempt to turn a seemingly uncontrollable situation into a more controllable one.

Conclusion

In the present chapter, we summarized and reviewed the literature on crying and personality, presented new data, and provided a theoretical basis for the continuation of this line of research. There is sufficient reason to hypothesize that personality is an important predictor of whether or not someone will cry in a certain situation, more specifically, whether one will cry in a less emotional situation or whether one will not shed tears in a highly emotional setting. On the other hand, it has to be admitted that research until now was highly a-theoretical, and as a consequence was lacking a clear rationale for examining certain personality attributes. In addition, crying variables were often poorly or ill-defined. Moreover, many studies failed to make a contribution to a better understanding of the nature of the relationship, because there was no attention for the different ways that might have led to the results.

Therefore, we would like to plead for more theoretically-driven research with proper attention to the different aspects of crying. In this chapter, we have shown that actual crying frequency and crying proneness both may be influenced by several personality, social psychological, and biological factors, operant conditioning processes, and the actual situation in which people find themselves, either because they purposefully seek it or in an attempt to avoid other more aversive conditions. A major step forward is Gross and Muñoz' (1995) distinction between antecedent-focused emotion regulation and response-focused emotion regulation (see also Gross & John, 1995, 1997). This model suggests that emotions may be regulated either by manipulating the input via the selection and avoidance of situations, situation modification, cognitive reappraisal or attentional strategies such as distraction and rumination, or by manipulating the output, using strategies to intensify, inhibit, prolong, or curtail ongoing emotional experience, expression, or physiological responding. We expect that a closer investigation of the more specific relation between personality and these two types of emotion regulation can deepen our understanding of the role of personality in the (non)expression of emotions in general and crying specifically.

References

Berry, D.S., & Pennebaker, J.W. (1993). Nonverbal and verbal emotional expression and health. *Psychotherapy & Psychosomatics*, *59*, 11–19.

Bolger, N., & Schilling, E.A. (1991). Personality and the problems of everyday life: The role of neuroticism in exposure and reactivity to daily stressors. *Journal of Personality, 59,* 355–386.

Buss, D.M. (1992). Manipulation in close relationships: Five personality factors in interactional context. *Journal of Personality, 60,* 477–499.

Chang, E.C. (1998). Dispositional optimism and primary and secondary appraisal of a stressor: Controlling for confounding influences and relations to coping and psychological and physical adjustment. *Journal of Personality and Social Psychology, 74,* 1109–1120.

Choti, S.E., Marston, A.R., Holston, S.G., & Hart, J.T. (1987). Gender and personality variables in film-induced sadness and crying. *Journal of Social and Clinical Psychology, 5,* 535–544.

De Fruyt, F. (1997). Gender and individual differences in adult crying. *Personality and Individual Differences, 22,* 937–940.

Frey, W.H. (1985). *Crying, the mystery of tears.* Minneapolis, MN: Winston Press.

Frey, W.H., Hoffman-Ahern, C., Johnson, R.A., Lykken, D.T., & Tuason, V.B. (1983). Crying behavior in the human adult. *Integrative Psychiatry, 1,* 94–100.

Frijda, N.H. (1986). *The emotions.* Cambridge: Cambridge University Press.

Frijda, N.H. (1997). On the functions of emotional expression. In: A.J.J.M. Vingerhoets, F.J. Van Bussel, & A.J.W. Boelhouwer (Eds.), *The (non)expression of emotions in health and disease* (pp. 1–14). Tilburg: Tilburg University Press.

Funk, S.C. (1992). Hardiness: A review of theory and research. *Health Psychology, 11,* 335–345.

Goldberg, J.R. (1987). Crying it out. *Health,* February, 64–66.

Gross, J.J., & John, O.P. (1995). Facets of emotional expressivity: Three self-report factors and their correlates. *Personality and Individual Differences, 19,* 555–568.

Gross, J.J., & John, O.P. (1997). Revealing feelings: Facets of emotional expressivity in self-reports, peer ratings, and behavior. *Journal of Personality and Social Psychology, 72,* 435–448.

Gross, J.J., & Muñoz, R.F. (1995). Emotion regulation and mental health. *Clinical Psychology: Science and Practice, 2,* 151–164.

Grossarth-Maticek, R., Bastiaans, J., & Kanazir, D.T. (1985). Psychosocial factors as strong predictors of mortality from cancer, ischaemic heart disease, and stroke: The Yugoslav prospective study. *Journal of Psychosomatic Research, 29,* 167–176.

Hemenover, S.H., & Dienstbier, R.A. (1996). Prediction of stress appraisals from mastery, extraversion, neuroticism, and general appraisal tendencies. *Motivation & Emotion, 20,* 299–317.

Hochschild, A.R. (1979). Emotion work, feeling rules, and social structure. *American Journal of Sociology, 85,* 551–575.

Keltner, D. (1996). Facial expressions of emotions and personality. In: C. Magai, & S. McFadden (Eds.), *Handbook of emotion, adult development, and aging* (pp. 385–401). San Diego, CA: Academic Press.

Kendell, R.E., Mackenzie, W.E., West, C., McGuire, R.J., & Cox, J.L. (1984). Day-to-day mood changes after childbirth: Further data. *British Journal of Psychiatry, 145*, 620–625.

Kennedy-Moore, E., & Watson, J.C. (1999). *Expressing emotion. Myths, realities, and therapeutic strategies.* New York: The Guilford Press.

King, A.M., Smith, D.A., & Neale, J.M. (1994). Individual differences in dispositional expressiveness: Development of the Emotional Expressivity Scale. *Journal of Personality and Social Psychology, 66*, 934–949.

King, L.A., & Emmons, R.A. (1990). Conflict over emotional expression: Psychological and physical correlates. *Journal of Personality and Social Psychology, 58*, 864–877.

Kottler, J.A. (1996). *The language of tears.* San Francisco, CA: Jossey-Bass.

Larsen, R.J., & Diener, E. (1987). Affect intensity as an individual difference characteristic: A review. *Journal of Research in Personality, 21*, 1–39.

Lensvelt, G., & Vingerhoets, A.J.J.M. (1998). Unpublished data.

Mattutini, A.T., & Cornelius, R.R. (1997). *Weeping as a function of self-monitoring.* Sixty-eighth Annual Meeting of the Eastern Psychological Association, Washington, DC.

Mayer, J.D., & Salovey, P. (1995). Emotional intelligence and the construction and regulation of feelings. *Applied and Preventive Psychology, 4*, 197–208.

Mischel, W. (1977). On the future of personality measurement. *American Psychologist, 32*, 246–254.

Peacock, E.J., & Wong, P.T.P. (1996). Anticipatory stress: The relation of locus of control, optimism, and control appraisals to coping. *Journal of Research in Personality, 30*, 204–222.

Peter, M., Vingerhoets, A.J.J.M., & Van Heck, G.L. (In press). Personality, gender and crying. *European Journal of Personality.*

Roger, D., & Najarian, B. (1989). The construction of a new scale for measuring emotion control. *Personality and Individual Differences, 10*, 845–853.

Snell, W.E., Miller, R.S., & Belk, S.S. (1988). Development of the Emotional Self-Disclosure Scale. *Sex Roles, 18*, 59–76.

Snyder, M. (1974). Self-monitoring of expressive behavior. *Journal of Personality and Social Psychology, 30*, 526–537.

Tamahori, L. (Director) (1995). *Once were warriors [film].*

Taylor, G.J, Bagby, R.M., & Parker, J.D.A. (1997). *Disorder of affect regulation: Alexithymia in medical and psychiatric illness.* Cambridge: Cambridge University Press.

Temoshok, L., & Dreher, H. (1992). *The Type C connection. The mind-body connection to cancer and your health.* New York: Random House.

Turvey, C., & Salovey, P. (1993/94). Measures of repression: Converging on the same construct? *Imagination, Cognition, and Personality, 13*, 279–289.

Unterberg, M.L. (1998). *Huilen en sexeverschillen bij adolescenten: Leeftijdtrends en de rol van menstruatie* [Gender differences in crying in adolescents:

Age trends and the role of menstruation]. Masters thesis. Tilburg: Department of Psychology, Tilburg University.

Van Gennip, M.A.M., & Vingerhoets, A.J.J.M. (1998). Zin en onzin van het begrip stressbestendigheid [Sense and nonsense of the concept stress resistance]. *Gedrag & Gezondheid, 26*, 201–209.

Vingerhoets, A.J.J.M., & Becht, M.C. (1997). *International Study on Adult Crying: Some first results.* Poster presented at the Annual Meeting of the American Psychosomatic Society, Santa Fe, NM (Abstract in *Psychosomatic Medicine, 59*, 85–86).

Vingerhoets, A.J.J.M., & Scheirs, J.G.M. (1999). Gender differences in crying: Empirical findings and possible explanations. In: A. Fischer (Ed.), *Gender and emotion. Social psychological perspectives* (pp. 143–165). Cambridge: Cambridge University Press.

Vingerhoets, A.J.J.M., Meerhof, L., & Van Heck, G.L. (1998). Gender differences in the relationship between repression and crying. *Psychosomatic Medicine, 60*, 98 (abstract).

Vingerhoets, A.J.J.M., Kop, P.F.M., & Labott, S. (1999). Huilen en psychotherapie [Crying and psychotherapy]. *Psychopraxis, 1*, 163–167.

Vingerhoets, A.J.J.M., Cornelius, R.R., Van Heck, G.L., & Becht, M.C. (2000). Adult crying: A model and review of the literature. *Review of General Psychology, 4*, 354–377.

Vingerhoets, A.J.J.M., Van den Berg, M.P., Kortekaas, R.T., Van Heck, G.L., & Croon, M.A. (1993). Weeping: Associations with personality, coping, and subjective health status. *Personality and Individual Differences, 14*, 185–190.

Vingerhoets, A.J.J.M., Van Geleuken, A.J.M.L., Van Tilburg, M.A.L., & Van Heck, G.L. (1997). The psychological context of crying episodes: Toward a model of adult crying. In: A.J.J.M. Vingerhoets, F.J. Van Bussel, & A.J.W. Boelhouwer (Eds.), *The (non)expression of emotions in health and disease* (pp. 323–336). Tilburg: Tilburg University Press.

Williams, D.G. (1982). Weeping by adults: Personality correlates and sex differences. *Journal of Psychology, 110*, 217–226.

Williams, D.G., & Morris, G.H. (1996). Crying, weeping or tearfulness in British and Israeli adults. *British Journal of Psychology, 87*, 479–505.

Williams, P.G., Wiebe, D.J., & Smith, T.W. (1992). Coping processes as mediators of the relationship between hardiness and health. *Journal of Behavioral Medicine, 15*, 237–255.

Zillmann, D. (1988a). Mood management through communication choices. *American Behavioral Scientist, 31*, 327–340.

Zillmann, D. (1988b). Mood management: Using entertainment to full advantage. In: L. Donohue & H.E. Sypher (Eds.), *Communication, social cognition, and affect* (pp. 147–171). Hillsdale, NJ: Erlbaum.

8 CRYING ACROSS COUNTRIES

Marleen C. Becht, Ype H. Poortinga, and Ad J.J.M. Vingerhoets

All healthy newborns cry, expressing hunger, pain, or other forms of distress. However, when growing up, crying behavior changes in important ways. Does culture affect this development, or do adults in various cultures cry for the same reasons and in the same situations? In other words, can adult crying best be seen as a universal phenomenon, or are there important differences among cultural populations? An early attempt to study cultural differences in crying behavior was reported by Borgquist (1906). He collected descriptions of crying episodes from missionaries and ethnologists in various regions of the world, and found the same crying inducing situations as in the reports of two hundred American colleagues who described their own experiences. However, he also noted cultural differences in the frequency of crying: "Tears are more frequently shed among the lower races of mankind than among civilized people" (p. 180). As evidence, Borgquist mentioned the many references to crying in writings about Latin races, Negroes, Indians, Japanese, Samoans, Sandwich Islanders and Maoris. Further he wrote: "Among civilized races there are [also] wide differences," and mentioned the English, who according to Darwin, shed tears much less freely than people from the continent. "Racial variations [in crying] are partly due to custom and, in part, to other causes" (p. 155). To Borgquist's credit it may be added that he emphasized the need for more research before definite conclusions could be reached.

It took more than eighty years before even a beginning was made to systematically compare crying behavior between cultural populations. Szabo and Frey (1991) asked Hungarian and American students to keep a crying diary and concluded that the American men cried more often than Hungarian men (1.5 versus 0.7 times per month). Hungarian women reported an average of 3.1 crying episodes per month and American women 5.3 episodes. In a comparison between British and Israeli students and faculty, Williams and Morris (1996) found substantial cultural differences, although sex was of more importance than the country the subjects lived in. Israelis estimated their number of crying episodes at 4.8 (men) and 17.4 (women) times a year, while the

English reported 8.4 and 31.7 times respectively. Comparing their results with American research by Kraemer and Hastrup (1986), Williams and Morris concluded that Israelis, especially women, reported a low crying frequency. In Kraemer and Hastrup's study, female American students estimated their annual crying frequency at 47.8 times per year, while the men reported only 6.5 episodes per year.

In this chapter, data will be presented on crying frequency, crying proneness and crying inducing situations in 29 countries. These data were collected in the context of the International Study on Adult Crying (ISAC; Vingerhoets & Becht, 1996). Because crying is generally considered an emotional expression, these data should be relevant to the current discussion on the relationship between culture and emotions (e.g., Kitayama & Markus, 1994; Mesquita et al., 1997). The interpretation of findings is dependent on the perspective one holds with respect to this relationship. Therefore, we first present a brief summary of the main viewpoints on the cross-cultural equivalence of emotions. Subsequently, crying will be described in a cultural context and international data will be presented. Finally, the interpretation of the differences found among countries will be elaborated on.

Emotions Across Cultures

Scientific discussions concerning emotion research often focus on the issue of to what extent empirical data can be compared validly across cultures. Two major perspectives on the relation between culture and human behavior can be distinguished: *relativism* and *universalism* (Berry et al., 1992; Scherer & Walbott, 1994). In *relativistic* approaches, human behavior is seen as a product of the culture in which it is found, and can only be understood within the context of that culture. The experience of emotions and emotional behavior is seen as largely dependent on the culture one lives in. For example, Kitayama and Markus (1994) wrote: "Specifically, we wish to establish that emotion can be fruitfully conceptualized as being social in nature." In strongly relativistic emotion research, it is emphasized that the emotions found in one society are not necessarily present in others. For instance, in her ethnographic analysis of the Ifaluk in the Pacific, C. Lutz (1988) describes the emotion "song" (translated as "justified anger") and argues that it is not found in the USA. Another example can be found in Wierzbicka (1994, 1998), who relies on linguistic analyses and interprets subtle differences in the meaning of emotion words as an indication of qualitatively different emotions. For crying, this would imply that people with different cultural backgrounds might cry for different reasons, with different accompanying emotions, or in different circumstances.

A *universalistic* approach emphasizes that humans in any culture have basically the same behavioral repertoire and range of feelings. In this viewpoint, emotions are seen as universal, internal states. Cultural variation is limited to differences in overt behavior reflecting basic emotions. For instance, "anger" and "rage" can be seen as different intensity levels of one and the same basic emotion. Often, the emphasis is on the identification of a limited number of biologically based emotion states. Well-known in culture-comparative research are the six (or seven) basic emotions (i.e., fear, anger, sadness, happiness, disgust, surprise and, recently added, contempt), distinguished by Ekman in terms of muscular patterns in the face (Ekman, 1994; Ekman & Friesen, 1971). Substantial empirical evidence shows that the facial expressions associated with each of Ekman's basic emotions are recognized across a wide range of societies, including illiterate populations, even though recognition tends to be higher in the case of emotions expressed by members of one's own ethnic group than by members of other ethnic groups. Another author arguing for a limited number of basic emotions is Izard (1977, 1994), who also includes the social emotions of guilt and shame, that have no characteristic muscular pattern of expression. As far as crying is concerned, fairly direct evidence comes from the field of human ethology. Because of its obvious physiological aspects, Eibl-Eibesfeldt (1997) posits that crying is fundamentally universal.

Various authors have made attempts to transcend the apparent dichotomy between relativistic and universalistic approaches. Berry et al. (1992) and Scherer and Wallbott (1994) consider cross-cultural differences in emotions as variations on common themes. Their argument is that all behavior has roots in psychological processes which are basically the same for people from any culture. Those basic processes may manifest themselves in quite different expressive behaviors or behavior patterns across cultures. Analogies are greeting rituals (Eibl-Eibesfeldt, 1997) and expressions of politeness (Brown & Levinson, 1987), which occur wherever people interact. Within the diversity of expression there are universal principles of social interaction. It should be noted that cross-cultural comparison always implies some similarity or identity that serves as a standard of comparison for culture-comparative research (Poortinga, 1997). In terms of such a standard, cultural differences can be quantified and perhaps valued. For crying, this means that we have to identify universal aspects across cultures before any differences can be interpreted.

Crying as Emotional Behavior in Cultural Context

Crying is closely related to the experience of emotions; emotions and emotional events can cause tears, emotions are felt during a crying spell, and crying may induce a change in mood (e.g., Cornelius, 1997; Frey, 1985). At least in Western societies, together with sadness, anger and frustration, powerlessness is among the most frequently reported feeling while crying (see also Chapter 5, this volume; Becht & Vingerhoets, 1997; Vingerhoets et al., 1997). However, more pleasant emotions are also associated with tearfulness. Being reunited with intimates may make many people brush away a tear. Having no appropriate behavioral response available seems to be an important determinant (Borgquist, 1906; Frijda, 1986; Vingerhoets et al., 1997, 2000).

Applying a cross-cultural approach to emotion research implies seeking a balance between universal and culture-specific aspects of emotions. Several authors consider emotions as multicomponential processes rather than as categorical events (e.g., Frijda, 1986; Mesquita et al., 1997; Russell, 1991; Scherer & Wallbott, 1994; Vingerhoets et al., 1997). The emotion process includes antecedent events and appraisals, physiological reactions, emotional behavior and regulation (Frijda, 1986). According to Mesquita et al. (1997), for each of these components there is empirical evidence of cultural invariance as well as evidence of cultural variation. Moreover, one finds arguments that the extent of similarity and variation depends on the level of generality (e.g., Mesquita & Frijda, 1992). In ethnographic research, culturally unique events and interpretations are given emphasis, while in a cross-cultural psychological analysis, the researcher looks more for common patterns at more abstract levels.

Antecedent events can differ in frequency of occurrence among cultures, and persons from different cultural populations may be affected differentially by particular kinds of events (Mesquita & Frijda, 1992). For example, there is a wide cultural variety in the frequency of exposure to still-births and the loss of newborns, which may considerably affect the appraisal and how people cope with such an event. In a study by Walbott and Scherer (1986), the reported frequencies of sadness being provoked by empathic experiences differed considerably among eight countries, with Great-Britain, Israel, Spain and Italy reporting high, and France, Switzerland and Belgium low frequencies. Relationship problems did not elicit sadness or grief in Israel as much as in the European countries, whereas bad news did. It is not clear to what extent these differences can be attributed to one particular event, namely that Israel was involved in the Lebanese war at the time of data collection. Asking for emotional events provoking sadness experiences in 27 countries, Scherer et al. (1988) found that the Japanese reported the death of an intimate in only 5% of all sadness antecedents, whereas for the European and American respondents more than 20% off all sadness was provoked by the death of someone. These examples show that there may be a differential exposure to situations, but that differences can also stem from a differential appraisal of the situations that are thought to be relevant for the experience of emotions. Ellsworth (1994) has emphasized that the appraisal of specific situations can be quite different, although dimensions of appraisal tend to be universal.

Obvious bodily features of crying are shedding tears, a lump in the throat, and sobbing, which are found in all groups investigated (Eibl-Eibesfeldt, 1997). Averaged over 37 countries on five continents, crying was reported in 55% of the sad experiences, and for a lump in the throat a similar percentage was reported (Scherer & Wallbott, 1994). In a previous study in eight European countries, it was remarkable that tears were only mentioned as a reaction to sadness, and never as a reaction to joy, fear or anger (Wallbott et al., 1986). Japanese subjects reported fewer physiological symptoms when feeling sad over the death of intimates than did USA and European subjects (Scherer et al., 1988). The question arises whether such differences are due to 'real' differences in the experience of emotion, or because it is socially less desirable in Japan to express sadness in such a way. A possible substantive explanation for the weaker Japanese reaction lies in the religious belief that a deceased person's soul always remains with the family, implying that the separation from a loved person is less final. This would indicate that it is in the *appraisal* of the event that the Japanese differed from the other cultural populations under investigation (Scherer et al., 1988). However, one should realize that this is a *post hoc* interpretation that needs further investigation.

Emotional behavior, including crying, cannot be studied separately from emotion regulation. The experience and (manifest) expression of emotions is censored by regulative rules, which may have strong cultural roots. Gross and Muñoz (1995) make a distinction between antecedent-focused emotion regulation, which is the avoidance or purposefully seeking of emotion inducing events, and response-focused emotion regulation, which refers to the restraint from or the facilitation of emotional expression. In any society there are prescriptions on *how* to behave at specific events, and *for whom* these prescriptions hold. For instance, it may be prescribed whose funerals one has to attend and how one should behave oneself.

To give some concrete examples: certain groups in Indonesia have been reported not to value intense emotions, like anger and sadness (Georges, 1995; Wellenkamp, 1992; 1995), except in specific situations. Accordingly, the expression of these emotions should be controlled. Still, in spite of the negative evaluation of crying among the Toraja in Indonesia, one is allowed and even encouraged to shed tears at funerals (Wellenkamp, 1992). On the Fiji Islands guests at a funeral are not allowed to cry until the body is buried; instead horns are blown (Brockhaus, in Stubbe, 1985). According to Rosenblatt et al. (1976), the Balinese are not supposed to cry during the whole period of mourning. Elsewhere one is obliged to shed tears and sometimes even official, paid mourners are hired. Eibl-Eibesfeldt (1997) describes how among the Yanomami in Venezuela withholding emotions can even result in punishment. Ekman and Friesen (1971) reported a well-known study on display rules, that clearly showed differences in the regulation of emotional expression between Japanese and American students. Japanese students hardly showed any facial expression of emotions in the presence of others. If they thought they were alone and nobody could see them, their emotional expressions corresponded to those of their American fellow students, suggesting that the difference in manifest behavior did not reflect a difference in emotional experience but rather a difference in regulation in a social context.

In conclusion, crying as an emotional process has several aspects such as the physiological concomitants which may be considered as being more universal, whereas others such as regulation (e.g., display rules) might be more culturally relative. Emotional experience and emotional expression cannot be taken as one and the same thing; regulative rules are likely to influence emotional experience and probably even more emotional expression. These rules and effects can differ across cultures, being a potential cause of cultural differences in research outcomes. In what follows, new cross-cultural data on crying behavior, collected in the ISAC study, will be presented.

The ISAC Project

From 1995 to 1998 data were collected on crying behavior among respondents from 35 countries, mainly students (Vingerhoets & Becht, 1996). Only countries with sample sizes of at least 30 subjects per sex within an age range of 16 to 28 years were included in the present analysis (Becht, 1998). This yielded a sample consisting of 1470 men (mean age 24 years) and 2100 women (mean age 23 years) from 29 countries.

The present chapter is limited to the following variables: (i) *estimated crying frequency:* "Can you estimate how often you have cried in the last four weeks?"; (ii) *estimated crying tendency:* "How would you rate your general tendency to cry?"; and (iii) *crying proneness.* This last variable was assessed by a checklist containing 54 items referring to events and emotions that may induce crying (Adult Crying Inventory—ACI; Vingerhoets, 1995; see Appendix).

Concerning crying frequency, women reported crying more often than men: on average, 2.7 times in four weeks, against 1.0 time for men. Less than three percent of all respondents, reportedly cried over ten times (men 0.6%, women 4.4%); their results were left out of the calculation of the mean scores. More than half of the men (55.4%) and 16.7% of the women reported not to have cried at all in the past four weeks. For women who did cry, the modal crying frequency was two times (21.4%), for men it was one time (20.8%). The mean frequencies per country are presented in Table 1.

The second question asked for a self-rating of the respondent's general crying tendency (in comparison to others) on a 10-point scale (1 = I hardly ever cry; 10 = I can cry very easily). This item yielded a mean score of 4.7. Also for this variable, a substantial sex difference was found; men rated themselves on average 3.3, the mean of the women was 5.9.

The participants rated the 54 ACI items on a Likert-scale (1 = I never cry; 7 = I always cry; see Appendix). The average score on this measure, further referred to as crying proneness, was 2.6 (Becht, 1998). In Table 1 the mean scores per country are shown. Figure 1 shows the mean scores per item, averaged over 29 countries, in descending order. It can be seen that women on all items obtained higher scores than men (overall means $M = 3.0$ and $M = 2.1$, respectively). To provide some impression of differences among countries, the scores of Indonesia and the USA are presented in Figure 2. Crying proneness correlated .44 with crying frequency and .66 with estimated crying tendency.

Table 1. Means per country of (four weeks) crying frequency, crying tendency and crying proneness

Country	Crying frequency		Crying tendency		Crying proneness	
	women	men	women	men	women	men
Australia	2.8	1.4	5.8	3.1	3.4	2.1
Belgium	3.0	1.1	5.8	3.0	3.0	2.0
Brazil	2.7	1.0	6.8	4.0	3.5	2.3
Bulgaria	1.8	0.4	5.5	3.6	2.7	2.0
Chile	3.6	1.2	6.2	3.6	3.1	2.1
Finland	3.1	1.4	6.5	3.1	3.3	2.1
Germany	3.2	1.5	6.3	3.9	3.1	2.2
Ghana	1.8	0.6	5.2	3.8	3.0	2.3
Greece	2.9	1.0	5.5	2.9	3.2	1.9
Iceland	1.9	0.6	5.9	2.5	2.6	1.5
India	2.3	0.9	5.0	3.2	2.8	2.2
Indonesia	2.1	0.8	5.6	3.1	3.2	2.2
Israel	2.7	1.3	6.0	2.8	3.0	1.7
Italy	3.2	1.7	6.2	4.4	3.0	2.0
Keya	2.1	1.3	5.3	3.8	3.3	2.7
Lithuania	3.1	0.8	5.9	2.8	2.8	1.9
Malaysia	2.1	0.6	5.4	2.3	3.0	1.8
Nepal	1.9	1.9	5.2	3.5	3.3	3.0
Netherlands	3.5	0.9	5.9	2.7	3.3	2.0
Nigeria	1.4	0.9	4.3	3.5	3.0	2.6
Peru	1.6	0.6	5.4	4.0	3.1	2.7
Poland	3.1	1.0	5.6	2.5	2.9	1.7
Portugal	2.3	0.6	6.3	2.8	2.8	2.0
Romania	2.4	0.9	5.2	3.2	2.6	1.8
Spain	2.8	0.6	6.5	3.3	3.3	2.4
Sweden	3.4	1.4	6.8	3.8	3.4	2.2
Switzerland	3.2	0.7	6.0	2.5	2.9	1.9
Turkey	3.5	1.1	6.2	4.1	3.2	2.3
USA	3.5	1.9	6.2	3.6	3.2	2.3

As has already been outlined before (Chapter 7, this volume), crying proneness and frequency of crying have been often confused in previous reports. Listing situations for which respondents have to indicate whether or not these make them cry, like the 54 items mentioned here, yields a score on crying proneness rather than on crying frequency; a high score does not necessarily mean that such an event often happens, or that the respondent actually cries often in such a situation. A proneness to cry will

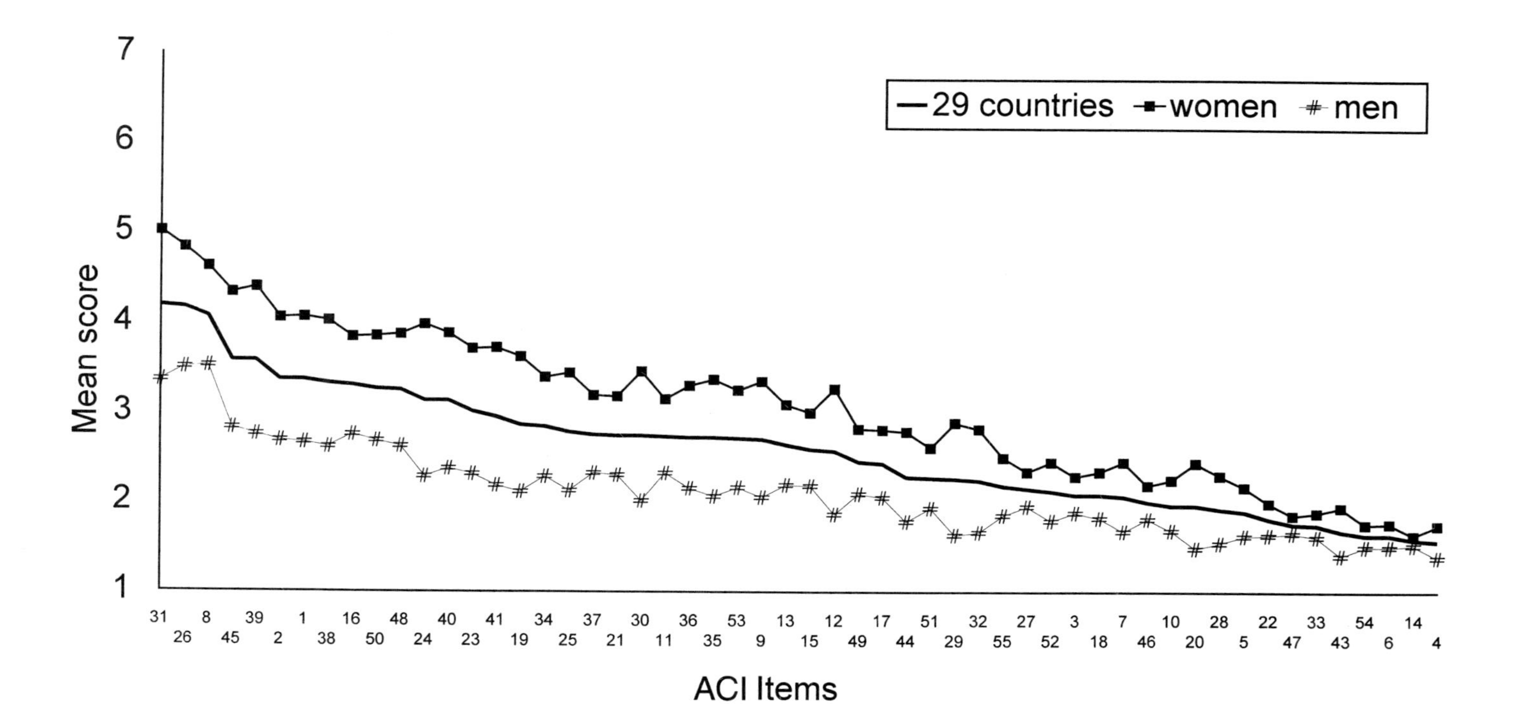

Figure 1. Adult Crying Inventory: Item means of the average scores per country for male and female respondents. The items are presented in descending order of the overall mean.

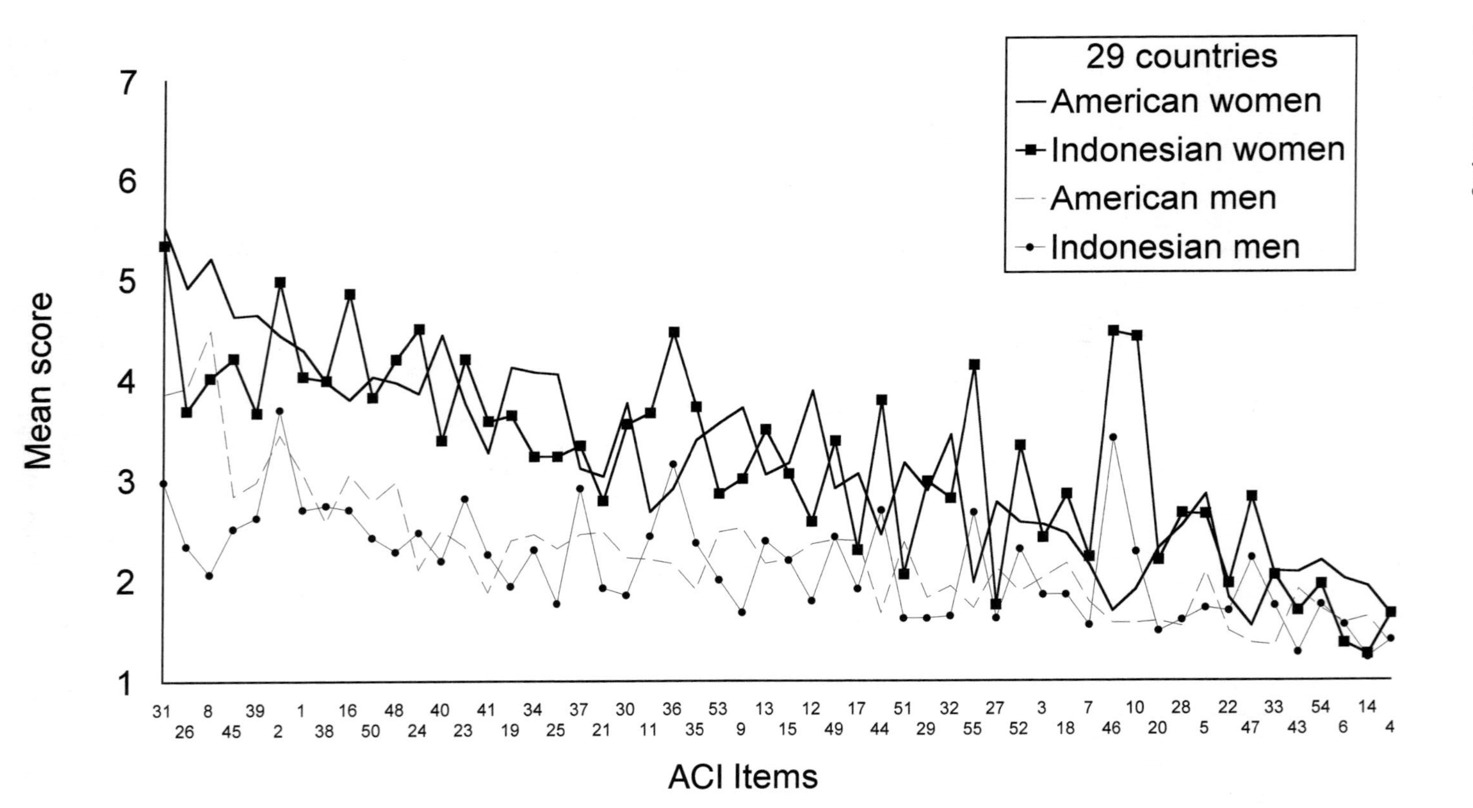

Figure 2. Adult Crying Inventory: Item means of the American and Indonesian respondents. The items are presented in descending order of the overall mean over 29 countries.

often not result in manifest crying behavior, depending on regulation strategies linked with person characteristics, sociodemographic factors and situational features (cf. Vingerhoets et al., 2000). This should be taken into consideration when interpreting data on crying. Crying frequency is often measured using a global estimate of the number of times a respondent has actually cried during a specific period. Depending on the period, this estimate can be more or less accurate. Frey (1985) found corresponding frequencies for a month's estimate and those obtained from keeping a diary, whereas estimates of crying in the past year often yielded much lower and presumably far less reliable values. Note that differences in crying frequency can also indicate differences in occurrence of the events, which may or may not be beyond the control of the individual; e.g., compare events like the death of an intimate with listening to beautiful music or watching a sad movie.

Cross-Cultural Research: Interpreting Differences

Among the women, Brazilians had the highest mean score on crying proneness (3.5), and Romanians and Icelanders the lowest (2.6). Among the men, Nepalese scored highest (3.0) and Icelanders lowest (1.5).

To what extent can we validly conclude from these results that Nepalese men indeed tend to cry somewhat more frequently than Romanian or Icelandic women? First, there are discrepant findings of the crying frequency data. Countries with high or low scores on crying proneness sometimes show non-matching findings with respect to the actual number of crying episodes. But there is more that should prevent us from drawing quick conclusions.

By interpreting cross-cultural differences at face value it is assumed that scores are directly comparable. However, in cross-cultural research, the equivalence of data should not be taken for granted. It is not always clear whether cross-cultural differences indeed represent valid (cultural) differences on a variable, or should rather be attributed to measurement artefacts. Numerous reasons for the existence of measurement artefacts have been identified in the literature (e.g., Van de Vijver & Leung, 1997). Items in questionnaires may differ in shades of meaning, even if the closest possible translation has been selected. Across societies, samples may differ in the tendency to create a good impression and to give more, or perhaps less, socially desirable answers. In the definition and operationalization of theoretical concepts, cultural elements may enter that are not shared by some of the cultural populations involved. For instance, the notion of numerical ability can hardly be expected to have the same meaning in both literate and illiterate societies. In all such instances, the same instrument used in different countries can produce non-comparable or "culturally biased" scores.

Detecting Bias in the ISAC Study

The term "bias" is generic for all nuisance factors threatening the validity of comparisons (Van de Vijver & Leung, 1997, p. 10). Inadequate translation of questions may affect the results of specific items and lead to *item bias*. General response tendencies, such as social desirability, are likely to influence most or all items in an instrument, hence the term *method bias*. Inequality in conceptual notions leads to a third form of bias, known as *construct bias*.

A major issue is to determine to what extent the ISAC data have been affected by bias, preventing the possibility of drawing firm conclusions concerning the validity of observed cultural differences in crying proneness.

Fortunately, there are psychometric techniques available to detect each of these three types of bias or inequivalence. Construct (in)equivalence is usually examined by checking the internal structure of a multiple-item instrument. If correlations between items are very different across cultures, it is difficult to maintain that an instrument pertains to one and the same psychological domain or construct across cultures. On the other hand, a strong correspondence in the pattern of inter-item correlations indicates structural equivalence. Since crying proneness is measured with 54 items, it is possible to compare the internal structure of this scale among countries. To this end, the following analyses were carried out. First, the general factor structure was analyzed using the data from all countries together. A two-factor solution with oblique rotation was most readily interpretable (Becht, 1998; see also Chapter 16, this volume). The first factor, which we will refer to as "Distress," represents unpleasant emotions or situations and contains items such as "I cry when things don't go as I want them to go," "I cry when I am in despair," and "I cry when I feel rejected by others." The second factor "Eustress" reflects more pleasant events, like "I cry when watching the awards ceremony at sporting events such as the Olympics," and "I cry when watching or hearing an admired person." These underlying factors can be seen as reflecting a shared concept of crying-inducing situations across cultures. Men and women did not differ in this respect. Thus, women generally do not tend to cry for different reasons than men.

Next, the results of the factor analyses of the individual countries were compared to the general solution. The factor solutions of Australia, Lithuania and the USA were closest to the common solution. Other countries showed clear differences, indicating that they did not fit the general pattern. These countries were Bulgaria, Finland, Malaysia, Peru, Poland and Turkey. These seven countries are from different parts of the world and it is not easy to identify any particular variable or complex of variables that distinguishes them from the other countries. It can be argued that each country has unique psychological features, but then the question of how the cultural uniqueness of samples with a better fit differs from samples with a poorer fit to the general structure must be answered. Removal of the data of the seven countries did not lead to noticeable change in general structure, demonstrating the consistency of this structure. For the time being, the observed lack of equivalence can tentatively be ascribed better to artefacts than to systematic cultural differences.

However, even perfect or near perfect identity in factorial structure does not necessarily exclude quantitative bias. After all, temperatures on a Fahrenheit and a Celsius scale correlate perfectly, but the scale values are quite different for any given temperature. Unfortunately, an analysis of method bias is not possible with the present data set, since no techniques are available to assess the impact of various sources of method bias with a single instrument. In particular, general response tendencies, such as the use of extreme response scale points, and possible effects of social norms and values (e.g., a "man" does not cry) may well have affected all or many items in a consistent, and hence with the available data, non-detectable fashion. The intercorrelations of the crying proneness scale score with crying frequency and estimated crying tendency may be considered as evidence of the validity of the country differences. However, since these various scores cannot be taken as independent measures (all three variables were part of the same instrument), caution is needed in the interpretation of these quite substantial intercorrelations. In short, the impossibility of ruling out method bias implies that any interpretation of overall cross-cultural differences in reported crying proneness of the various samples can only be tentative.

For the 22 countries that showed an acceptable factorial fit, explorations of item bias yielded significant cultural differences for all

Table 2. The ten most biased items and the countries explaining a proportion of variance larger than one percent

08	I cry over the loss of a love relationship (Indonesia)
10	I cry when I experience disgust or contempt for something or somebody (Brazil, Indonesia, Kenya)
23	I cry when I feel powerless (Lithuania)
39	I cry when I am in despair (Indonesia, Kenya, Nepal, Nigeria)
44	I cry when I am ill (Indonesia, Kenya, Nepal, Nigeria)
46	I cry when practicing religious activities such as praying, listening to preachers reading holy books (Ghana, Indonesia, Nigeria)
47	I cry when I hear the national anthem and/or see the national flag rise (Indonesia, Nepal)
51	I cry when I attend or witness memorial meetings (Brazil, Iceland)
52	I cry when I am reuniting with friends or family members (No specific country)
55	I cry when I have achieved success (Indonesia, Nepal)

items (for details see Becht, 1998), indicating that they could not be taken as culturally unbiased. Further analysis identified ten items which were biased to a substantial extent (see Table 2).

Most biased was "I cry when practicing religious activities such as praying, listening to preachers, reading holy books" (item 46). In the graph for Indonesia (Figure 2), this item shows a clear peak. It is not surprising that being moved by religious experiences is a source of cross-cultural differences among samples of which some are largely agnostic, whereas others are extremely religious. Such incidental differences demonstrate that specific items can elicit quite different reactions in different cultures and, thus, do not represent a similar aspect of the domain of situations that elicit crying across countries. Hence, cross-cultural data obtained with questionnaires including such items cannot be interpreted at face value. The elimination of highly biased items should lead to an item set that better represents crying cross-culturally. At the same time, our findings show that in a study involving many countries it is impossible to find an item set that represents crying in each country in exactly the same way. On the other hand, the identification of biased items can generate ideas for further research on very specific cross-cultural differences in the potential of situations and psychological states to induce crying. Religious experiences are one such example.

Results of a Reduced Data Set

A reduced data set was prepared, deleting the data of the seven countries and ten items for which substantial bias was found (see Figure 3). When descriptive statistics were calculated, the overall mean score on the crying proneness scale differed less than .1 from the results of the initial data set, which means that there was no systematic relationship between the level of item scores and item bias. Moreover, the values of the single item variables (crying frequency and crying tendency), and the intercorrelations between the proneness score and these single item variables did not change to any appreciable extent. In short, the removal of 7 countries and 10 items on the basis of suspicion of inequivalence did not affect the results systematically, indicating that the general results are fairly representative. Since we did not remove countries when only a single item was biased, or items when they were biased in only one country, one should be aware that the remaining items may still be biased for certain countries. In Figure 4 the results of Indonesia and the USA on the reduced data set with 44 items are shown: The order of the items is much the same as in the previous figure (the general structure remained by and large the same), a number of the high and low peaks, especially in the Indonesian graphs, have disappeared (elimination of item bias), and some peaks remained (possible country-specific bias).

Tragic events (item 31) and funerals (item 26) received the highest mean score, similarly so for men (M = 3.45 and 3.57) and women (M = 5.13 and 4.88). Watching sad movies or television programs (item 45) had the third highest mean for women. Making love had the lowest average score of all the situations included in the questionnaire (M = 1.36 and 1.72, for men and women), although for women hearing a happy song rated even lower (M = 1.64). Generally, "distress" items show higher mean scores than "eustress" items for both men and women (M = 2.9 and 2.3 respectively).

Crying and Country Characteristics

The available findings do not provide clear leads about systematic differences between clusters of societies that have certain characteristics in common. It has been suggested that in so-called collectivistic societies the expression of intense emotions is regarded as less appropriate than in individualistic societies (e.g., Matsumoto, 1989). A somewhat

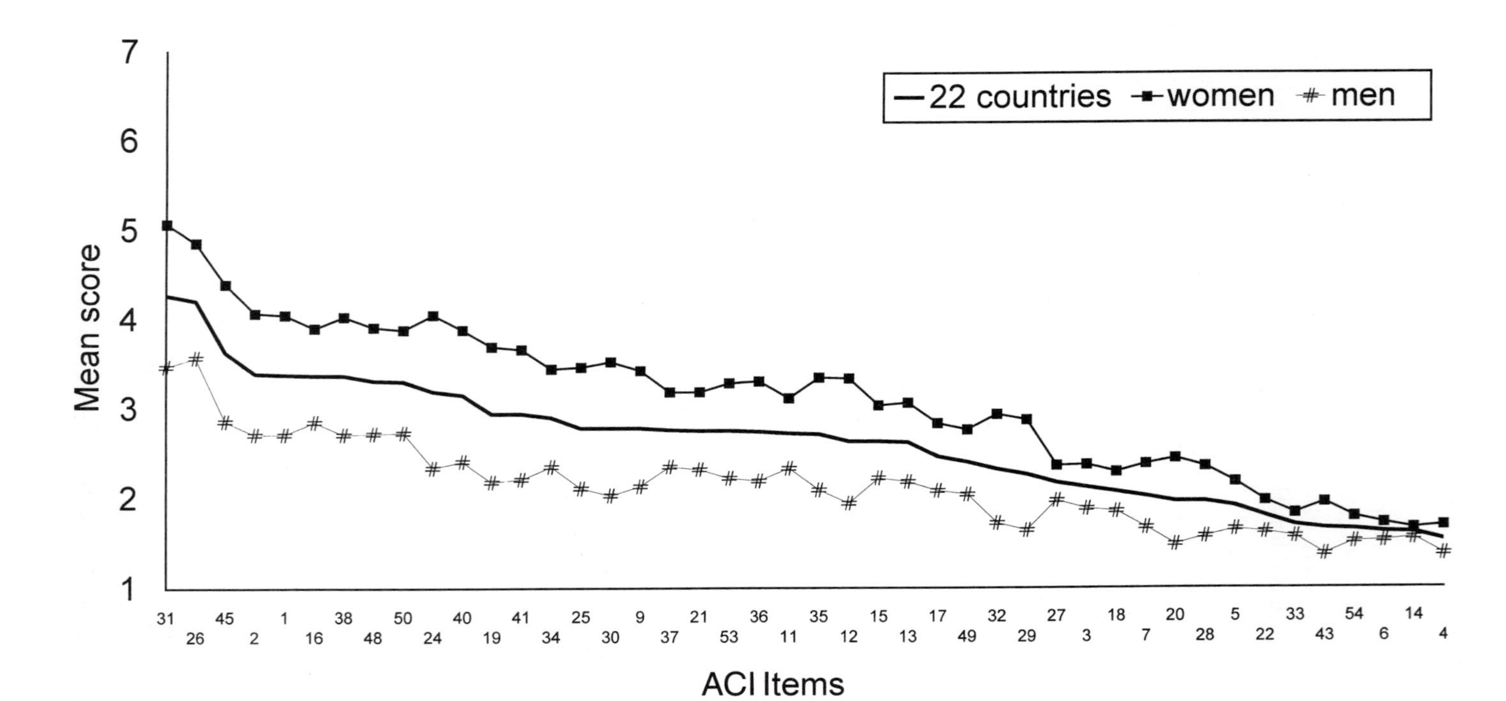

Figure 3. Reduced Adult Crying Inventory: Item means of the average scores per country for male and female respondents. The items are presented in descending order of the overall mean.

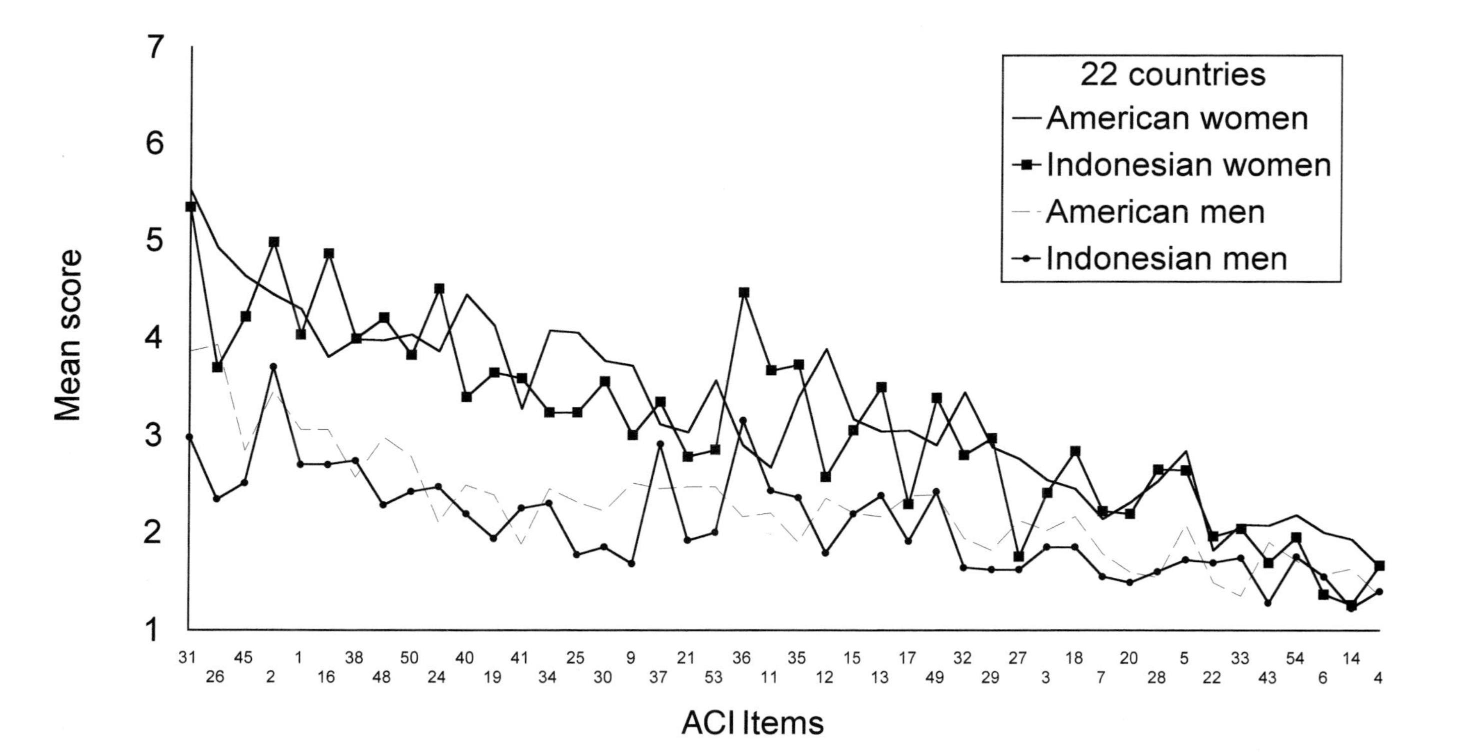

Figure 4. Reduced Adult Crying Inventory: Item means of the American and Indonesian respondents. The items are presented in descending order of the overall mean over 22 countries.

contrasting conjecture has been put forward by Pennebaker et al. (1996), namely, that people in warm climates, where societies generally tend to belong to the collectivistic category, are more emotionally expressive than inhabitants of colder areas. The reason for this might be that living in warm climates takes place more often outside, resulting in more frequent and intense social contact. In order to empirically check on these two hypotheses, we investigated whether the ISAC data supported any of the postulated relationships.

For this purpose, we calculated Pearson's correlations between our crying scores, on the one hand, and, on the other hand, Hofstede's (1983) country scores on the Individualism-Collectivism dimension and temperature indices. Only one significant correlation was found: crying frequency was highly negatively correlated with the average annual temperature ($r = -.75$, for 20 countries), indicating that people living in cold areas cry more often than those living in warm regions, which goes against the prediction of Pennebaker et al. (1996). Crying tendency and general crying proneness failed to show any relationship with these country characteristics. The notion that individuals in so-called collectivistic countries would cry less than those in individualistic countries is not supported by these findings.

It should be noted that studying cross-cultural differences with samples from only a few countries can easily yield misleading outcomes, as may be illustrated when we compare only the Indonesian and US data of the present study. In Indonesia, often considered a collectivistic country, the crying frequency was half the number of times reported in the USA, suggesting that shedding tears is more restrained in the collectivistic country. At the same time, however, no significant differences were found for the other two variables, i.e., crying tendency and crying proneness. Similarly, in a comparison with, for example, Chile (fairly high on collectivism and close to the overall mean tendency) one could easily find a more individualistic country with a higher mean (e.g., Sweden), but also one with a considerably lower mean (e.g., Switzerland). This once more emphasizes the necessity of a sufficiently large sample of countries when one aims to examine universal relationships.

Discussion and Conclusions

In the present chapter we presented cross-cultural data on crying frequency, crying tendency and crying proneness which in some cases seem to yield impressive cross-cultural differences. A crucial question, however, is to what extent these data can be considered as valid and indeed reflective of cultural differences in crying. With respect to crying frequency, it may be noted that retrospective estimates of the number of crying episodes in the last month have been shown to be rather accurate in a North-American sample (Frey, 1985). Assuming this to be valid for all countries included, one has to conclude that apart from some more extreme scores, the overall differences among countries are rather small (Table 1). The same holds for crying tendency.

One should be aware of the limitations of the highly educated samples used in this study, which in many cultures hardly can be seen as representative for the total population. On the other hand, because of the choice of the student samples, we were able to compare reasonably homogeneous groups across cultures.

Since actual crying will be influenced by regulative social norms and rules, which differ considerably among cultures, it is important to consider crying proneness across a range of different situations. Our analyses do not allow an answer to the intriguing questions of whether and to what extent cross-cultural differences in emotional crying reflect differences in the experience of emotion, and to what extent they merely reflect display rules.

Questionnaire data, both overall scores (method bias) and separate items (item bias) are vulnerable to misrepresentation of cross-cultural differences. However, after examination of the effects of bias in ISAC, consistent patterns of findings still emerged. Psychometric techniques were used to control for structural bias. Seven countries differed on a structural level: Bulgaria, Finland, Malaysia, Peru, Poland and Turkey, but we are not able to come up with an explanation for these differences. There appears to be not a single factor or complex of factors that could distinguish them from the other countries in the study. The stable structure of the crying proneness scale in 22 of the 29 countries with an adequate data set suggests that crying proneness covaries among different situations in a similar fashion in many cultures. This finding fits our everyday experience that we can understand people in other societies emotionally, exemplified by the fact that movies and TV soaps, in which the communication of emotions is essential, transfer well across cultural boundaries. On some items there was substantial cross-cultural variance. If a distinction is made between western and non-western countries, there were more seriously biased items in non-western countries. Assuming other artefacts (e.g., translation errors) being equal, the larger extent of bias in non-western samples can be interpreted in two ways, either as due to the overrepresentation of Western countries in the data set, or as a western bias in questionnaire construction. Unfortunately, questionnaire studies by themselves are not suitable for a comparison of overall cross-cultural differences in score levels. Observations, interviews, and other ways of assessment should be added to validate scores. Also, to accurately interpret differences, emotion regulation strategies as described by Gross and Muñoz (1995) should be taken into account in future research.

There can be little doubt that in some societies more than in others, crying, especially by men, is seen as an undesirable sign of weakness and this may well have affected subjects' ratings. The ethnocentric comments of an early researcher like Borquist, quoted in the introduction, reflect a standard of desirable behavior, probably more than anything else. Still, this does not preclude cross-cultural differences in meaning and functionality (Kottler, 1996). Thus, it remains likely that students in Iceland are less prone to cry than those in Brazil. In search for universal features, bias should be avoided or eliminated, as best as possible. However, bias can also point to interesting differences between cultures and hence guide future research. For instance, there were some interesting differences between countries and between items in the extent of bias. Ghana, Indonesia and Nigeria contributed most to the variation of the item about crying during religious practices (item 44), suggesting a possible link with the Islam religion. At the same time, there are other countries where the Muslim religion is widely practiced but that did not show the same effect. Such differences in the experience of religion or religious regulations may be an interesting topic for further research. In this respect, it is interesting that T. Lutz (1999) shows the relevance of crying in a religious context in previous times in western cultures.

No support was found for two hypotheses derived from the literature that posit relationships between the expression of emotions and sociocultural (individualism-collectivism) or ecocultural characteristics (average annual temperature). Previous findings were based on data from fewer countries than the present study, and thus are more likely to have yielded incidental findings. An alternative interpretation for the contradictory findings can be that the crying and emotion processes are not in all respects equivalent, despite their close association. For example, it may be argued that crying is more of an ultimate response, which is *less* likely to occur when emotional expression is socially permissible, in this way possibly preventing the escalation of situations to the point of crying.

Finally, our data once more show that women report higher crying frequencies and a higher crying proneness than men, although the differences in some countries (e.g., Nepal, Nigeria, Kenya) are clearly smaller than in others. At the same time, our findings indicate that men and women by and large cry for the same reasons.

In conclusion, the ISAC study has resulted in a large cross-cultural data set that offers a glimpse at cross-cultural aspects of crying. One should be aware of the major limitations of the included samples (students), and the applied method (self-reports). Nevertheless, some interesting findings have emerged that may be helpful in formulating hypotheses that can be tested in future research. We hope that the ISAC study further stimulates the study on crying in order to come to a better understanding of the cultural dimension of this intriguing, and typical human, emotional expression.

Acknowledgement

Ad Vingerhoets and Marleen Becht want to express their gratitude to the following colleagues who participated as collaborators in the ISAC study and provided data: Cindy Gallois and Matthew Jones (Australia), Harald Wallbott (Austria), Filip de Fruijt, Cathrin Nilsen, Bernard Rimé, Pierre Philippot and Emanuelle Zech (Belgium), Ana Cristina Limongi and Juan Perez-Ramos (Brasil), Elka Todorova (Bulgary), Eugenia Vinet (Chile), Sun Yuming and Zheng Xue (China), Samir Farag and Ahmed El Azayem (Egypt), Jukka Tontti (Finland), Irmela Florin and Stefanie Glaessel (Germany), Charity Akotia (Ghana), Tanya Anagnostopolou (Greece), Suzanne Kulcsar, Imre Janszky and Janos Nagy (Hungary), Fridrik Jonsson and Jakob Smari (Iceland), Y. Bhushan, Sandhya Karpe, and H.L. Kaila (India), Yulia Airyza (Indonesia), Shulamith Kreitler and Shiri Nussbaum (Israel), Pio Enrico Ricci Bitti (Italy), Mary Richardson and Susan Anderson (Jamaica), Josiah Oketch-Oboth (Kenya), Rytis Orinta (Lithuania), Cynthia Joseph (Malaysia), Shishir Subba (Nepal), Mary-Anne Mace (New Zealand), Akinsola Olowu (Nigeria), Alegria Majluf (Peru), Joanna Kossewska (Poland), Felix Neto (Portugal), Adriana Baban (Romania), Debo Akande (South Africa), Chon Kyum Koo (South-Korea), Jose Bermudez (Spain), Töres Theorell and Ulf Lundberg (Sweden), Janique Sangsue (Swiss), Ferda Aysan (Turkey), Moira Macquire (UK) and Randolph Cornelius (USA).

References

Becht, M.C. (1998). *Crying across countries: A comparative study of the tendency and frequency of crying in 35 countries*. Master's thesis, Department of Psychology, Tilburg University, The Netherlands.

Becht, M.C., & Vingerhoets, A.J.J.M. (1997, March). *Why we cry and how it affects mood.* Annual Meeting of the American Psychosomatic Society, Santa Fe, NM (Abstracted in *Psychosomatic Medicine, 59*, 92).
Berry, J.W., Poortinga, Y.H., Segall, M.H., & Dasen, P.R. (1992). *Cross-cultural psychology: Research and applications.* Cambridge: Cambridge University.
Borgquist, A. (1906). Crying. *American Journal of Psychology, 17*, 149–205.
Brown, P., & Levinson, S.C. (1987). *Politeness: Some universals in language use.* Cambridge: Cambridge University Press.
Cornelius, R.R. (1997). Toward a new understanding of weeping and catharsis? In: A.J.J.M. Vingerhoets, F.J. van Bussel, & A.J.W. Boelhouwer (Eds.), *The (non)expression of emotions in health and disease* (pp. 303–321). Tilburg: Tilburg University Press.
Eibl-Eibesfeldt, I. (1997). *Die Biologie des menschlichen Verhaltens* [The biology of human behavior]. Weyarn: Seehamer Verlag.
Ekman, P. (1994). Strong evidence for universals in facial expressions: A reply to Russell's mistaken critique. *Psychological Bulletin, 115*, 268–287.
Ekman, P. & Friesen, W.V. (1971). Constants across cultures in the face and emotion. *Journal of Personality and Social Psychology, 17*, 124–129.
Ellsworth, P.C. (1994) Sense, culture, and sensibility. In: S. Kitayama, & H. Markus (Eds.), *Emotion and culture: Empirical studies of mutual influence* (pp. 23–50). Washington, DC: American Psychological Association.
Frey, W.H. (1985). *Crying: The mystery of tears.* Minneapolis, MN: Winston Press.
Frijda, N.H. (1986). *The emotions.* New York: Cambridge University Press.
Georges, E. (1995). A cultural and historical perspective on confession. In: J.W. Pennebaker (Ed.), *Emotion, disclosure and health* (pp. 11–22). Washington, DC: American Psychological Association.
Gross, J.J., & Muñoz, R.F. (1995). Emotion regulation and mental health. *Clinical Psychology: Science and Practice, 2*, 151–164.
Hofstede, G. (1983). National cultures in four dimensions: A research-based theory of cultural differences among nations. *International Studies of Management and Organisations, 8*, 46–74.
Izard, C.E. (1977). *Human emotions.* London: Plenum.
Izard, C.E. (1994). Innate and universal facial expressions: Evidence from developmental and cross-cultural research. *Psychological Bulletin, 115*, 288–299.
Kitayama, S., & Markus, H.R. (Eds.) (1994). *Emotion and culture: Empirical studies of mutual influence.* Washington, DC: American Psychological Association.
Kottler, J.A. (1996). *The language of tears.* San Francisco, CA: Jossey-Bass.
Kraemer, D.L., & Hastrup, J.L. (1986). Crying in natural settings: Global estimates, self-monitored frequencies, depression and sex differences in an undergraduate population. *Behaviour Research and Therapy, 24*, 371–373.

Lutz, C. (1988). *Unnatural emotions*. Chicago: University of Chicago Press.
Lutz, T. (1999). *Crying. The natural and cultural history of tears*. New York: Norton.
Matsumoto, D. (1989). Cultural influences on the perception of emotion. *Journal of Cross-Cultural Psychology, 20*, 92–105.
Mesquita, B., & Frijda, N.H. (1992). Cultural variations in emotions: A review. *Psychological Bulletin, 112*, 179–204.
Mesquita, B., Frijda, N.H., & Scherer K.R. (1997). Culture and emotion. In: J.W. Berry, P.R. Dasen, & T. Saraswathi (Eds.), *Handbook of cross-cultural psychology: Vol. 2. Basic processes and human development* (pp. 255–297). Boston, MA: Allyn and Bacon.
Pennebaker, J.W., Rimé, B., & Blankenship, V.E. (1996). Stereotypes of emotional expressiveness of Northerners and Southerners: A cross-cultural test of Montesquieu's hypotheses. *Journal of Personality and Social Psychology, 70*, 372–380.
Poortinga, Y.H. (1997). Towards convergence. In: J.W Berry, Y.H Poortinga, & J. Pandey (Eds.), *Handbook of cross-cultural psychology: Vol. 1. Theory and Method* (pp. 347–387). Boston, MA: Allyn and Bacon.
Rosenblatt, P.C., Walsh, R.P., & Jackson, D.A. (1976). *Grief and mourning in cross-cultural perspective*. New Haven, CN: HRAF Press.
Russell, J.A. (1991). Culture and the categorization of emotions. *Psychological Bulletin, 110*, 426–450.
Scherer, K.R., & Wallbott, H.G. (1994). Evidence for universality and cultural variation of differential emotion response patterning. *Journal of Personality and Social Psychology, 66*, 310–328.
Scherer, K.R., Matsumoto, D., Wallbott, H.G, & Kudoh, T. (1988). Emotional experience in cultural context: A comparison between Europe, Japan, and the United States. In: K.R. Scherer (Ed.), *Facets of emotion: Recent research* (pp. 5–30). Hillsdale, NJ: Erlbaum.
Stubbe, H. (1985). *Formen der Trauer: Eine kulturanthropologische Untersuchung* [Manifestations of sorrow: A culture-anthropological study]. Berlin: Dietrich Reimer Verlag.
Szabo, P., & Frey, W.H. (1991, July). *Emotional crying: A cross-cultural study*. Second European Congress of Psychology, Budapest, Hungary.
Van de Vijver, F.J.R., & Leung, K. (1997). *Methods and data analysis for cross-cultural research*. Thousand Oaks, CA: Sage.
Vingerhoets, A.J.J.M. (1995). *Adult Crying Inventory*. Department of Psychology, Tilburg University, the Netherlands.
Vingerhoets, A.J.J.M., & Becht, M.C. (1996, August). *The ISAC study: Some preliminary findings*. International conference on The (Non)Expression of Emotions in Health and Disease. Tilburg University, The Netherlands.
Vingerhoets, A.J.J.M., Van Geleuken, A.J.M.L., Van Tilburg, M.A.L., & Van Heck, G.L. (1997). The psychological context of crying: Towards a model of adult crying. In: A.J.J.M. Vingerhoets, F.J. van Bussel, & A.J.W.

Boelhouwer (Eds.), *The (non)expression of emotions in health and disease* (pp. 323–336). Tilburg: Tilburg University Press.
Vingerhoets, A.J.J.M., Cornelius, R.R., Van Heck, G.L., & Becht, M.C. (2000). Crying: a review of the literature. *Review of General Psychology*, *4*, 354–377.
Wallbott, H.G. & Scherer, K.R. (1986). The antecedents of emotional experiences. In: K.R. Scherer, H.G. Wallbott, & A.B. Summerfield (Eds.), *Experiencing emotion: A cross-cultural study* (pp. 69–83). Cambridge: Cambridge University Press.
Wallbott, H.G., Ricci-Bitti, P., & Bänninger-Huber, E. (1986). Non-verbal reactions to emotional experiences. In: K.R. Scherer, H.G. Wallbott, & A.B. Summerfield (Eds.), *Experiencing emotion: A cross-cultural study* (pp. 98–116). Cambridge: Cambridge University Press.
Wellenkamp, J. (1992). Variation in the social and cultural organization of emotions: The meaning of crying and the importance of compassion in Toraja, Indonesia. In: D.D. Franks, & V. Gecas (Eds.), *Social perspectives on emotion: Vol. 1. A research annual* (pp. 189–216). Greenwich, CT: Jai Press.
Wellenkamp, J. (1995). Cultural similarities and differences regarding emotional disclosure: Some examples from Indonesia and the Pacific. In: J.W. Pennebaker (Ed.), *Emotion, disclosure and health* (pp. 293–311). Washington, DC: American Psychological Association.
Wierzbicka, A. (1994). Emotion, language and cultural scripts. In: S. Kitayama, & H.R. Markus (Eds.), *Emotion and culture: Empirical studies of mutual influence* (pp. 133–196). Washington, DC: American Psychological Association.
Wierzbicka, A. (1998). Angst [Anxiety]. *Culture and Psychology, 4,* 161–188.
Williams, D.G., & Morris, G.H. (1996). Crying, weeping or tearfulness in British and Israeli adults. *British Journal of Psychology, 87*, 479–505.

9 THE SOCIAL PSYCHOLOGICAL ASPECTS OF CRYING

Randolph R. Cornelius and Susan M. Labott

Social psychology may be defined as the "scientific study of how people's thoughts, feelings, and actions are affected by others" (Feldman, 1998, p. 4), where "others" may be "real or imagined, present or absent." Captured within the scope of this very general definition is an enormous and ever widening field of inquiry. Within the diversity of phenomena studied by social psychologists, however, one may discern two very broad domains of inquiry, the study of the influence of the situation or, more broadly, *social context* (and the explanatory/methodological framework that accompanies it, often called "situationism"), and the study of *social cognition*. The study of social context encompasses everything from the influence of social norms and imagined others on individuals' thoughts, feelings, and actions to what happens during face-to-face interaction. Social cognition is concerned with how people make sense of their social environments and, more specifically, how they perceive and make judgments about themselves and others (Fiske & Taylor, 1991).

Social Perception, Social Context and the Study of Crying

While it is possible to separate these two domains of inquiry analytically, in the context of actual social psychological research, of course, the two are often intertwined. Thus, those who study social context cannot study how people are influenced by the situation without considering how people interpret it, and those who study social cognition are often concerned with the ways in which the situation influences peoples' judgments about themselves and others. The distinction is nevertheless a useful one in that it draws attention to the relative emphasis placed on the situation or on cognition in explanations of social behavior and it highlights the different kinds of questions different social psychologists tend to ask.

In the context of the present examination of the social psychology of crying, crying will be considered from within each of the two domains. From the perspective of social cognition we will ask, How are crying persons perceived by others? Considering the social context, we will ask, How does the situation influence crying? How is the experience of crying by the one who cries influenced by the situation? In asking these questions, we must be prepared to acknowledge the reciprocal effects of social context and social cognition in that, for example, social context effects may depend on how crying is perceived by others and how the crying person is perceived by others may depend on the social context within which the crying occurs. Although consideration of the kinds of stimuli that elicit crying falls within the purview of social psychology, discussion of this aspect of social context is reserved for its own chapter (see Chapter 5 in this volume). The present review will also focus only on perceptions of adult crying and the ways in which adult crying is influenced by its social context. Adequate in-depth reviews of the factors that influence infant crying and of perceptions of infant crying are available elsewhere (see, for example, Michelsson et al., 1990; Boukydis, 1985; Lester & Boukydis, 1985; see also Chapter 3 in this volume).

Social Psychology and Emotion

To anticipate one of our conclusions, we would ultimately like to argue strongly for considering crying as a form of *social communication*. As such, crying may be seen as a means by which individuals actively engage their social environments and communicate important aspects of how they perceive themselves, the situation in which they find themselves, and the relationships they share with others in the situation. Our view of crying as a communicative act is consonant with Averill's (1980) definition of emotions as "transitory social roles." Generated by an individual's appraisal of the situation and structured by both biological constraints and learned social rules, emotions, while cognitively mediated, are experienced as beyond the individual's control, yet are based on his or her understanding of the social implications of his or her behavior (see Averill, 1980, p. 312). Similarly, Sarbin (1989) describes emotions as rhetorical acts designed to maintain one's moral identity. The purpose of the emotional enactment is to modify a social relationship. Therefore, while experienced as beyond our control, emotional expressions are actually instrumental in nature.

In a view quite similar to our own, Sadoff (1966) specifically discusses the communicative or help-seeking functions of crying. In his view, adults cry to communicate that a significant aspect of the self is in danger; tears signal the need for others to help manage the situation.

How is Crying Perceived by Others?

General social norms. Any social psychological analysis of how crying persons are perceived by others must begin with an account of the normative context within which crying takes place. General social norms for crying have never really been studied systematically, however. We know that there is considerable cultural variation in the kinds of situations in which people cry and we have some indication that, even though crying appears to be a universal form of emotional expression, cultures vary in the extent to which they value and countenance crying, at least in public (Kottler, 1996; Rosenblatt et al., 1976; Scherer et al., 1988; Wallbott & Scherer, 1988). While Kottler (1996) is certainly correct in his assertion that, "crying exists in a cultural context that is affected by the way events are defined (good, bad, or indifferent), and by the rules for the way feelings should be expressed (stoic restraint, howls, wails, or silent tears)" (p. 108), our knowledge of the cultural norms for crying is based primarily on inference: we assume that there are variations in cultural norms for crying and variations in crying behavior across cultures (see Chapter 8 in this volume) and between social groups within cultures, although we only have reliable information on differences in the norms for male and female crying in the United States (see below).

Nevertheless, it is probably safe to say, based on the level of interest in crying and the way it is portrayed in the popular media, at least in the United States, that most adult crying is generally regarded quite positively (see, for example, Heiman, 1997). In a systematic study of the content of 70 popular articles on crying and weeping published in English-speaking countries between 1848 and 1985, Cornelius (1986) found that most presented crying as a positive, even desirable, form of emotional expression. Indeed, the majority contended that holding back tears in situations that elicit them may have adverse consequences for psychological and physical health (see Chapter 13 in this volume). There is no reason to believe that popular articles on crying published since 1985 portray crying any less positively. Indeed, as Kottler (1996) asserts, attitudes toward crying, especially the crying of men, may be becoming more positive (see, for example, Mason, 1997).

Differences in the Perception of Crying Women and Men

Most of what we know about how crying is perceived by others comes from studies of differences in the evaluations of crying men and women. In a questionnaire study of a sample of university students, Jesser (1982; 1989) found that men were more likely than women to report feeling "confused or irritated" in the presence of another person who was crying. Men also evaluated the tears of another man much more negatively than those of a woman. In contrast, women's tears were seen much more positively. Interestingly, both men and women indicated that it was easier to cry in the presence of a woman than in the presence of a man, possibly because of the more positive attitudes toward crying persons attributed to women.

Cretser et al. (1982) obtained similar results from a large sample of university students in a study that compared how participants thought *other people* regarded crying with how they themselves regarded it. Cretser et al. found that both women and men saw people in general as having basically positive attitudes toward crying women. Their participants indicated that people would probably consider the crying of a woman to be acceptable, would probably sympathize with her, and would feel like helping her. Both women and men saw people in general as having basically negative attitudes toward crying men, with participants indicating that people would probably look down on a crying man and see his crying as inappropriate and a sign of weakness. Men and women differed, however, with regard to how they personally viewed crying men.

Men and women did not differ appreciably in their personal attitudes toward crying women; both indicated acceptance of and sympathy for a crying woman. Men's attitudes toward the crying of other men, however, mirrored what both men and women saw as the general attitude toward crying men. That is, men indicated that a man's crying would be less acceptable to them than would a woman's crying and that they would be more sympathetic toward and feel more like helping a crying woman than a crying man. Men more often reported that they would look down on a crying man and consider his crying inappropriate and a sign of weakness than did women. Women did not differ in their responses to crying men and women. They indicated they would feel acceptance and sympathy toward a crying man and would feel like helping him, just as they would a crying woman.

Plas and Hoover-Dempsey (1988), in a questionnaire/interview study of the employees of a large company (presumably in the United States), found that the majority of those in their sample indicated that they would feel like comforting and would be "moved" by the tears of a colleague crying in the office, whether male or female. Men, however, were more likely to report that they would feel awkward, confused, and manipulated by another's crying, while women were more likely to report that they would feel helpless.

In a laboratory study involving university students conducted by Staebler Tardino (1996), participants read a script of an employee expressing sadness, anger, or no emotion in a conversation with a co-worker in the same office. Sadness was described in the script as being expressed through speech content, vocal tone, and tears. In this situation, type of emotional expression significantly influenced how the employee was evaluated. Specifically, individuals expressing sadness were judged to be more passive, lacking in control, and experiencing more negative work relationship consequences relative to those who expressed no emotion. Overall, those expressing either sadness or anger in the workplace were perceived more negatively than individuals who expressed no emotion. In this study, the gender of the participant had little impact on the ratings of the protagonist in the script.

Results dramatically different from these, as well as those of Cretser et al. (1982), were obtained by Labott et al. (1991) in a laboratory study that examined university students' reactions to the crying (or laughing) of another person with whom they viewed a sad film. The other person with whom participants viewed the film was actually a confederate of the experimenters who had been trained to either laugh, act as if he or she were crying, or show no emotional response to the film. After the film was over, participants completed a number of measures designed to assess, among other things, what they thought of the confederate. Confederates who cried during the film, regardless of their gender, were seen as significantly more depressed and emotional than confederates who laughed or showed no emotional response. Most importantly, in contrast to the findings of earlier research, both male and female participants rated male confederates who cried and female confederates who were non-emotional highest in likeability. Female confederates who either laughed or cried and male confederates who showed no emotion were liked the least. Significantly, there was no indication that male confederates who cried were seen as more feminine than male confederates who laughed or showed no emotion.

To explain why male confederates were liked more when they cried, Labott et al. (1991) conjecture that perhaps because crying in men is such an infrequent event, participants "may believe that something truly important must have occurred for a man to engage in this behavior" (p. 412). Along with the finding that nonexpressive men were liked the least, this may signal an important change in social expectations about gender and emotional expressiveness, at least with regard to male expressiveness (since it is unclear why nonexpressive women were liked the least). Labott et al. conclude on the basis of these findings that, "social perceptions of those who engage in emotional expressions (especially crying) may have changed in the past several years. Crying, no longer seen as a predominantly 'feminine' behaviour, may currently even be more encouraged in men than in women" (p. 412).

Hill and Martin (1997), in a laboratory study involving a sample of female university students that was designed to assess, among other things, the relationship between empathic crying and cognitive dissonance, found that crying confederates (in this study, always females) elicited greater sympathy than non-crying confederates. Consistent with Labott et al.'s findings, confederates who cried were not evaluated more positively than noncrying confederates.

Psychotherapists' Perceptions of Crying Clients

Crying in psychotherapy is examined in detail elsewhere in this volume (see Chapter 12), but it seems appropriate to consider at least briefly in the present context the perceptions of psychotherapists toward the crying of clients during therapy. Given its aim, the kinds of events that are discussed during it, and the kinds of emotions it may be expected to elicit, it would not be surprising to discover that crying is a common occurrence during psychotherapy sessions (although see Vingerhoets & Becht, 1997, Table 4b). Trezza et al. (1988) asked a sample of clinical psychologists in private practice in the United States to describe their experiences with and attitudes toward crying during psychotherapy (see also Chapter 12). The clinicans in their sample, on average, regarded crying as a "generally healthy" behavior of potential help in decreasing depression.

It is clear from the writings of psychotherapists associated with psychodynamic or cathartic therapies (e.g., Greenacre, 1965; Groen, 1957; Löfgren, 1966; McCrank, 1983) that crying in the therapy session is highly valued by other therapists as well. Indeed, Arthur Janov (1992), the developer of so-called Primal Therapy, has said that, "I do not believe that anyone can get well in psychotherapy without it" (p. 317). Nevertheless, Mills and Wooster (1987) speculate that some psychotherapists may feel that crying is an "interruption" and a "hindrance" to their work with a client and may place either implicit or explicit demands on the client to suppress his or her tears. Mills and Wooster see the kind of emotional catharsis that crying purportedly brings about (see Chapter 11 in this volume) as crucial to psychotherapeutic success and counsel therapists that it is important to give clients "permission" to cry during therapy.

Health Care Professionals' Perceptions of Crying Patients

Although conspicuously absent from participants' reports of places in which they cry (Kraemer & Hastrup, 1986; Vingerhoets et al., 1997), possibly because of the way they are asked about such matters (see Chapter 5 in this volume), hospitals are probably sites of frequent crying. Health care professionals often find themselves in situations in which they must deal with the crying of patients and patients' relatives, the crying of coworkers, as well as their own crying (Krauser, 1989; Wagner et al., 1997). Forster and Forster (1971), attempted to systematically assess how nurses respond to crying patients by presenting a sample of nursing students in the United States with tape recordings of five "typical crying patient situations." Half of the participants listened to a crying patient describe various complaints. The other half listened to a non-crying patient describe the same complaints. (The tapes were recorded by professional actresses.) After listening to each of the five situations, participants described how they would respond to the patient's complaints. Participants also rated how helpless, irritated, anxious, and embarrassed they would feel in each situation. Forster and Forster had predicted, on the basis of earlier reports of how nurses interact with crying patients (Knowles, 1959; Norris, 1957), that participants' responses to crying patients would be "less effective" than those toward the non-crying patients and that they would express more negative emotions toward the former than toward the latter. After scaling their participants' responses to the patient scenarios from least to most effective, Forster and Forster discovered that, contrary to what they had expected, nurses who interacted with crying patients were somewhat more effective in communicating with their patients—more willing to listen and more effective at finding solutions—than nurses who interacted with noncrying patients. The difference between the two groups, however, was not statistically significant. Although also not significant, nurses who listened to crying patients reported feeling more helpless, irritated, anxious and embarrassed.

Wagner et al. (1997) distributed a questionnaire to a sample of doctors, nurses and final year medical students in Australia in which, among other things, participants were asked about their attitudes and reactions to crying patients and co-workers. The responses of all three groups toward crying patients were fairly positive. Participants reported that their most likely reactions toward a crying patient were holding his/her hand, becoming personally affected, and trying to soothe the patient's distress with words. Participants' least likely reactions were remaining unmoved and unemotional and leaving the room. Nurses were more likely to hold a patient's hand than were doctors or students, and nurses and students were more likely to start crying themselves. Students were more likely to regard crying patients as "lacking in will" than were doctors or nurses. All three groups indicated that patients and relatives should be allowed to cry in the hospital. Nurses were more likely than doctors and students to indicate that hospital personnel should be allowed to cry as well.

In summary, while the data are not nearly as complete or unambiguous as they could be, it may be concluded that individuals' perceptions of those who cry are on the whole positive. People report accepting and being emotionally moved by others' tears and say they feel sympathy and acceptance toward a person who is crying. Judging from both the popular and scientific literature of the past two decades, perceptions of men's crying may be undergoing a profound change. Men who cry are apparently not seen as feminine and their tears are no longer seen as a sign of weakness. Finally, there appear to be contexts in which crying is especially accepted. Crying, of course, has always been an expected and accepted part of funerals and grieving (Rosenblatt et al., 1976). It appears as if crying during psychotherapy and in hospitals (by patients, family members, and, perhaps, staff) is also acceptable, if not outright welcome.

How does the Situation Influence Crying and the Experience of Crying?

The major precipitants of crying tend to be interpersonal, although media events are also frequent antecedents to crying (Frey et al., 1983). In this section, we consider only the contextual features of the situations in which crying takes place; the specific elicitors of crying are discussed elsewhere in this volume (see Chapter 5).

It is clear that context greatly influences the tendency to cry. Crying very obviously occurs much more frequently in some contexts, e.g., funerals, than in others, e.g., at the office. Vingerhoets et al. (1997), in a study of Dutch and Flemish women who responded to a questionnaire on crying in a popular magazine, found that the majority of self-reported crying episodes occurred at home. Similar findings were obtained in the International Study on Adult Crying (ISAC; Vingerhoets & Becht, 1997), a questionnaire study involving close to 4000 college and university students in 30 different countries around the world. Interestingly, while 71% of men and 78% of women reported crying most often at home, a sizable fraction (13% vs. 8%, respectively) reported crying in public places.

Instructional Set Effects

Kraemer and Hastrup (1988), in a laboratory study of self control of crying, instructed a sample of female university students to try to cry or try to inhibit their crying while they viewed a sad film. Compared with those who were instructed to try to inhibit their crying, a significantly higher percentage of participants who were instructed to try to cry during the film indeed cried. Participants instructed to cry also cried for a significantly longer duration than did participants instructed not to cry.

Labott and Teleha (1996) examined the interaction of inhibition and expression instructions with participants' self-reported crying propensity on skin conductance and self-reported distress during a sad film. To ascertain crying propensity, women who were students at a university were asked to estimate the number of times they had cried in the month prior to the study. A median split determined those with a high and low propensity to cry. In the expression condition, participants were told to try to express any emotions elicited by the film as fully as they could, while in the inhibition condition, participants were told to inhibit their emotions as strongly as they could. Labott and Teleha did not report if there were significant effects of the instructions on participants' crying in response to the film. There was, however, a significant interaction between instructional set and crying propensity on the level of stress participants reported during the film such that participants who were instructed to behave in a manner opposite to their natural crying propensities and who did so (expression for low frequency criers and inhibition for high frequency criers) reported higher stress levels than participants who were instructed to behave in a manner consistent with their natural crying propensities and who did so.

In the study described earlier, Hill and Martin (1997) placed participants in a situation in which it was conveyed to them that emotional expression or emotional restraint reflected support and empathy for another person in distress (a confederate). They found that participants cried longer and more intensely in the former than in the latter situation, and interpreted these findings as indicating that crying may communicate emotional support and empathy in some situations.

Effects of the Presence of Others Who are Crying

Cornelius (1981), in an interview/questionnaire study of episodes of crying in a sample of university students, found that the most frequently reported event preceding his participants' crying was crying by another person. Jesser's (1982) participants also reported crying in response to the crying of another. These results suggest that adults may be prone to contagious crying or "empathic" crying in the same way that infants apparently are (Martin & Clark, 1982). Labott et al. (1991), in the study of reactions to a crying or laughing confederate described earlier, however, found no evidence of crying contagion. Recall that Labott et al. (1991) had participants watch a sad film in the presence of a confederate who either laughed or acted as if he or she were crying during the film. While the confederate's emotional expression had a definite effect on how participants' perceived him or her and the film, crying by the confederate did not appear to elicit crying by participants. Participants did appear to laugh more when they were paired with a laughing confederate, however.

Employing a very similar methodology but a different film, Hill and Martin (1997), did obtain evidence for crying contagion. They found that participants who watched a sad film in the presence of a crying confederate cried more than did participants who watched the film in the presence of a confederate who did not cry. As described above, because participants tended to cry more when it was conveyed to them that crying was a form of emotional support, Hill and Martin interpret these results as evidence of empathic crying.

What might account for the difference in the outcomes of Labott et al. (1991) and Hill and Martin's (1997) studies? In addition to the different films used in the two studies, participants were led to believe in Hill and Martin's study that the confederate was depressed, while nothing about the confederate's emotional state was conveyed to participants in Labott et al.'s study. It may be that contagious or empathic crying requires knowing something about the meaning of another person's tears. More generally, contagious crying may depend on the relationship between the person who cries and the one who witnesses his or her crying. The kind and intensity of crying will probably also make a difference, as will a number of other contextual factors. Wagner et al.'s (1997) study of hospital personnel, for example, suggests that role requirements may have a strong influence on whether or not one responds to another's crying with tears of one's own. Given the artificiality of most laboratory situations in which participants have the opportunity to observe another person crying and the low intensity of the observed crying (which usually is not directly observed anyway), perhaps it is not too surprising that contagious or empathic crying has not been more regularly observed.

Reactions of Others

It seems reasonable to assume that the reactions of others to one's crying would have a significant effect on the duration and intensity of crying. Anecdotally, we know that people will often suppress or attempt to suppress their tears if they know others will disapprove of them (see Plas & Hoover-Dempsey, 1988). Data collected in the context of the ISAC-project (Vingerhoets & Becht, 1997) suggest, on the other hand, that disapproval of one's crying by others is rarely seen as a reason to stop crying. It may be the case, however, that social feedback has an effect on whether or not one will *begin* crying but little effect on crying once it has begun. Mattutini and Cornelius (1997), in a study involving a sample of college students, found that high self-monitors, who are generally thought of as being highly sensitive to situational cues (Snyder, 1974), indicated that they had significantly more control over whether or not they began crying in a particular situation than did low self-monitors. High and low self-monitors did not differ in whether or not they thought they could stop themselves once they had started crying.

Cornelius et al. (1997; see also Cornelius, 1997), in a questionnaire study of college and university students in which participants were asked to describe an episode of crying in which they felt better after crying and one in which they did not feel better, found that the two types of episodes were distinguished by how positive participants perceived the reactions of others to their crying to be. In the ISAC-project, participants' reports of feeling better after crying were also associated with positive changes in the situation in which they cried and in the relationship they shared with whomever was present when they cried (see Chapter 11 this volume).

Although the data are not conclusive and are far from complete, it appears as if the reactions of others may influence the timing of the onset of crying and perhaps other characteristics of crying bouts as well. The reactions of others may also influence how individuals feel after they cry.

Other Situational Effects

In a test of some aspects of their cognitive restructuring model of crying, Martin and Labott (1991), presented university students with a sad, crying-inducing film that was followed by one of three different conditions. A third of the participants listened to a humorous audio tape and then watched a humorous videotape after the film, a third were treated to a repeat of the final crying-eliciting scenes of the film, and a third simply waited an equivalent amount of time. Martin and Labott found that the humorous audio and videotapes eliminated crying after the film, waiting reduced but did not eliminate crying, and repeating the crying-eliciting scenes from the film maintained crying at about the same level it was at the end of the film. Participants self-reported depressed mood followed the same pattern. Thus, as might be expected, crying appears to be influenced by the kinds of contextual events that follow it.

To summarize, while there is still much to be learned about situational influences on crying, it is clear that when people cry they are often quite sensitive to the context within which their crying takes place. Although not yet conclusively demonstrated in the laboratory, there is also evidence that crying by another person may be an important elicitor of crying. Studies involving instructional set manipulations, in addition, indicate not only that the initiation and duration of crying may be influenced by the situation, but that the meaning of crying may depend on social contextual cues as well.

Implications of Research on the Social Psychology of Crying for a Theory of Crying

Whatever the specific content of individuals' perceptions of others' crying, it is clear that crying is a compelling form of emotional expression that is difficult for others to ignore. Tears demand our attention and demand a response (Cornelius, 1981), although we may not always know how to respond to them. Given the powerful effects that crying may have on observers' perceptions of and behavior toward those who cry, it seems reasonable to consider whether or not crying may be something more than simply the expression of deeply felt emotion, that is, whether or not crying may be a form of social communication (Cornelius, 1981; Hill & Martin, 1997; Jakobs, 1998). This conclusion is strengthened by the findings that indicate that crying is context sensitive and, especially, by the findings that suggest that the meaning of tears, i.e., whether they are experienced as conveying empathy or not, may depend on situational factors.

Before exploring the implications of this notion, it is important to note that thinking of crying as a form of social communication does not mean excluding solitary crying from consideration. To do so would be ill-advised, as a great deal of crying by adults (perhaps as much as 50%) appears to be a solitary experience (Vingerhoets & Becht, 1997; Vingerhoets et al., 1997). Fridlund (1994) cogently argues, however, that emotional expressions may be considered communicative even when they take place alone and in private (see pp. 160–166). According to Fridlund (1994), the presence of others is "one of the *least* important criteria for ascertaining the sociality of facial displays" (p. 160). What applies to other facial displays may apply to crying as well.

Crying as a Communicative Act

Conceptualizing crying as a communicative act allows us to see the parallels between the distress and alarm calls of non-human animals (Marler & Evans, 1997), the crying of infants (see Chapter 3 in this volume), and the crying of adults in terms of their ecological functions and the selection pressures that shaped them. From the perspective of the behavioral ecology of emotional expression (Fridlund, 1994), crying may be seen as a *display* that, whatever else it may do for the person who cries, communicates something important about him or her to others. Following Bowlby (1973), Ainsworth et al. (1978) and others (e.g., Barr, 1990) consider infant crying as an attachment behavior or signal *par excellence* that ensures that an infant will be attended to by its caretaker. Crying as a communicative act that signals that one is in need may be seen as developing out of a simple, reflex-like distress response (Bell & Ainsworth, 1972; Hill & Martin, 1997; Lester, 1984). Cornelius (1981) has argued that adult crying also involves a plea for support in situations in which attachment bonds are broken or are in danger of being broken and hence may be seen as a kind of adult analog of the infant attachment signal. Recent studies of social support seeking indicate that crying by adults is indeed an effective elicitor of solace from others (see Barbee et al., 1998).

In addition to communicating that one is in need of succor, crying may be seen as a display that alerts others to danger and that may curtail aggression by others (Shaver et al., 1987). As Hill and Martin's (1997) findings reveal, crying may also convey emotional support or empathy. At a more specific level, crying may communicate many different meanings depending on the social context within which it occurs and the goals and desires of the person who cries. The behavioral ecology perspective also allows us to consider the use of crying by individuals to achieve particular interaction goals as more than mere manipulation (cf. Krebs & Dawkins, 1984) and may thus be seen as complementary to the impression management perspective in social psychology (Baumeister, 1982). From this perspective, then, knowing how crying is perceived by others and the ways in which it is sensitive to contextual cues are important but only become interesting when we examine the role that crying plays in ongoing episodes of social interaction, what it communicates to others, how others respond to it, how it may be used as a tool of self-presentation, and so on. We are still a long way from understanding the interpersonal dynamics of episodes of crying in terms of what and how crying communicates what it does. Adopting the perspective that crying is a social communicative display may not only help us order the available data on crying but may reveal new questions to ask about crying because it places crying more centrally within its ecological context, namely, social interaction and social relationships.

References

Ainsworth, M.D.S., Blehar, M.C., Waters, E., & Wall, S. (1978). *Patterns of attachment: A psychological study of the strange situation*. Hillsdale, NJ: Lawrence Erlbaum Associates.

Averill, J.R. (1980). A constructivist view of emotion. In: R. Plutchik & H. Kellerman (Eds.), *Emotion: Theory, research and experience, vol. 1* (pp. 305–339). New York: Academic Press.

Barbee, A.P., Rowatt, T.L., & Cunningham, M.R. (1998). When a friend is in need: Feelings about seeking, giving, and receiving social support. In: P.A. Andersen & L.K. Guerrero (Eds.), *Handbook of communication and emotion: Research, theory, applications, and contexts* (pp. 281–301). San Diego: Academic Press.

Barr, R.G. (1990). The early crying paradox: A modest proposal. *Human Nature*, *1*, 355–389.

Baumeister, R.F. (1982). A self-presentational view of social phenomena. *Psychological Bulletin*, *91*, 3–26.

Bell, S.M., & Ainsworth, M.D. (1972). Infant crying and maternal responsiveness. *Child Development*, *43*, 1171–1190.

Boukydis, C.F.Z.(1985). Perception of infant crying as an interpersonal event. In: B.M. Lester & C.F.Z. Boukydis (Eds.), *Infant crying. Theoretical and research perspectives* (pp. 187–215). New York: Plenum.

Bowlby, J. (1973). *Attachment and loss: Vol. 2. Separation: Anxiety and anger*. New York: Basic Books.

Cornelius, R.R. (1981) *Weeping as social interaction: The interpersonal logic of the moist eye*. Unpublished Doctoral Dissertation, University of Massachusetts, Amherst, MA.

Cornelius, R.R. (1986). *Prescience in the prescientific study of weeping? A history of weeping in the popular press from the mid-1800's to the present*. Fifty-seventh Annual Meeting of the Eastern Psychological Association, New York, NY.

Cornelius, R.R. (1997). Toward a new understanding of weeping and catharsis? In: A.J.J.M. Vingerhoets, F.J. Van Bussel, & A.J.W. Boelhouwer (Eds.), *The (non)expression of emotions in health and disease* (pp. 303–321). Tilburg: Tilburg University Press.

Cornelius, R.R., DeSteno, D., Labott, S., Oken, J., & Armm, J. (1997). *Weeping and catharsis: A new look*. Unpublished manuscript, Vassar College.

Cretser, G.A., Lombardo, W.K., Lombardo, B., & Mathis, S. (1982). Reactions to men and women who cry: A study of sex differences in perceived societal attitudes versus personal attitudes. *Perceptual and Motor Skills*, *55*, 479–486.

Feldman, R.S. (1998). *Social psychology*. Upper Saddle River, NJ: Prentice-Hall.

Fiske, S.T., & Taylor, S.E. (Ed.) (1991). *Social cognition*. New York: McGraw-Hill.

Forster, B., & Forster, F. (1971). Nursing students' reactions to the crying patient. *Nursing Research*, *20*, 265–268.

Frey, W.H., Hoffman-Ahern, C., Johnson, R.A., Lykken, D.T., & Tuason, V.B. (1983). Crying behavior in the human adult. *Integrative Psychiatry*, *1*, 94–98.

Fridlund, A.J. (1994). *Human facial expression. An evolutionary view*. San Diego: Academic Press.

Greenacre, P. (1965). On the development and function of tears. *Psychoanalytic Study of the Child*, *20*, 249–259.

Groen, J. (1957). Psychosomatic disturbances as a form of substituted behavior. *Journal of Psychosomatic Research*, 2, 85–96.

Heiman, J.D. (1997, July 6). Baby, don't stop crying. *New York Daily News*, p. 24–26.

Hill, P., & Martin, R.B. (1997). Empathic weeping, social communication, and cognitive dissonance. *Journal of Social and Clinical Psychology*, *16*, 299–322.

Hoover-Dempsey, K.V., Plas, J.M., & Wallston, B.S. (1986). Tears and weeping among professional women: In search of new understanding. *Psychology of Women Quarterly*, *10*, 19–43.

Jakobs, E. (1998). *Faces and feelings in social context*. Enschede, Netherlands: Print Partners Ipskamp.

Janov, A. (1992). *The new primal scream: Primal therapy 20 years on*. New York: Enterprise Publishers.

Jesser, C.J. (1982). *Gender and crying among college students*. Annual meeting of the Midwest Sociological Society, Des Moines, IA.

Jesser, C.J. (1989, Winter/spring). Men and crying. *Changing Men*, p. 12, 15.

Knowles, L.N. (1959). How can we reassure patients? *American Journal of Nursing*, *59*, 832–835.

Kottler, J.A. (1996). *The language of tears*. San Francisco: Jossey-Bass.

Kraemer, D.L., & Hastrup, J.L. (1986). Crying in natural settings. *Behaviour Research and Therapy*, *24*, 371–373.

Kraemer, D.L., & Hastrup, J.L. (1988). Crying in adults: Self-control and autonomic correlates. *Journal of Social and Clinical Psychology*, *6*, 53–68.

Krauser, P.S. (1989). Tears. *JAMA: Journal of the American Medical Association*, *261*, 3612.

Krebs, J.R., & Dawkins, R. (1984). Animal signals: Mind-reading and manipulation. In: J.R. Krebs & N.B. Davies (Eds.), *Behavioural ecology. An evolutionary approach* (pp. 380–402). Oxford: Blackwell.

Labott, S.M., Martin, R.B., Eason, P.S., & Berkey, E.Y. (1991). Social reactions to the expression of emotion. *Cognition and Emotion*, *5*, 397–417.

Labott, S.M., & Teleha, M.K. (1996). Weeping propensity and the effects of laboratory expression or inhibition. *Motivation and Emotion*, *20*, 273–284.

Lester, B.M. (1984). A biosocial model of infant crying. In: L. Lipsitt & C. Rovee-Collier (Eds.), *Advances in infancy research* (pp. 167–212). Norwood, NJ: Ablex.

Lester, B.M., & Boukydis, C.F.Z. (Ed.). (1985). *Infant crying. Theoretical and research perspectives*. New York: Plenum.
Löfgren, L. (1966). On weeping. *International Journal of Psychoanalysis*, *47*, 345–383.
Marler, P., & Evans, C.S. (1997). Communication signals of animals: Contributions of emotion and reference. In: U.C. Segerstrale & P. Molnar (Eds.), *Nonverbal communication: Where nature meets culture* (pp. 151–170). Mahwah, NJ: Lawrence Erlbaum Associates, Inc.
Martin, G.B., & Clark, R.D. (1982). Distress crying in neonates: Species and peer specificity. *Developmental Psychology*, *18*, 3–9.
Martin, R.B., & Labott, S.M. (1991). Mood following emotional crying: Effects of the situation. *Journal of Research in Personality*, *25*, 218–244.
Mason, D. (1997, September). The wet look is in. *New Choices*, pp. 40–43.
Mattutini, A.T., & Cornelius, R.R. (1997). *Weeping as a function of self-monitoring*. Sixty-eighth Annual Meeting of the Eastern Psychological Association, Washington, DC.
McCrank, E.W. (1983). Crying behavior in the human adult: Commentary. *Integrative Psychiatry*, *1*, 98–99.
Michelsson, K., Paajanen, S., Rinne, A., Tervo, H., & al., e. (1990). Mothers' perceptions of and feelings towards their babies' crying. *Early Child Development and Care*, *65*, 109–116.
Mills, C.K., & Wooster, A.D. (1987). Crying in the counseling situation. *British Journal of Guidance and Counseling*, *15*, 125–131.
Norris, C.M. (1957). Nurse and the crying patient. *American Journal of Nursing*, *57*, 323–327.
Plas, J.M., & Hoover-Dempsey, K.V. (1988). *Working up a storm: Anger, anxiety, joy, and tears on the job*. New York: W.W. Norton.
Rosenblatt, P.C., Walsh, R.P., & Jackson, D.A. (1976). *Grief and mourning in cross-cultural perspective*. New Haven, CT: Human Relations Area Files.
Sadoff, R. (1966). On the nature of crying and weeping. *Psychiatric Quarterly*, *40*, 138–146.
Sarbin, T.R. (1989). Emotions as narrative employments. In: M.J. Packer & R.B. Addison (Eds.), *Entering the circle: Hermeneutic investigations in psychology* (pp. 185–201). New York: State University of New York Press.
Scherer, K.R., Wallbott, H.G., Matsumoto, D., & Kudoh, T. (1988). Emotional experience in cultural context: A comparison between Europe, Japan, and the United States. In: K.R. Scherer (Eds.), *Facets of emotion* (pp. 5–30). Hillsdale, NJ: Lawrence Erlbaum.
Shaver, P., Schwartz, J., Kirson, D., & O'Connor, C. (1987). Emotion knowledge: Further explorations of a prototype approach. *Journal of Personality and Social Psychology*, *52*, 1061–1086.
Snyder, M. (1974). Self-monitoring of expressive behavior. *Journal of Personality and Social Psychology*, *30*, 526–537.

Staebler Tardino, V.M. (1996) *Violating norms for expression: Perceptions of expressors of negative emotions in the workplace.* Unpublished Masters' Thesis, Southern Illinois University, Carbondale, IL.

Trezza, G.R., Hastrup, J.L., & Kim, S.E. (1988). *Clinicians' attitudes and beliefs about crying behavior.* Fifty-ninth Annual Meeting of the Eastern Psychological Association, Buffalo, NY.

Vingerhoets, A.J.J.M., & Becht, M. (1997). *The ISAC-project: Some preliminary findings.* Unpublished manuscript, Tilburg University.

Vingerhoets, A.J.J.M., Van Geleuken, A.J.M.L., Van Tilburg, M.A.L., & Van Heck, G.L. (1997). The psychological context of crying episodes: Toward a model of adult crying. In: A.J.J.M. Vingerhoets, F.J. Van Bussel, & A.J.W. Boelhouwer (Eds.), *The (non)expression of emotions in health and disease* (pp. 323–336). Tilburg, NL: Tilburg University Press.

Wagner, R.E., Hexel, M., Bauer, W.W., & Kropiunigg, U. (1997). Crying in hospitals: A survey of doctors', nurses', and medical students' experience and attitudes. *Medical Journal of Australia, 166*, 13–16.

Wallbott, H.G., & Scherer, K.R. (1988). How universal and specific is emotional experience? Evidence from 27 countries on five continents. In: K.R. Scherer (Eds.), *Facets of emotion* (pp. 31–56). Hillsdale, NJ: Lawrence Erlbaum.

10 MENSTRUAL CYCLE, PREGNANCY, AND CRYING

Antje Eugster, Myriam Horsten, and Ad J.J.M. Vingerhoets

Women cry more frequently and more intensely than men (see chapter 6, this volume; Vingerhoets and Scheirs, 2000). Several factors explaining these gender differences in crying have been proposed, mainly focusing on differences in socialization and a possible differential exposure to stressful experiences, both unwanted (e.g., because of a greater emotional involvement with others) and wanted (e.g., a high interest in watching sad or melodramatic movies). The conclusion that these factors are most relevant is often based without justification on findings that show that the sex difference in crying is not there from birth. We will consider that the lack of sex differences in crying in newborns does not necessarily refute the role of biological factors as at least a partial explanation for the observed sex differences in adult crying.

In the present chapter, we will summarize the scarce literature on crying and the following four related factors in women: (1) prolactin levels; (2) the menstrual cycle; (3) pregnancy; and (4) the post-partum period. Since only a small number of publications on these issues are available, we will also discuss the relationship between these conditions and negative mood or depression, under the as yet unproven assumption that a depressed mood may be often accompanied by a higher crying proneness. Since it has been established that, in particular during the fertile years, women also have a higher prevalence of depression than men, it seems reasonable to have a closer look at phenomena linked with fertility and their possible association with mood disturbance and, in particular, with crying.

Crying and Prolactin

According to Frey (1985), three hormones are important in human tears: prolactin, the adrenocorticotropic hormone (ACHT), and the endorphin leucine-enkephalin. Here we will focus particularly on prolactin, since

Frey postulates that this hormone may lower the threshold to cry. Before going into the arguments put forth by Frey, some brief introductory statements about this hormone. Prolactin is released by the anterior part of the endocrine master gland, the pituitary, also referred to as the adenohypophysis. Its secretion is mainly under the control of inhibiting factors, released by the hypothalamus, which are under the influence of the dopaminergic system. The biological functions of this hormone have in particular to do with fertility, reproduction, and caring behavior. Increased levels of prolactin may cause menstrual disorders and infertility. Prolactin increases during pregnancy and reaches its physiological peak levels in the post-partum period, when it stimulates lactation (Albrecht, 1980; Frey, 1985 Greenspan, 1991).

As already said, Frey (1985) postulated that prolactin facilitates crying, probably because higher levels of prolactin lower the threshold for tearing. This hypothesis was based on the following observations. First, Frey based his arguments on data suggesting that the sex difference in crying becomes manifest at the age of thirteen, when puberty and the development of the secondary sex characteristics start (Hastrup, cited in Frey, 1985). According to Frey, this sex difference in crying frequency may be partly accounted for by the fact that during that period, girls develop higher prolactin levels than boys. Frey further points out that so called dry eye symptoms occur in particular during periods when prolactin levels are low, such as following menopause or as a side effect of drugs that inhibit prolactin secretion. In addition, Frey reports a case of a patient in whom administering drugs to reduce prolactin secretion was effective in the control of pathological crying. A final argument was the observation that administering prolactin to marine ducks increased the secretory activity of the salt glands, which are similar in location and innervation to the human lacrimal gland.

We think that there are some more theoretical arguments that make an association between prolactin and crying likely and warrant further exploration. As already indicated, high levels of prolactin are found in women just after childbirth, when the mothers start breast feeding. At about the same time, many women experience the so-called maternity blues, with crying and tearfulness as the main characteristics. Later on, we will expand more on this topic, here we limit ourselves to the mere fact of a coincidence of greater crying proneness and high plasma prolactin levels. Moreover, there are some additional interesting coincidences. For example, there is the conjecture of Theorell (1992) that prolactin is a hormone reflecting passivity and feelings of powerlessness in crisis situations. Theorell based his hypothesis on the results of four longitudinal real-life studies, all showing an association between passive coping and helplessness, on the one hand, and increased plasma prolactin on the other hand, whereas active coping was accompanied by unchanged or even decreased levels of prolactin. It is striking that just this same psychological state of powerlessness and a lack of adequate responses has been hypothesized as particularly relevant for crying (Bindra, 1972; Frijda, 1986; Vingerhoets et al., 1997). In addition, Jacobs et al. (1986) showed that individuals who had lost their spouse or were threatened with loss had marked prolactin responses during a stressful interview focusing on separation anxiety and depression. More specifically, their data suggested that both depression and separation anxiety, each in conjunction with high levels of the other but not independently, rendered these specific individuals more physiologically sensitive or, alternatively, the global distress had to pass a certain threshold before triggering the prolactin response. Prolactin has further been hypothesized to be involved in the development of mood disorders (i.e., depression and premenstrual tension; see De La Fuente & Rosenbaum, 1981). Remarkably, the connection between high prolactin levels and crying proneness may also hold in more pleasurable conditions, like when making love and experiencing an orgasm (the "love cry", see Fraenkel, 1932; Offit, 1995). Indeed, also in those conditions, prolactin levels have been found to increase considerably (Exton et al., 1999). Finally, there is some evidence that drunk individuals can cry rather easily and experience crying spells (Whitters et al., 1985). Alcohol use has also been shown to increase prolactin levels (Soyka et al., 1991).

In conclusion, these observations suggest an association between prolactin and crying proneness. However, one should be aware that in all of the above mentioned studies, the evidence is circumstantial at best. There are no examples of studies in which both prolactin levels and crying (proneness) have been measured directly, except for Vingerhoets et al. (1992), who compared crying proneness of 13 endocrinologically established hyperprolactineamic women and healthy controls, but failed to find a difference in crying behavior between these groups. However, it should be emphasized that the prolactin levels in these patients, although sufficiently elevated to produce menstrual cycle disturbances, were rather low as compared to levels found in pregnancy and after delivery. It might be that the less dramatic (but perhaps more chronic) elevations of prolactin suffice to cause disruptions in the menstrual cycle, whereas higher (more temporarily) increases are needed in order to influence tearing and crying proneness.

In conclusion, we feel that Frey's (1985) prolactin hypothesis is still worth consideration. However, more research is needed to investigate direct associations between prolactin levels and crying behavior. We want to emphasize that this prolactin hypothesis should not and does not mean discounting social and cultural influences on crying behavior. It merely suggests that hormonal influences may play a role in addition to environmental factors. Prolactin may be hypothesized to lower the threshold for shedding tears, but for a crying episode to occur there is still the need for an external or internal trigger.

The Menstrual Cycle, Mood, and Crying

It has been established that some women recurrently experience not only physical (e.g. edema, lower abdominal pain, breast tenderness, headache, abdominal bloating, fatigue) but also psychological (e.g., depression, irritability, tension, anxiety) symptoms in the pre-menstrual period. Does that also imply that women cry more often during specific phases of the menstrual cycle? In a large-scale questionnaire study by Hamburg et al. (1968), approximately 1100 women described their experience of 47 symptoms four times (during their most recent menstrual flow, during the week before their most recent menstrual flow, during the remainder of their most recent cycle, and during their worst menstrual cycle). Factor analyses of menstrual, premenstrual, intermenstrual and worst menstrual symptoms resulted in eight symptom groups; (1) pain; (2) concentration; (3) behavioral change; (4) dizziness; (5) water retention; (6) negative affect; (7) positive arousal; and (8) control symptoms. Concerning the negative affect scale, 20% of the women had moderate to strong and severe symptoms of irritability, mood swings, depression and/or tension in the premenstrual phase of their most recent menstrual cycle. The researchers further found that women not using oral contraceptives (OC) complained of greater severity of symptoms in both the menstrual and premenstrual phase. These women complained, amongst other things, of greater severity of restlessness, irritability, mood swings, depression, and tension, compared with the OC women, suggesting influence by the sex hormones. According to Golub (1992) and Scambler and Scambler (1993), fertile women generally notice some emotional, physical, or behavioral symptoms in the week before menstruation, referred to as premenstrual tension. In about 3% to 5% of women, the symptoms are so severe that they interfere with work or interpersonal relationships.

The presence of a link between premenstrual syndrome (PMS) and depression is well known. Halbreich and Endicott (1985) found that about two–thirds of women with a history of major affective disorder (such as depression) experience significant pre-menstrual dysphoria, and most women with marked pre-menstrual depression have a history of psychiatric illness. Pre-menstrual dysphoria is also associated with increased risk of future major depressive disorder (Golub, 1992).

Although few studies have paid specific attention to crying during the pre-menstrual period, it seems reasonable to expect that crying occurs more often during the pre-menstrual period, be it either a symptom of PMS or in reaction to the pain and discomfort or the interference with normal social and work activities. So, it would be important to know not only *whether* women cry more often during certain periods of their menstrual cycle, but also *why* they cry during that specific period. Below we will summarize the results of the studies that have reported data on crying or crying proneness in relation to the menstrual cycle.

In a retrospective study by Moos (1968) with 839 women rating forty-seven symptoms associated with their most recent and their worst menstrual cycles, it was found that the percentage of women who reported enhanced crying proneness increased five times during the premenstrual period and four times during menstruation, compared to the intermenstrual period. However, this study was retrospective and measured self-reported crying proneness instead of actual crying behavior, thereby limiting any strong conclusions.

In a study by Frey and coworkers (cited in Frey, 1985, and Frey et al., 1986), among 85 normal female subjects who were not on birth control or other hormone medication, the number and length of crying episodes during the course of the menstrual cycle were examined. Data on actual crying behavior were collected by having subjects keep a crying diary. Increased crying was observed during the following periods: (i) four to six days *before* the beginning of the menstrual period, (ii) three to five days *after* the onset of menstruation and (iii) around ovulation. These three peaks of crying did not correlate with marked fluctuations of any sex hormones such as progesterone or estrogen. Remarkably, crying was quite low during the three days prior to onset of menstruation, days that are generally considered as the most problematic.

In a retrospective self-report study by Lee and Rittenhouse (1991), 594 registered nurses indicated the frequency with which they experienced a list of symptoms just before or at the beginning of their menstrual cycle. Results revealed that 44% of single women (never married, separated, divorced, and widowed) reported experiencing depression and crying frequently or every month, in comparison to 30% of married and partnered women. When dividing the women into age categories, the women over 30 years of age had a significantly lower prevalence of depression and crying than the women between 21 and 30 years of age.

Horsten et al. (1997) specifically asked the 2018 female participants in their retrospective cross-cultural study whether they felt that there was a relation between phase of the menstrual cycle and crying proneness and if so, to indicate on which days during the menstrual cycle they felt more prone to cry. Forty-five per cent of the participants agreed that their crying proneness fluctuated as a function of the phase of their cycle. Surprisingly, striking differences were found between the samples of different cultures. For example, whereas in countries like China, Peru and Romania only 15 to 20% admitted such a relationship, in other countries (e.g., Australia, Chile, Finland, Kenya, The Netherlands, Spain, and Turkey) these percentages were between 60 and 70%. A significant increase in self-reported crying proneness was reported from the seventh day before menstruation until the second day of the periods. In addition, slight increases were found on the first day after menstruation and around the ovulation. In addition to the remarkable cultural differences in percentages of women confirming the requested relationship, a further finding casting doubt on a merely biological underlying basis is that OC users more or less indicated a similar pattern of days with increased crying proneness as non-OC users. One may thus wonder to what extent these findings represent implicit knowledge and theories on the requested relationships rather than the actual state of affairs (e.g., Marván & Escobedo, 1999). In addition, there is the more general question whether it is justified to equate self-reported crying proneness to actual crying behavior.

Van Tilburg et al. (1999) tackled the issue of the role of menstruation in crying with a different approach when they examined crying proneness and crying frequency in girls in the age groups 12, 13 and 14 years. No significant differences were found for crying proneness and crying frequencies between same age menstruating girls and non-menstruating girls, suggesting that menarche does not systematically influence crying related behavior.

Recently, Vingerhoets[1] performed a pilot-study on 21 OC users and 21 non-OC users, who completed a crying diary very similar to the one

---

1 The investigator is indebted to Mariëlle van de Rijt, Angelina Klokgieters, Stannie Simonis, Inge Stuifbergen, and Colinda Boers for their help in data collection.

applied by Frey and coworkers (cited in Frey, 1985, and Frey et al., 1986). Although the samples differed too much in terms of age, education and marital status to allow for definitive conclusions concerning the role OCs play in determining mood, some interesting observations were made. OC users failed to report differences in crying frequencies during the different phases of menstruation, whereas, for non-OC users, significantly more frequent crying was reported during the first days of the menstruation, which seems in correspondence with Hamburg et al. (1968) findings.

Van Tilburg and Vingerhoets (2000) collected data on mood, crying proneness, and actual crying applying a diary method in 82 female students, during two full menstrual cycles. Whereas significant mood changes were detected, in particular, during the menstruation, no accompanying differences in actual crying behavior were observed. More specifically, increased scores on items like bad tempered, sad, fearful and irritable and decreased scores on relaxed and emotionally stable showed very weak associations with actual crying behavior. In addition, surprisingly, a significant difference was found in actual crying frequency between the OC and the non-OC group, with the OC group crying more frequently. This study thus demonstrates that care is warranted when generalizing from concepts like bad mood and emotional lability to crying proneness if not actual crying.

In conclusion, the scarce data available on the possible role of menstruation in crying do not allow any definite conclusion. In some studies, support was found for increased crying during the premenstrual phase and the first days of the menstruation, but other studies failed to find such an association. The basis of a possible increase in crying during some menstrual phase is relatively unexplored; could it be biological factors affecting crying proneness, or might it be a response to other symptoms experienced (e.g., pain and discomfort) or to a changed interaction with the social and work environment? We would like to recommend future concurrent studies which not only focus on crying frequency, but which also try to obtain insight into the specific reasons why women cry in order to be able to compare these motives for different menstrual phases.

Pregnancy, Mood, and Crying

It has long been common knowledge that pregnant women not only undergo bodily changes, but also mental changes when carrying their child (Ellis, 1936). Similar to the results of studies on mood and menstrual cycle, the literature on this issue is characterized by a wide variety of study designs, applied measures, and research findings (Elliott et al., 1983).

In order to determine whether there are any systematic fluctuations in degree of tension over the 40 weeks of pregnancy, Grimm (1961) studied 200 women cross-sectionally, using a balanced design in which a single evaluation was made of several carefully balanced groups of women seen at different periods of their pregnancy. Results showed that the index of tension was greatest in the last half of the third trimester and differed significantly with the first, second and first half of the third trimester. This author wondered to what extent generalizations could be made over women, because not all women reacted in the same way, thereby emphasizing the great individual variability in the amount of tension experienced at different times during pregnancy.

Murai and Murai (1975) studied mood in 128 pregnant women and 38 controls, thereby testing the women at two different points during pregnancy. This study showed that the pregnant group scored higher on surgency, fatigue, and social affection and lower on sadness than the control group. Most emotional stability was observed in the middle phase of pregnancy. In addition, slight differences in emotional stability were observed between nulliparous and multiparous women, with the latter showing the greatest stability.

In a study by Lubin et al. (1975), anxiety, depression and somatic symptoms were assessed in 93 women during the second, fifth and eighth month of pregnancy. Results revealed that anxiety decreased significantly during the second trimester and reverted back to the initial level in the third trimester. No significant variation for depression was found over trimesters. Variation for somatic symptoms over trimesters depended on the previous pregnancy history of the women. Nulliparous women and women who had experienced previous live birth and no terminated pregnancy did not differ significantly on somatic symptoms over the three trimesters. However, women with previous live birth and previous terminated pregnancy scored significantly higher on somatic symptoms during the second trimester, as compared to the other two trimesters.

Although the main focus of the study of Gorsuch and Key (1974) was to investigate the relationship between anxiety and abnormalities of pregnancy prospectively by comparing two subgroups (women whose pregnancy and delivery were problem free and women whose pregnancy and delivery had one or more abnormalities), a closer look at the graph provided in their report reveals a slight increase in average state-anxiety in the second trimester of normal pregnancy (see also Elliot et al., 1983), which is in contrast to the results of Lubin et al. (1975).

In a study by Kumar and Robson (1984), 119 first-time mothers were interviewed and clinically screened repeatedly during their pregnancy and until one year after the delivery, using the General Health Questionnaire (GHQ; Goldberg, 1972). Results revealed that the incidence of depression significantly increased in the first trimester. In many cases previous depressions had remitted before the subject's second interview (around the 24th week of pregnancy).

Ballinger (1982) followed 47 women through pregnancy to 10 days after delivery. During pregnancy, measures were taken at around 10–16 weeks, then again at 32 and 38 weeks. No significant changes were found in anxiety and hostility during pregnancy. However, depression increased significantly during pregnancy.

In a concurrent study by Elliot et al. (1983) with 128 pregnant women, psychological changes over the course of pregnancy and in the first postpartum year were measured. Compared with available norms, the pregnant women scored *below* the norms on depression and significantly *higher* on anxiety in early pregnancy. However, the authors noted that the available norms were not divided by age group, thereby limiting the comparison with a pregnant sample. Apart from an increase in ratings of discomfort and scores on somatic symptoms in late pregnancy, no significant changes in depression, anxiety or tension were found over the course of pregnancy. The authors concluded that their findings contradicted the results of the study of Lubin et al. (1975), who found a U-shape for anxiety over trimesters. Possible explanations for the conflicting results may be the different measures that were applied and differences in the study samples (socially homogeneous or heterogeneous). The authors concluded that their results supported the view of Grimm (1961), in that reactions to pregnancy are so highly individual that no generalizations could be made.

In a cross-sectional study by Helregel and Weaver (1989) with 112 pregnant women and 18 non-pregnant women, it was found that pregnant women scored highest on negative affect around the 22nd week of pregnancy and immediately preceding delivery. Remarkably, during these two stages of pregnancy women expressed a strong preference for comedy programs on television. Women 25 to 28 weeks pregnant, associated with predominantly positive affective states, were most interested in action adventure programs. These findings support the notion that individuals often employ television as a mood management strategy and that during pregnancy a different pattern of TV program preferences can be expected as a result of changed affect. This result reveals the complex relationships between basic mood, the possible (un)successful application of mood management strategies, and finally what may be called "distress behavior," including crying.

Smith et al. (1990) studied 97 pregnant women concurrently in order to examine the relationship between mood disturbance, psychosocial factors, obstetric events and alternations in the hypothalamic pituitary adrenal axis. The women were interviewed and completed questionnaires during the 28th and 38th week of pregnancy and again on the second postpartum day and three months after delivery. Results revealed that mood disturbance peaked at about 38 weeks of pregnancy and slowly returned to its initial level.

In a concurrent questionnaire survey by Sugawara et al. (1997), 1329 pregnant women were assessed three times during pregnancy (in the early, middle, and late phase of pregnancy) and 5 days, 1 month, and 6 months after delivery. In order to be able to compare a premenstrual mood changing group with a non-changing group, premenstrual mood was measured retrospectively in early pregnancy. Women who reported experiencing premenstrual irritability scored significantly higher on depression across all measure points during pregnancy than the non-irritable group. In the middle of pregnancy, the irritable group reported higher levels of anxiety about pregnancy or delivery than the non-irritable group.

In a concurrent study by Paarlberg et al. (1996), 369 nulliparous women were assessed in each trimester of their pregnancy on, amongst other things, depression, anxiety and somatic complaints. While depression and anxiety did not change significantly throughout pregnancy, somatic complaints were reported less frequently in the second trimester as compared with the first and the third trimester.

To summarize, contradictory results have been found in these studies of psychological changes during the course of pregnancy, which may be partly accounted for by differences in methodology (questionnaires or clinical interview, retrospective versus concurrent designs), specific mood indicators, composition of the samples (e.g., with respect to the factor parity and other relevant psychosocial factors related to the pregnancy, e.g. whether the child was planned, availability of necessary resources, etc).

Very few studies have specifically focused on crying proneness or crying frequency during pregnancy. Vingerhoets[2] examined the response of 396

---

2 The researcher would like to thank Marieke Paarlberg for her willingness to provide him with the data of the pregnant women and Nienke Bosma, Marieke Brouwer, Evan Essen, Nadine Lommers, Aafke Seebregts, and Sheila Vermaas for their help in further data collection and analysis.

primiparous pregnant women of the above mentioned Paarlberg et al. (1996) study to two crying related items of a symptom checklist ("lump in throat" and "prone to cry") during the first, second and third trimester of their pregnancy and 275 age-matched non-pregnant controls. The non-pregnant control group was assessed only once. Results showed that the pregnant women reported a higher crying propensity, with no significant changes in crying propensity over the course of pregnancy. Frey and coworkers (unpublished data) examined crying frequency and duration as a function of the number of weeks pregnancy (16 to 40 weeks of pregnancy) in a diary study. Data were obtained from 43 pregnant women and a control group of 77 non–pregnant women. No significant differences in crying frequency or average duration of crying episodes between pregnant and non–pregnant women were found. In addition, no significant variations in crying frequency as a function of numbers of weeks pregnant were found. Eugster reanalyzed these data[3], pooling them in two trimesters (no data were available for the first trimester). No significant differences in crying episodes were observed between the second and third trimester.

Lutjens (1998) examined crying frequency and crying proneness during and after pregnancy in 76 women who had given birth one to four months earlier. Crying frequency and proneness were measured retrospectively. Results revealed that 56% of the women reported that they cried more often during their pregnancy than before that state. In addition, 62% reported greater crying proneness during pregnancy than before. It was notable that crying frequency and proneness differed over months. The women reported lower crying behavior during the second trimester than the first (minus month 1) and the third trimester, thereby showing a U-shape pattern, similar to Lubin et al.'s (1975) findings. A U-shape was also found for uncomfortable moods, while for positive moods like relaxed and joyful a reversed U-shape was found.

In sum, little data are available to answer the question whether there are any systematic variations in crying during pregnancy. There are some data suggesting an inverted U-shape over trimesters in emotional stability, but other studies fail to find such a relationship. Very similar to the situation with menstrual cycle, it seems not justified to extrapolate negative mood and/or emotional lability to actual crying. Some authors emphasize the individuality of the reactions to pregnancy, which may be understood easily, given the huge impact of pregnancy and approaching

---

3 The researcher would like to thank W.H. Frey, C. Ahern, D. Nelson, T. Peterson, L. Erickson, L. Edwards, and M. Koszalka for making their data set available for her.

motherhood on women. Pregnancy is a complex life event, possibly associated with many other major or minor life changes and several psychosocial aspects. As stated before, in order to come to an adequate understanding of emotional experience and emotional expression in such complex biopsychosocial conditions, one should have adequate attention for all these dimensions. The application of mood management strategies as described by Helregel and Weaver (1989) makes things even more complex. As yet, there is little empirical evidence in support of a close association between crying during pregnancy and plasma levels of prolactin, which increase steadily from the fifth week in pregnancy until delivery (Albrecht, 1980). However, only empirical data, preferably collected in a concurrent research design, can yield more insight into these complex relationships. Here again, we would like to stress that it is important not only to pay attention to crying frequency, but also to focus on the triggering factors.

Postpartum Period, Mood, and Crying

During the puerperium, three categories of emotional disturbances can be distinguished: at one extreme is maternity blues, at the other is postpartum psychosis and located in between these two is postpartum depression. While postpartum psychoses are severe illnesses that commonly require admission to the hospital, postpartum depression is less severe and occurs more frequently (Beck, 1991). Postpartum depression, which has its onset in the first year after delivery and may last months, is characterized by, amongst other things, anxiety, despair and feelings of helplessness. Maternity blues, a transitory phenomenon of mood changes that may begin within the first few days after delivery and lasts anywhere from one day through the first 10 days postpartum or longer, is characterized by symptoms of tearfulness or crying, depression, anxiety, irritability, headache, and lability of mood (Beck, 1991; Kennerley & Gath, 1989a; Pitt, 1973; Yalom et al., 1968). All measures aimed at assessing maternity blues include items referring specifically to crying or tearfulness (the Dutch term for maternity blues "kraamtranen" literally refers to tears). In contrast to the above discussed issues, crying here takes a more central place.

Incidence estimates of maternity blues typically range from 40% to 75%, although Japanese studies typically yield a lower incidence (15–20%, see Kumar, 1994). These large differences in estimates may partly result from differences in the precise definition and/or measurement of maternity blues. In addition, in a critical review of maternity blues research, Beck (1991) found inconsistent results with respect to which days the blues symptoms occurred most frequently or were the most severe, although the period from day 3 to day 5 postpartum is most frequently identified as such (e.g., Pitt, 1973, Davidson, 1972). There is evidence that maternity blues occurs in many cultures (except for Japan) with the same frequency (cf. Harris, 1981; Khalid, 1989; Kumar, 1994; Morsbach et al., 1983; Rohde et al., 1997; Sutter et al., 1995; York, 1990; see, however, for some critical comments Macy, 1983). Kumar (1994) claims that social and cultural factors play a minor role in this phenomenon. On the other hand, there are some studies suggesting that women scoring highly on neuroticism (e.g., Kendell et al., 1984), with a previous neurotic depression (Stein et al., 1981), or with a poor family and/or marital relationship generally have a higher risk. No clear or conflicting associations have been reported for age, parity, social class, social support, sense of disappointment, and previous life events (cf. Condon & Watson; 1987; Kennerley & Gath, 1989b).

In a study by Hamburg et al. (1968), thirty-nine pregnant women were studied 1–3 weeks before delivery and for 10 days after delivery in order to examine the degree of postpartum depression. They were measured by interviews, behavioral observations and psychological tests. Results revealed that in the first 10 days after delivery, 67% of the women had episodes of crying lasting at least 5 minutes, 13% cried continuously for over 2 hours, 15% between 1 and 2 hours and 38% between 5 minutes and 1 hour. From the interviews it became apparent that although crying was usually associated with sadness, some women cried for emotional reasons other than sadness while other women experienced sadness and hopelessness without showing this by crying. For comparison, during a 10 day span late in the third trimester of pregnancy, 20% of the women cried, with no episode lasting more than 60 minutes. Eight months postpartum, 22% women had cried in the previous days, with no episodes lasting more than 30 minutes. In addition to crying, insomnia, restlessness and undue concern about the health of the newborn were found during the 10 days postpartum.

Stein (1980) investigated postpartum mental change and body weight change in women with normal pregnancies. He found a temporal relationship between the peak day of mental change and onset of weight loss. He identified a peak on postpartum days 4 and 5 specifically for crying. In a later study, Stein et al. (1981) established that weight nor electrolyte excretion differed between blues and non-blues cases. However, among women with a clear cut onset of mood disturbance, the start of the weight change and the rise in sodium excretion occurred at around the time of the mood swing. In other studies, a relationship with breast compression (Stein et al., 1981), platelet MAO activity, maternal B-endorphin/B-lipotrophin, and, indeed, prolactin (cf. Wilson, 1985) has been suggested. Some authors wondered whether the maternity blues is peculiar to the puerperium or whether it may also occur after surgery. This again has yielded some conflicting conclusions. Whereas Levy (1987) concluded that many post-operative dysphoric symptoms peaked in a similar way to those in the maternity group, both Kendell et al. (1984) and Iles et al. (1989) failed to find what they call a 'day five' phenomenon after gynecological surgery.

Stein et al. (1981) provides some more details on the clinical aspects of the maternity blues. Specifically with respect to crying, the authors note that there is always an association with mood disturbance, although not necessarily depression, but also elation, acute anxiety, or occasionally depersonalization. Many women could provide a reason for their tears, but also recognized the frailty of their explanation. Most reported reasons were minor problems with the baby, feeding difficulties, jaundice, minor conflicts with the husband or the nursing staff, or, especially in multigravid women, concerns about the other child(ren). Generally, there was an overwhelming need to rationalize their tears, although the women seemed to have more insight into their mood disturbance some days later.

Ballinger (1982) assessed women at day 1 to day 5 after delivery and again at day 10 after delivery. In women with spontaneous vaginal deliveries, no significant change occurred in anxiety and hostility between postpartum days 1 and 5. In this group, depression was significantly lower on the fifth day after delivery than at 38 weeks during pregnancy. For women who had a caesarean section, anxiety scores between days 1 and 5 differed significantly, with a decrease in scores from day 1 to day 5. Specifically, in this group, a significant fall in the score on depression was found between days 5 and 10 after delivery.

Ruchala and Halstead (1994) studied 50 postpartum women who were released within the normative postpartum hospital stay (defined as 2 days for vaginal deliveries and 4 days for Caesarean deliveries), on day 10 and 14 post discharge. Women were asked to describe their early postpartum experiences. Interview data revealed that fatigue was an underlying theme for both primiparous and multiparous women, in addition to issues like physical concerns, concerns about their social network and emotional concerns. Crying was the most frequently reported emotional symptom. Interview data revealed that 24% of the primiparous and 32% of the multiparous women reported frequent episodes of crying in the first two weeks post discharge from the hospital. The women appeared to rationalize their emotional changes by relating them to physical discomfort, the physical condition of their bodies and the impact of the baby on their personal freedom. Many of the women identified fatigue as a possible underlying reason for feeling depressed and irritable.

Meares et al. (1976) asked women 6 to 18 months following the birth of their child to indicate retrospectively how they felt after the birth of their baby. Twenty of the 49 (41%) women reported having experienced maternity blues. Crying was remembered as a more frequent occurrence than depression, although this difference did not reach statistical significance.

Results of the retrospective study by Lutjens (1998) revealed that in the first month postpartum, 82% of the 76 women actually cried and 54% reported they had maternity blues tears. The incidence of reported crying, lump in throat and maternity blues tears were highest on the fifth day postpartum. In addition, parity and crying frequency during pregnancy predicted postpartum actual crying and crying proneness. Parity also appeared to be a predictor of maternity blues. Nulliparous women reported a higher crying proneness, actual crying and tears in the first month postpartum than multiparous women. However, because of the retrospective study design, these results have to be interpreted with caution.

The longitudinal questionnaire survey of Sugawara et al. (1997) revealed that women who reported experiencing premenstrual irritability, scored significantly higher on depression than the non–irritable group 5 days after delivery, 1 month after delivery, and 6 months after delivery. In addition, five days after the delivery, the irritable group reported more tiredness than the non-irritable group. These data thus suggest that premenstrual mood change might be an important risk factor for the occurrence of both prenatal and postpartum depression, which raises the question whether they share a similar (physiological?) background. However, as the authors noted, the results have to be interpreted with caution, since premenstrual mood was measured retrospectively with a single item, measuring only one aspect of the premenstrual syndrome, casting doubt on the validity of the applied measure.

To summarize, maternity blues with crying and tearfulness as most characteristic features is a commonly seen condition in the period after childbirth. Although there might be some association with personality traits (i.e. neuroticism), evidence suggests that the phenomenon shows hardly any relationship with cultural or psychosocial factors, which suggests that it is a mainly biologically determined state. However, one should be aware that these women will nevertheless often try to attribute their crying bouts to specific events or worries. Future research should be focused on the possible underlying biological mechanisms, including serum prolactin levels. Data suggesting that there is a correspondence with the psychological response to a major surgical operation are not convincing, although one might argue that surgery is not a proper control condition, but that one should rather focus on other life events with a strong impact on one's life, like marriage.

Conclusion

In the present chapter we have reviewed the literature on crying and mood in relation to prolactin, menstrual cycle, pregnancy, and the post-partum period. Except for the maternity blues, little data are available with respect to the relation with crying. In addition, studies which did focus on crying often used a retrospective design, which prevents us from drawing any definitive conclusions. Although it is tempting to assume a close relationship between crying and concepts such as emotional lability and mood swings, the data presented in this chapter clearly demonstrate that one has to be careful in making generalizations from such concepts to crying proneness, that is not to say, actually crying too easily (see Figure 1 as an illustration of this statement).

There is some circumstantial evidence supporting Frey's (1985) prolactin hypothesis, although it is not possible to draw definite conclusions. Studies in which crying proneness and frequency as well as prolactin are measured are urgently needed to get a more reliable picture of such an association. In addition, the effects of societal conditioning on crying behavior and other psychosocial factors should not be neglected. Recent evidence has linked PMS reports with negative socialization about menstruation (Taylor et al., 1991), and beliefs about whether one is pre-menstrual have powerful effects on symptoms (Ruble, 1977). Not only is emotional crying in general more accepted in women, the pre-menstrual and

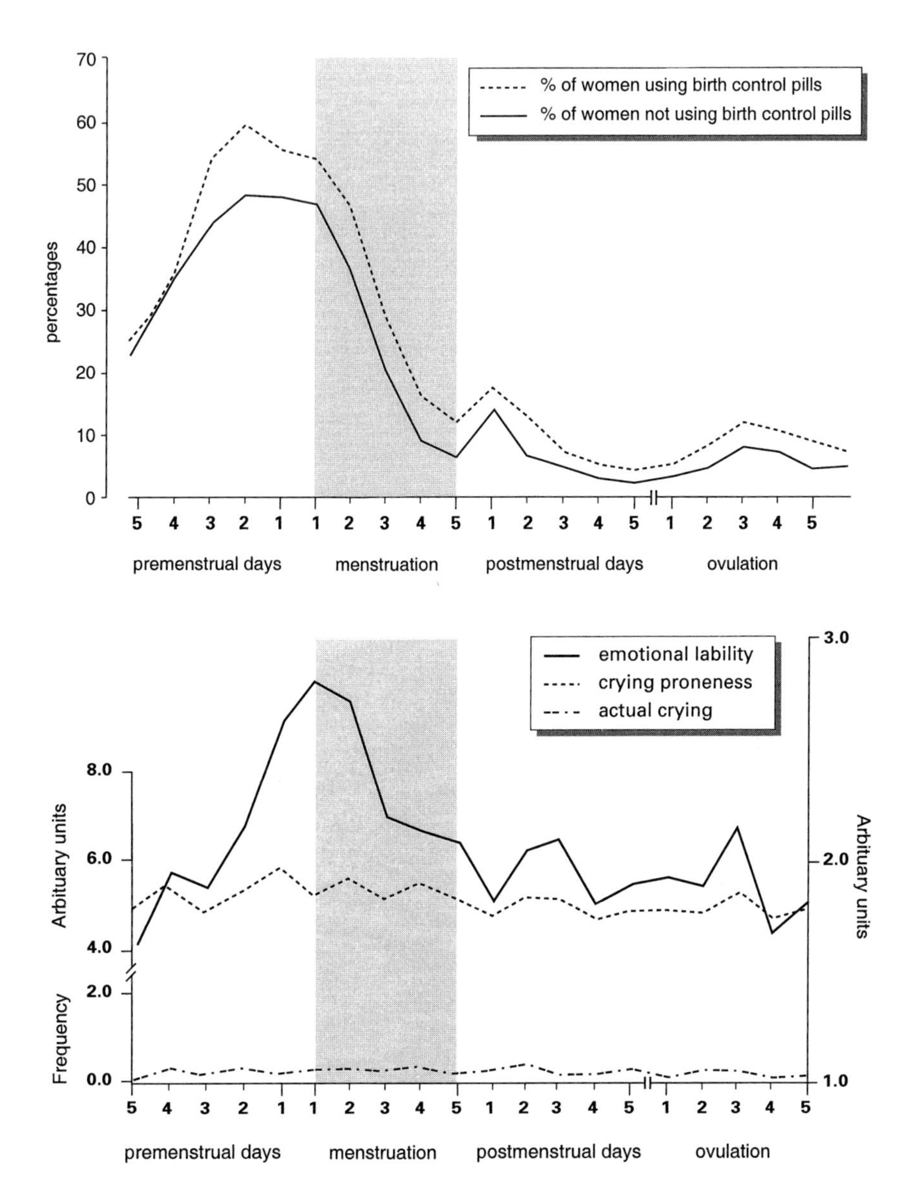

Figure 1. Data on self-reported crying propensity as a function of phase of mestrual cycle collected in two studies applying a different methodology. Above, women retrospectively indicated during which days of their menstrual cycle they felt more prone to cry (Horsten et al., 1997). Below, data from a diary study on emotions and menstrual cycle are presented (Van Tilburg & Vingerhoets, 2000). No changes were reported for actual crying and crying proneness. However, emotional lability showed a remarkable increase more or less parallelling the self-reported crying propensity in the retrospective study.

postnatal period are conditions in which it is 'acceptable' to let go and have emotional outbursts. It has been suggested that PMS and its symptoms are a survival strategy for women that allows culturally unacceptable behaviors to be labeled as "sick" (i.e., a result of illness) (Laws, 1983).

There is no clear evidence from concurrent studies supporting the existence of fluctuations in crying frequency during the menstrual cycle. In retrospective studies it has been found that crying behavior and self-reported crying proneness increase before menstruation, a few days after the onset of menstruation, and around ovulation. More women report increased crying proneness during the premenstrual period and during menstruation, compared to the intermenstrual period. However, in concurrent research designs, these findings could not be supported. OC use, age, and having a partner are further factors that may play a role. In addition, one should consider that crying may occur in response to PMS symptoms (e.g., pain and discomfort) rather than being a symptom. Moreover, the possibility that the impact of these symptoms on availability for sex and work performance can give rise to interpersonal conflicts, which in turn comprise a major reason for crying, should be considered.

Like studies concerning changes in depression and anxiety over the course of pregnancy, studies on crying during pregnancy have also yielded contradictory results. While significant changes in crying frequency and propensity during the course of pregnancy or differences in crying propensity between pregnant and non-pregnant women have been reported, other studies have failed to yield evidence in support of a greater emotional lability and/or crying proneness. Again, the contradictory results may at least partially be attributed to differences in the methodology and composition of the samples, in terms of age, socioeconomic status, but also other relevant psychosocial factors related to pregnancy. Nonetheless, results do suggest that a considerable proportion of women cry more often during the first 10 days after delivery.

In conclusion, more research on relations between menstrual cycle, pregnancy and crying is definitely needed to enable more definitive conclusions. The possible biological co-determinants of crying, in particular, the hormone prolactin, deserve further attention. Real progress can only be made if adequate attention is given to the wide variety of relevant factors. The findings described in this chapter strongly suggest that crying indeed can be best considered a biopsychosocial phenomenon.

References

Albrecht, B.H. (1980). The maternal adenohypophysis. In: D. Tulchinsky & K.J. Ryan (Eds.), *Maternal-fetal endocrinology* (pp. 97–114). Philadelphia/London/Toronto: W.B. Saunders Company.

Ballinger, C.B. (1982). Emotional disturbance during pregnancy and following delivery. *Journal of Psychosomatic Research, 26,* 629–634.

Bindra, D. (1972). Weeping: A problem of many facets. *Bulletin of the British Psychological Society, 25,* 281–284.

Beck, C.T. (1991). Maternity blues research: A critical review. *Issues in Mental Health Nursing, 12,* 291–300.

Condon, J.T., & Watson, T.L. (1987). The maternity blues: Exploration of a psychological hypothesis. *Acta Psychiatrica Scandinavica, 76,* 164–171.

Davidson, J. (1972). Postpartum mood change in Jamaican women: A description and discussion on its significance. *British Journal of Psychiatry, 121,* 659–663.

De La Fuenete, J., & Rosenbaum, A. (1981). Prolactin in psychiatry. *American Journal of Psychiatry, 138,* 1154–1160.

Elliott, S.A., Rugg, A.J., Watson, J.P., & Brough, D.I. (1983). Mood changes during pregnancy and after the birth of a child. *British Journal of Clinical Psychology, 22,* 295–308.

Ellis, H. (1936). The psychic state during pregnancy. In: H. Ellis (Ed.), *Studies in the psychology of sex* (Vol 2. pp 201–229). New York: Modern Library (original publication dates from 1906).

Exton, M.S., Bindert, A., Krüger, T., Scheller, F., Hartmann, U., & Schedlowski, M. (1999). Cardiovascular and endocrine alterations after masturbation-induced orgasm in women. *Psychosomatic Medicine, 61,* 280–289.

Fraenkel, E. (1932). Das "koitale Wort." [Coital speech]. *Zeitschrift für Sexualwissenschaft und Sexualpolitik, 18,* 436–452.

Frey, W.H. (1985). *Crying: The mystery of tears.* Minneapolis, MN: Winston Press.

Frey, W.H., Ahern, C., Gunderson, B.D., Tuason, V.B. (1986). Biochemical, behavioral and genetic aspects of psychogenic lacrimation: The unknown function of emotional tears. In: F.J. Holly (Ed.), *The preocular tear film in health, disease, and contact lens wear* (pp. 543–551). Lubbock, TX: Dry Eye Institute.

Frijda, N.H. (1986). *The emotions.* Cambridge: Cambridge University Press.

Goldberg, D. (1972). *The detection of psychiatric illness by questionnaire.* London: Oxford University Press.

Golub, S. (1992). *Periods. From menarche to menopause.* Sage: Newbury Park.

Gorsuch, R.L., & Key, M.K. (1974). Abnormalities of pregnancy as a function of anxiety and life stress. *Psychosomatic Medicine, 36,* 352–362.

Greenspan, F.S. (1991). *Basic and clinical endocrinology.* East Norwalk, CT: Appleton & Lange.

Grimm, E.R. (1961). Psychological tension in pregnancy. *Psychosomatic Medicine, 23,* 520–527.

Halbreich, U., & Endicott, J. (1985). Relationship of dysphoric premenstrual changes to depressive disorders. *Acta Psychiatrica Scandinavica, 71,* 331–338.

Hamburg, D.A., Moos, R.H., & Yalom, I.D. (1968), Studies of distress in the menstrual cycle and the postpartum period. In: R.P. Michael (Ed.), *Endocrinology and human behaviour* (pp. 94–116). London: Oxford University Press.

Harris, B. (1981). "Maternity blues" in East African clinic attenders. *Archives of General Psychiatry, 38,* 1293–1295.

Helregel, B.K., & Weaver, J.B. (1989). Mood-management during pregnancy through selective exposure to television. *Journal of Broadcasting & Electronic Media, 33,* 15–33.

Horsten, M., Becht, M., & Vingerhoets, A.J.J.M. (1997). Crying and the menstrual cycle. *Psychosomatic Medicine, 59,* 102–103.

Iles, S., Gath, D., & Kennerley, H. (1989). Maternity blues: II. A comparison between post-operative women and post-natal women. *British Journal of Psychiatry, 155,* 363–366.

Jacobs, S., Brown, S.A., Mason, J., Wahby, V., Kasl, S., & Ostfeld, A. (1986). Psychological distress, depression, and prolactin response in stressed persons. *Journal of Human Stress, 12,* 113–118.

Kendell, R.E., Mackenzie, W.E., West, C., McGuire, R.J., & Cox, J.L. (1984). Day-to-day mood changes after childbirth: Further data. *British Journal of Psychiatry, 145,* 620–625.

Kennerley, H., & Gath, D. (1989a). Maternity blues: I. Detection and measurement by questionnaire. *British Journal of Psychiatry, 155,* 356–362.

Kennerley, H., & Gath, D. (1989b). Maternity blues: III. Associations with obstetric, psychological, and psychiatric factors. *British Journal of Psychiatry, 155,* 367–373.

Khalid, R. (1989). Maternity blues and puerperal depression in Pakistani women. *Pakistan Journal of Psychological Research, 4,* 71–80.

Kumar, R. (1994). Postnatal mental illness: A transcultural perspective. *Social Psychiatry and Psychiatric Epidemiology, 29,* 250–264.

Kumar, R., & Robson, K.M. (1984). A prospective study of emotional disorders in childbearing women. *British Journal of Psychiatry, 144,* 35–47.

Laws, S. (1983). The sexual politics of pre-menstrual tension. *Women's Studies International Forum, 6,* 19–31.

Lee, K.A., & Rittenhouse, C.A. (1991). Prevalence of perimenstrual symptoms in employed women. *Women & Health, 17,* 17–32.

Levy, V. (1987). The maternity blues in post-partum and post-operative women. *British Journal of Psychiatry, 151,* 368–372.

Lubin, B., Gardener, S.H., & Roth, A. (1975). Mood and somatic symptoms during pregnancy. *Psychosomatic Medicine, 37,* 136–146.

Lutjens, C. (1998). *Mood and crying during pregnancy and postpartum: A pilot study.* Master's Thesis, Department of Psychology, Tilburg University.

Macy, C. (1983). The occurrence of "maternity blues" in Scottish and Japanese women: A meta-analysis. *Journal of Reproductive and Infant Psychology, 1,* 61–62.

Marván, M.L., & Escobedo, C. (1999). Prementrual symptomatology: Role of prior knowledge about presmenstrual syndrome. *Psychosomatic Medicine, 61,* 163–167.

Meares, R., Grimwade, J., & Wood, C. (1976). A possible relationship between anxiety in pregnancy and puerperal depression. *Journal of Psychosomatic Research, 20,* 605–610.

Moos, R.H. (1968). The development of a Menstrual Distress Questionnaire. *Psychosomatic Medicine, 30,* 853–867.

Morsbach, G., Sawaragi, I., Riddell, C., & Carswell, A. (1983). The occurrence of "maternity blues" in Scottish and Japanese mothers. *Journal of Reproductive and Infant Psychology, 1,* 29–35.

Murai, N., & Murai, N. (1975). A study of moods in pregnant women. *Tohoku Psychologia Folia, 34,* 10–16.

Offit, A.K. (1995). *Night thoughts: Reflections of a sex therapist.* Northvale, NJ: Jason Aronson.

Paarlberg, K.M., Vingerhoets, A.J.J.M., Passchier, J., Heinen, A.G.J.J., Dekker, G.A., & van Geijn, H.P. (1996). Psychosocial factors as predictors of maternal well-being and pregnancy-related complaints. *Journal of Psychosomatic Obstetrics and Gynaecology, 17,* 93–102.

Pitt, B. (1973). Maternity blues. *British Journal of Psychiatry, 122,* 431–433.

Rohde, L.A., Busnello, E., Wolf, A., Zomer, A., Shansis, F., Martins, S., & Tramontina, S. (1997). Maternity blues in Brazilian women. *Acta Psychiatrica Scandinavica, 95,* 231–235.

Ruble, D.N. (1977). Premenstrual symptoms: A reinterpretation. *Science, 197,* 291–292.

Ruchala, P.L., & Halstead, L. (1994). The postpartum experience of low-risk women: A time of adjustment and change. *Maternal-Child Nursing Journal, 22,* 83–89.

Scambler, A., & Scambler, G. (1993). *Menstrual disorders.* London: Tavistock/Routledge.

Smith, R., Cubis, J. Brinsmead, M., Lewin, T., Singh, B., Owens, P., Chan, E., Hall, C., Adler, R., Lovelock, M., Hurt, D., Rowley, M., & Nolan, M. (1990). Mood changes, obstetric experience and alternations in plasma cortisol, beta-endorphin and corticotrophin releasing hormone during pregnancy and the puerperium. *Journal of Psychosomatic Research, 34,* 53–69.

Soyka, M., Goerig, E., & Naber, D. (1991). Serum prolactin increase induced by ethanol: A dose-dependent effect not related to stress. *Psychoneuroendocrinology, 16,* 441–446.

Stein, G. (1980). The pattern of mental change and body weight change in the first post-partum week. *Journal of Psychosomatic Research, 24,* 165–171.

Stein, G., Marsh, A., & Morton, J. (1981). Mental symptoms, weight changes, and electrolyte excretion in the first post partum week. *Journal of Psychosomatic Research, 25,* 395–408.

Sugawara, M., Toda, M.A., Shima, S., Mukai, T., Sakakura, K., & Kitamura, T. (1997). Premenstrual mood changes and maternal mental health in pregnancy and the postpartum period. *Journal of Clinical Psychology, 53,* 225–232.

Sutter, A.L., Leroy, V., Dallay, D., & Bourgeois, M.L. (1995). Post partum blues et depression post-natale. Etude d'un echantillon de 104 accouchees [Post-partum blues and after-birth depression in a sample of 104 mothers]. *Annales Medico-Psychologiques, 153,* 414–417.

Taylor, D., Woods, N.F., Lentz, M.J., Mitchell, E.S., & Lee, K.A. (1991). Perimenstrual negative affect: Development and testing of an explanatory model. In: D.L. Taylor & N.F. Woods (Eds.), *Menstruation, health, and illness* (pp. 103–118). Washington DC: Hemisphere.

Theorell, T. (1992). Prolactin—a hormone that mirrors passiveness in crisis situations. *Integrative Physiological and Behavioral Sciences, 27,* 32–38.

Van Tilburg, M.A.L., & Vingerhoets, A.J.J.M. (2000). Menstrual cycle, mood, and crying. Poster presented at the Annual Meeting of the American Psychosomatic Society, Savannah, GA.

Van Tilburg, M.A.L., Unterberg, M.L., & Vingerhoets, A.J.J.M. (1999). *Crying during adolescence: The role of gender, menstruation, and empathy.* Poster presented at the second international conference on The (Non)expression of Emotions in Health and Disease, Tilburg, Netherlands.

Vingerhoets, A.J.J.M., & Scheirs, J. (2000). Gender differences in crying: Empirical findings and possible explanations. In: A. Fischer (Ed.). *Gender and emotion. Social psychological perspectives* (pp. 143–165). Cambridge: Cambridge University Press.

Vingerhoets, A.J.J.M., Assies, J., & Poppelaars, K. (1992). Weeping and prolactin. *International Journal of Psychosomatics, 39,* 81–82.

Vingerhoets A.J.J.M., Van Geleuken, A.J.M.L., Van Tilburg, M.A.L., & Van Heck, G.L. (1997). The psychological context of crying: Towards a model of adult crying. In: A.J.J.M. Vingerhoets, F.J. van Bussel, & A.J.W. Boelhouwer (Eds.), *The (non)expression of emotions in health and disease* (pp. 323–336). Tilburg: Tilburg University Press.

Whitters, A.C., Cadoret, R.J., & Widmer, R.B. (1985). Factors associated with suicide attempts in alcohol abusers. *Journal of Affective Disorders, 9,* 19–23.

Wilson, K.C.M. (1985). Mood changes after child birth. *British Journal of Psychiatry, 146,* 215–216.

Yalom, I.D., Lunde, D.T., Moos, R.H., & Hamburg, D.A. (1968). The 'Postpartum Blues' syndrome: Description and related variables. *Archives of General Psychiatry, 18,* 16–27.

York, R. (1990). Pattern of postpartum blues. *Journal of Reproductive and Infant Psychology, 8,* 67–73.

11 CRYING AND CATHARSIS

Randolph R. Cornelius

Among both scientists and the lay public, crying, or more specifically, the shedding of emotional tears, is often assumed to result in some form of tension-reduction or emotional catharsis. Catharsis is a term derived from the Greek word *katharsis*, meaning "purification," and a related word meaning "to cleanse and purify" (Jackson, 1994) and the idea is that, by crying, one may effectively purge oneself of negative affect. Standing behind this idea is another, namely, that *not* crying when it is called for may result in ill-health because the emotions that we have failed to release have the power to essentially poison our bodies, resulting in the formation of a variety of physical and psychological symptoms (see Groen, 1957).

The notion that crying results in tension-reduction or emotional relief typically takes the following form: Crying of a sufficient intensity, duration and form (see Scheff, 1979) is assumed to more or less automatically result in tension-reduction. The failure to cry in the appropriate manner in situations calling for it is assumed to result in a "discharge" of the emotional tension through other means, whether physical or symbolic. This redirected discharge is almost always seen as being destructive in one way or another, as when the failure to cry during bereavement results in the formation of skin eruptions or some other somatic manifestation (cf. Groen, 1957).

A basic assumption of the cathartic model of crying is that crying is the outcome of a hydraulic-like process in which pain, anger, sadness or other related negative affects must be given some more or less direct expression (see Nichols & Zax, 1977, pp. 3–12; Scheff, 1979, pp. 12–25). Crying is thus seen as a kind of "safety valve," for it allows the "draining off" of the energy accompanying negative affect (Sadoff, 1966). According to Sadoff (1966), "Crying seems to have the function of washing away pain and painful affects from the body" (p. 493). If this washing away cannot be accomplished directly, it will be accomplished indirectly and will take the form of a variety of physical symptoms from urticaria to asthma (see Sadoff, 1966, for a review).

Catharsis, in the context of the "model" described in this chapter, appears to have two meanings. It may refer to a set of *physiological* changes, e.g., decreases in sympathetic arousal, muscle tension, etc., or to a set of *emotional* changes, e.g., decreases in negative affect (cf. Nichols & Zax, 1977). Those who write about crying and catharsis do not always take care to distinguish between these two meanings but it is clear that many assume that both occur. In what follows, catharsis will be defined broadly to mean either.

The cathartic model of crying is often associated with psychoanalytic commentaries and case histories involving crying and studies of the ill effects of not crying published in the psychosomatic medicine literature from the 1940's to the 1960's. The model may also be found, in a slightly modified form, however, in more recently published work on crying (Crepeau, 1980/1981; Frey, 1985; Frey et al., 1981). Efran and Spangler (1979), for example, have argued that, "*all* psychologically induced tears signify relief from tension" (p. 71, emphasis added). In spite of the ubiquity of the assumption that crying may bring about cathartic relief and the vehemence with which it is asserted that crying has such positive effects (cf. Janov, 1992), what evidence is there that crying leads to catharsis?

In this chapter, I review the evidence for catharsis or tension reduction after crying. There are many case studies that describe emotional relief following crying in the psychotherapy literature, especially that with a psychoanalytic orientation (see, for example, Heilbrun, 1955; Sadoff, 1966). However, because of the unsystematic nature of such studies, the presumption that crying is followed by catharsis in many of them, the possible existence of strong demand in the psychotherapeutic relationship to report feeling better after crying, and the fact that none of the studies explicitly tested the hypothesis that crying results in emotional relief, such studies are not entirely reliable and so are not included in my review. Instead, I focus on a series of more rigorous empirical studies of various aspects of crying. To be sure, only the most recent of these explicitly tested the cathartic functions of crying. However, the inclusion in many of them of before and after crying measures of affect and, in some, autonomic activity, allows for at least a tentative evaluation of the hypothesis that crying brings about emotional relief.

After reviewing the evidence for emotional catharsis after crying, I briefly examine what little evidence there is that is relevant to the question of whether crying is associated with positive changes in health status (see also Chapter 13 in this volume). Finding that self-report and laboratory studies of crying provide contradictory evidence of crying-induced catharsis, I conclude by suggesting some of the conditions under which crying will and will not lead to catharsis.

Does Crying Lead to Catharsis?

When people are asked to describe how they feel after crying, most indicate that they feel relieved or more relaxed and report experiencing an increase in positive affect or decrease in negative affect. In his classic study of crying, Borgquist (1906) asked a sample of university students to describe their experiences with crying by means of a self-report questionnaire. Borquist reported that the majority indicated that "the effect of crying is good" (p. 181). Bindra (1972) administered a self-report questionnaire to a sample of Canadian university students on which they were asked to describe a recent occasion on which they cried. He reported, without giving exact numbers, that "some subjects" said they "felt better" after they had cried. Crepeau (1980) obtained the responses of a sample of healthy people, people suffering from ulcers, and people suffering from colitis to a questionnaire on crying as part of a study on the relationship between crying and health status. A large percentage of the participants in her study described their crying as "a release of tension or some form of letting pent up emotions out" (p. 73).

By means of a structured interview, Cornelius (1981) asked a sample of university students to describe the most recent occasion on which they cried in the presence of another person. His participants reported that, on average, their moods and overall experience of the situation were much more positive after crying than before. Frey et al. (1983) asked a large and heterogeneous sample of adults to keep a diary of their crying experiences for 30 days. The majority of men and women in the sample indicated that they "generally felt better" and experienced a decrease in the intensity of their emotions after they had cried. Lombardo et al. (1983) administered a self-report questionnaire to a sample of college students and found that the majority of women and a sizable percentage of men in the sample reported feeling "relieved" after crying. Somewhat smaller, but still appreciable, percentages of both women and men indicated that they felt more "relaxed" after crying. In another diary study of crying among students at a large university, Kraemer and Hastrup (1986) reported that crying was accompanied by significant decreases in MAACL depression scores. Finally, data collected as part of the International Study on Adult Crying from samples of college and university students in 29 countries around the world (Vingerhoets & Becht, 1997) revealed that the majority of respondents experienced marked positive changes in mood following crying. Results were similar for respondents' descriptions of their experiences with crying in general and for a specific, recent episode of crying.

The results of these studies suggest that people do indeed often feel better after they have cried in the sense that they experience decreases in negative affect and increases in positive affect. Such positive effects of crying have not always been obtained in self-report questionnaire studies, however, and have never been observed when the effects of crying are assessed under controlled laboratory conditions.

Labott and Martin (1987), in a questionnaire study on the use of crying as a coping device among university students, found that those who cried often in response to negative life events were *more* likely to report high levels of mood disturbance (tension-anxiety, depression-dejection, anger-hostility, fatigue, and confusion) than were those who rarely cried in the face of such events. These results led Labott and Martin to conclude that crying is not an effective coping device in that it "may enhance distress, instead of releasing or reducing it" (p. 162). Similar effects have been consistently found in laboratory studies of crying.

In laboratory studies of crying, self-report and, sometimes, physiological measures of affect are typically taken before, during, and after participants are shown a sad film. Since women cry at higher frequencies than do men under such circumstances, such studies often only employ women as participants, a factor that could restrict their generalizability. Such studies, however, provide strong evidence that, at least in the short run, crying does not lead to anything resembling emotional catharsis.

Marston et al. (1984), in a study representative of laboratory investigations of crying, collected self-reports of mood and bodily tension and monitored heart rate while university students watched a film called *The Champ* (Lovell & Zeffirelli, 1979). Marston et al. found that the participants in their study showed significant decreases in happiness and significant increases in heart rate and self-reported bodily tension, sadness and anger regardless of whether they cried or not in response to the film. In another study, Choti et al. (1987) found significant increases in self-reported sadness, frustration, depression, anger and muscle tension, and decreases in happiness in university students who cried in response to the film *Peege* (Knapp et al., 1973). Kraemer and Hastrup (1988), using a similar methodology and the same film, found significant increases in self-reported depression, a non-significant decrease in skin conductance level and no change in heart rate in university students who cried in response to the film compared to those who did not cry. Labott and Martin (1988) found that, even though participants who cried in response to the film *Brian's Song* (Junger-Witt & Kulik, 1971) rated themselves as somewhat less stressed than participants who did not cry, the difference was not significant. This led Labott and Martin to conclude, "there is no evidence that tears resulted in any greater reduction of subsequent stress" (p. 213).

Similar findings were obtained by Silverstein et al. (1986), who found no difference in MAACL sadness scores between those who cried in response to the film *Peege* and those who did not. In a study of the psychological and physiological effects of expressing or inhibiting positive and negative emotions, Labott et al. (1990) asked university students to either inhibit or express freely any emotions they felt while viewing a sad and a happy film. They found that participants' moods were, overall, more negative in the expression condition than in the inhibition condition. Martin and Labott (1991) found that crying during a sad film (*Brian's Song*) was associated with higher levels of depressed mood among the university students who watched the film.

In the most comprehensive psychophysiological study of crying, Gross et al. (1994) obtained self-reports of affect before and after participants, all students at a large university, watched brief excerpts from the film *Steel Magnolias* (Stark & Ross, 1989). Gross et al. also continuously monitored nine physiological responses (heart rate, skin conductance level, finger temperature, pulse transmission time to the ear and to the finger, finger pulse amplitude, respiration period, respiration depth, and general somatic activity) while participants watched the excerpts. The physiological responses of participants who cried during the film excerpts were compared with those who did not by calculating a set of change scores for each measure. Each participant who did not cry was randomly matched with one who did and the difference between a pre-film baseline and the period during which crying occurred was determined. The results clearly failed to support the hypothesis that crying leads to decreases in negative affect and physiological arousal in that crying during the film excerpts was associated with increases in self-reported sadness, embarrassment, and pain after the film, and increases in heart rate, skin conductance level, finger temperature, and general somatic activity while crying.

These studies indicate that when people cry in response to a sad film and careful measures are taken of self-reported affect and physiological arousal in a controlled setting, crying does not appear to lead to decreases in arousal, tension, or negative affect. Quite the contrary, crying is associated with increases in arousal, tension, and negative affect. One could raise the objection here that the major difference between the outcomes of these studies and the self-report studies in which subjects describe crying episodes they have chosen themselves is that the former are too artificial and represent situations that do not occur in "real life." Kraemer and Hastrup (1988) have argued, however, that, given the frequency with which subjects report crying in response to a sad film in everyday life and the number of subjects who report using films as a stimulus to cry, such a criticism is not warranted. Also, this criticism does not apply to Labott and Martin's (1987) self-report questionnaire study of the use of crying as a coping device. Clearly, crying does not always lead to catharsis. Crying does also not appear to be necessarily beneficial to one's health, as the cathartic model of crying would predict.

Is Crying Associated with Positive Changes in Health Status?

The assumption that crying is cathartic is often coupled with the assumption that crying should be associated with positive increases in health status. To date, however, only one non-case history has provided evidence for this. Crepeau (1980), in her study comparing the crying behavior of patients suffering from ulcers and colitis with healthy controls, found that those in her healthy sample tended to cry more often and had more positive attitudes toward crying than did her ulcer and colitis patients, although the differences were quite small.

Very different results were obtained in a study by Labott and Martin (1990) of the use of crying as a coping device and the prevalence of physical disorders of various kinds. Labott and Martin (1990) asked a large sample of adults ($N = 510$, 225 men, 285 women) from a city in the mid-western United States contacted via telephone to report the prevalence of 11 physical disorders (ulcers, high blood pressure, asthma, thyroid disease, skin disorders, colitis, headaches, arthritis, diabetes, heart disease, and jaw-joint disorders) and their tendency to use crying as a means of coping with stress. Labott and Martin found no relationship between the use of crying as a coping strategy and physical disorder for men, but for women, as age increased, crying was associated with an *increased* frequency of the 11 disorders. In a Dutch study, Vingerhoets et al. (1993) also found that self-rated health status was not related to crying frequency.

The studies by Labott and Martin (1990) and Vingerhoets et al. (1993) clearly fail to support the view that crying is associated with positive health outcomes. These two studies, as well as the one carried out by Crepeau (1980), however, only assessed the relationship between health outcomes and what might be called trait crying or crying proneness. It could be argued that such an assessment provides only an indirect measure of the relationship between crying and health. A much more direct assessment of the effects of crying on health, at least in the very short run, was obtained by Labott et al. (1990) in the study described above on the effects of the inhibition or expression of crying.

Recall that Labott et al. (1990) asked participants in their study to either inhibit or express their emotions while they watched a sad and a happy film. In addition to self-report measures of affect taken before and after viewing each film, Labott et al. obtained saliva samples from participants and assessed secretory immunoglobulin A (S-IgA) levels from them. Labott et al. found that participants who cried in response to the sad film had significant decreases in S-IgA levels, an indication of immunosuppression, relative to subjects who did not cry.

Thus, there is some indication that, rather than being beneficial to health, crying may be accompanied by physiological changes associated with compromises in health status.

Predicting When Crying Will and Will Not Have Cathartic Effects

Taken together, the laboratory studies of the physiological and emotional consequences of crying suggest that crying may not have the beneficial effects that it is assumed by many to have. Rather, crying appears to be associated with increased negative affect and the physiological markers of increased sympathetic nervous system arousal and somatic activity, precisely the opposite of what the cathartic model would predict. It would be a mistake, however, to ignore the many self-report studies of crying that suggest that people *do* sometimes experience cathartic effects when they cry, or at least *believe* that they do. How may we account for the discrepancies in the two types of studies? I would like to suggest two possibilities.

First, it may be the case that crying does indeed have cathartic effects but that these appear over a period of time longer than that assessed in most laboratory studies (Gross et al., 1994; J.W. Pennebaker, personal communication, August 1996). Most laboratory studies of crying to date have assessed the effects of crying only in the few minutes that immediately follow it or, at most, a couple of hours afterward (e.g., Kraemer & Hastrup, 1988). Gross et al. (1994), for example, found that crying during their stimulus film was associated with increases in negative affect immediately after the film, a period of only a few minutes. Moreover, because they obtained no post-crying measures of the physiological systems they also monitored, their evaluation of whether or not crying leads to any kind of physiological recovery was based only on a comparison of what the participants in their study experienced shortly before or while they were crying with what they experienced during the few minutes immediately preceding their crying. It could be quite reasonably argued that these conditions do not allow for a fair test of the cathartic model of crying. It may be the case that when people are asked how they usually feel after crying or how they felt after a particular episode of crying they are basing their judgments on a much longer span of time, possibly hours, or even days.

This raises the possibility that the putative cathartic effects of crying may be due simply to the passage of time and are nothing more than artifacts or contrast effects (N.H. Frijda, personal communication, August 1996), since it is likely that people are not comparing how they felt after crying with how they felt in a similar situation when they did not cry. To be sure, highlighting the different time-frames involved in laboratory and self-report assessments of the cathartic effects of crying does not account for the failure to find any significant health benefits associated with crying. However, although some theoretical accounts of catharsis and crying do explicitly tie the health benefits of crying to its tension-reducing functions (cf. Crepeau, 1980), the influence of crying on health status does not necessarily depend on crying having immediate positive effects.

The second explanation for why laboratory and self-report studies have produced conflicting pictures of the cathartic effects of crying involves the social context within which crying occurs. Perhaps people do not feel better after crying in the laboratory because their crying is in response to something about which they can do nothing and nothing in the situation changes as a result of their crying. If one is moved to tears during an argument with one's romantic partner but the issues that led one to cry have not been resolved, it is not likely that one will feel better after crying. Similarly, if one's tears are met by the disregard or disapproval by others, crying in the situation will probably not make one feel better. Evidence that my colleagues and I have gathered strongly suggests that self-reports of affect change after crying may be influenced by whether or not the issues and events that lead one to cry are resolved, by the social impact of one's crying, and by the feedback one receives for one's crying (see Cornelius et al., 1997; Cornelius, 1997).

In one study (Cornelius et al. 1997, Study 1; see also Cornelius, 1997), male and female college and university students were asked to describe an episode of crying in which they felt better after crying and one in which they did not feel better after crying. The two episodes did not differ in terms of the kinds of events involved (e.g., dissolution of a romantic relationship, death of a loved one), crying intensity, or a number of other variables (e.g., how typical the episode was).

Content analysis of participants' descriptions of the features of the episodes that led them to either feel better or not feel better after they cried, however, indicated that the two episodes did differ in terms of the apparent resolution of the issues that led participants to cry. Episodes in which participants reported feeling better after they cried were ones in which the issues that led them to cry achieved some kind of resolution (e.g., "Hearing her say that we'd work out the financial situation"). Episodes in which participants reported not feeling better after they cried were ones in which the issues that led them to cry did not achieve resolution by the time they cried or could not be resolved (e.g., "The situation was still unchanged. Even though I had cried, I couldn't resolve either my emotions or the original conflict"). The two episodes also differed in the extent to which participants reported that their crying itself helped to resolve the situation: Ratings of how much their crying helped resolve the situation were higher for the episode in which they felt better after crying than for the episode in which they did not feel better after crying.

For episodes in which participants described crying in the presence of another person, participants were significantly more pleased with the other person's reactions to their crying in the episode in which they felt better after crying than in the episode in which they did not feel better. Participants also rated their crying in the episode in which they felt better after crying as more positive, helpful, and constructive.

The two episodes also differed in terms of the duration of participants' crying: Participants reported crying for a longer period of time in the episode in which they did not feel better after crying. Since crying duration might have been related to the other variables that distinguished the two episodes, this finding hinders any clear interpretation of these results. A second study (Cornelius et al., 1997, Study 2) was thus designed to more directly test the hypothesis that people will not report feeling better after crying if the events that led them to cry remain unresolved.

As a laboratory analog to situations in which the events that lead a person to cry are resolved or not resolved, participants were given one of three types of information intended to influence their expectations about the outcome of the events depicted in a sad film. Immediately before viewing the film (*Peege*), participants, adult women residing or working on a college campus, were given a brief paragraph to read that was designed to lead them to believe that the events in the film reached either a positive, negative, or ambiguous (no) resolution. Supporting the hypothesis that outcome resolution may influence self-reports of affect change after crying, participants who received ambiguous outcome information about the events in the film and who cried during the film reported higher levels of sadness after the film compared with those who received ambiguous information but who did not cry. Participants who received negative outcome information and who cried reported lower levels of sadness after the film compared to those in the same information condition who did not cry. Contrary to expectation, participants who received positive outcome information and who cried reported higher levels of sadness relative to those in the same information condition who did not cry, although their level of sadness was below that of participants in the ambiguous information condition. Participants in the positive information condition, however, also reported feeling the least sense of resolution after the film and so these results may not be that surprising. Indeed, several participants reported that they did not find the positive outcome information believable.

The results of these two studies may be taken as tentative support for the hypothesis that a person's reports of affect change after crying may be influenced by how he or she interprets the outcome of the events that led him or her to cry in the first place. Data collected as part of the International Study on Adult Crying (Vingerhoets & Becht, 1997) provide support for the contention that self-reports of affect change after an episode of crying may depend on the impact of one's crying (Cornelius, 1997). Participants in the study on the average indicated that they generally felt better after they cried and that they felt better after crying in a recent episode of crying they were asked to describe. Their reports of how they felt after crying in the specific episode they described, however, were significantly related to how the episode turned out. Specifically, participants who indicated that their crying changed the situation they were in for the better or changed for the better the relationship they shared with someone who was present when they cried reported feeling "mentally" better after they cried relative to those whose crying did not have such positive consequences. Interestingly, there were no effects for the influence of participants' crying on how they felt "physically" after crying, suggesting that it may be important to distinguish between the physical and cognitive/emotional effects of crying (cf. Nichols & Zax, 1977).

The results of these studies suggest some of the conditions under which cathartic crying might take place. Positive affect change and tension reduction may indeed follow crying but it may take considerable time for such effects to appear. More research needs to be conducted on the emotional and physiological changes that follow crying with careful attention paid to the time course of such changes. It remains to be seen whether the critical variable is crying itself or simply the passage of time. Careful attention also needs to be paid to the social context of crying and the effects that social context has on the experience of crying (see Chapter 9 in this volume). Crying, except perhaps when due to organic brain damage, is always about something and always has a social context. The outcome of the events that lead a person to cry, the impact that his or her tears have on others, and the responses of others to his or her tears may be important influences on what a person feels before, during, and after he or she cries. Crying may not have cathartic effects if the issues that have moved one to tears have not been resolved or if one receives negative feedback from others about one's tears (cf. Plas & Hoover-Dempsey, 1988).

Whether or not one feels better after crying may also depend crucially on how one feels before one begins to cry. We know from laboratory studies of crying that crying is more likely to be elicited by a sad film from those with higher initial levels of sadness, but we really have no idea what effects baseline levels of various emotions or physiological arousal have on the emotional and physiological consequences of crying. Future research should examine this issue more closely.

References

Bindra, D. (1972). Weeping, a problem of many facets. *Bulletin of the British Psychological Society, 25*, 281–284.

Borgquist, A. (1906). Crying. *American Journal of Psychology, 17*, 149–205.

Choti, S.E., Marston, A.R., Holston, S.G., & Hart, J.T. (1987). Gender and personality variables in film induced sadness and crying. *Journal of Social and Clinical Psychology, 5*, 535–544.

Cornelius, R.R. (1981). Weeping as social interaction: The interpersonal logic of the moist eye. (Doctoral dissertation, University of Massachusetts, Amherst, 1981). *Dissertation Abstracts International, 42,* 3491B–3492B.

Cornelius, R.R. (1997). Toward a new understanding of weeping and catharsis? In: A.J.J.M. Vingerhoets, F.J. Van Bussel, & A.J.W. Boelhouwer (Eds.), *The (non)expression of emotions in health and disease* (pp. 303–321). Tilburg, The Netherlands: Tilburg University Press.

Cornelius, R.R., DeSteno, D., Labott, S., Oken, J., & Armm, J. (1997). *Weeping and catharsis: A new look*. Unpublished manuscript, Vassar College.

Crepeau, M.T. (1981). A comparison of the behavior patterns and meanings of weeping among adult men and women across three health conditions. (Doctoral dissertation, University of Pittsburgh, 1980). *Dissertation Abstracts International, 42,* 137B–138B.

Efran, J.S., & Spangler, T.J. (1979). Why grown-ups cry: A two-factor theory and evidence from *The miracle worker. Motivation and Emotion, 3,* 63–72.

Frey, W.H. (1985). *Crying: The mystery of tears.* Minneapolis: Winston Press.

Frey, W.H., DeSota-Johnson, D., & Hoffman, C. (1981). Effect of stimulus on the chemical composition of human tears. *American Journal of Ophthalmology, 92, 559–567.*

Frey, W.H., Hoffman-Ahern, C., Johnson, R.A., Lykken, D.T., & Tuason, V.B. (1983). Crying behavior in the human adult. *Integrative Psychiatry, 3,* 94–98.

Groen, J. (1957). Psychosomatic disturbances as a form of substituted behavior. *Journal of Psychosomatic Research, 2,* 85–96.

Gross, J.J., Fredrickson, B.L., & Levenson, R.W. (1994). The psychophysiology of crying. *Psychophysiology, 31,* 460–468.

Heilbrun, G. (1955). On weeping. *Psychoanalytic Quarterly, 27,* 245–255.

Jackson, S.W. (1994). Catharsis and abreaction in the history of psychological healing. *History of Psychiatry, 17,* 471–491.

Janov, A. (1992). *The new primal scream: Primal Therapy 20 years on.* New York: Enterprise Publishers.

Junger-Witt, P. (Producer), & Kulik, B. (Director). (1971). *Brian's song* [Film]. Burbank, CA: Columbia TriStar.

Knapp, D., Berman, L.S. (Producers), & Kleiser, R. (Director). (1973). *Peege* [Film]. New York: Phoenix/BFA.

Kraemer, D.L., & Hastrup, J.L. (1986). Crying in natural settings: Global estimates, self-monitored frequencies, depression and sex differences in an undergraduate population. *Behavior Research and Therapy, 24,* 371–373.

Kraemer, D.L., & Hastrup, J.L. (1988). Crying in adults: Self-control and autonomic correlates. *Journal of Consulting and Clinical Psychology, 6,* 53–68.

Labott, S.M., & Martin, R.B. (1987). The stress-moderating effects of weeping and humor. *Journal of Human Stress, 13,* 159–164.

Labott, S.M., & Martin, R.B. (1988). Weeping: Evidence for a cognitive theory. *Motivation and Emotion, 12,* 205–216.

Labott, S.M., & Martin, R.B. (1990). Emotional coping, age, and physical disorder. *Behavioral Medicine, 16*, 53–61.

Labott, S.M., Ahleman, S., Wolever, M.E., & Martin, R.B. (1990). The physiological and psychological effects of the expression and inhibition of emotion. *Behavioral Medicine, 16*, 182–189.

Lombardo, W.K., Cretser, G.A., Lombardo, B., & Mathis, S.L. (1983). Fer cryin' out loud—There's a sex difference. *Sex Roles, 9*, 987–995.

Lovell, D. (Producer), & Zeffirelli, F. (Director). (1979). *The champ* [Film]. Hollywood, CA: MGM.

Martin, R.B., & Labott, S.M. (1991). Mood following emotional crying: Effects of the situation. *Journal of Research in Personality, 25*, 218–244.

Marston, A., Hart, J., Hilleman, C., & Faunce, W. (1984). Toward the laboratory study of crying. *American Journal of Psychology, 97*, 127–131.

Nichols, M.P., & Zax, M. (1977). *Catharsis in psychotherapy*. New York: Gardner Press.

Plas, J.M., & Hoover-Dempsey, K.V. (1988). *Working up a storm: Anger, anxiety, joy, and tears on the job*. New York: W.W. Norton.

Sadoff, R.L. (1966). On the nature of crying and weeping. *Psychiatric Quarterly, 40*, 490–503.

Scheff, T. (1979). *Catharsis in healing, ritual and drama*. Berkeley: University of California Press.

Silverstein, S.M., Hastrup, J.L., & Kraemer, D.L. (1986). Individual differences in crying in a laboratory setting. Paper presented at the Fifty-seventh Annual Meeting on the Eastern Psychological Association. Buffalo, NY.

Stark, R. (Producer), & Ross, H. (Director). (1989). *Steel magnolias* [Film]. Burbank, CA: Columbia/TriStar.

Vingerhoets, A.J.J.M., & Becht, M. (1997). *The ISAC study: Some preliminary findings*. Unpublished manuscript, Tilburg University.

Vingerhoets, A.J.J.M., Van den Berg, M., Kortekaas, R.Th., Van Heck, G.L., & Croon, M. (1993). Weeping: Associations with personality, coping, and subjective health status. *Personality and Individual differences, 14*, 185–190.

12 CRYING IN PSYCHOTHERAPY

Susan M. Labott

In this chapter, I will first provide a definition of adult crying, followed by what little data we have on the frequency of crying in psychotherapy. Next, the major theoretical views on crying in psychotherapy will be presented, including psychodynamic, experiential, cognitive-behavioral, and primal therapy perspectives. Ideas about crying in response to happy events will also be described. The therapist's stance toward crying in treatment will be explored, as will case studies in which crying frequency was changed. Finally, the consequences of crying in psychotherapy and empirical data on the efficacy of crying in psychotherapy will be presented. Directions for research on the role of crying in psychotherapy will also be discussed.

Crying Defined

Crying in adults is generally defined as a strong emotional experience (either personal or empathic) which occurs in the context of tears. Theorists and researchers have studied various intensities of crying, from tears in the eyes for a few seconds to sobbing for long time periods. The term weeping has sometimes been used synonymously with crying; the term crying will be generally used in this chapter to refer to emotional reactions involving tears which occur in psychotherapy.

Whereas a number of empirical and theoretical studies of crying have appeared in the last 10–15 years (e.g., Cornelius, 1997; Kraemer & Hastrup, 1988; Marston et al., 1984), few have studied crying in the context of psychotherapy. Therefore, we know little about the frequency, intensity, and duration of crying in this context—certainly much less than the information we have accumulated about crying in the laboratory. In the only study which directly asked therapists about their experiences with crying clients in treatment, Trezza et al. (1988) focused on the frequency of crying as a percentage of total therapy sessions. Of the 227 clinical psychologists and social workers they surveyed, 21% of sessions involved clients with watery eyes, 15%

included some tears, 9% involved many tears, and 3% included sobbing. We have no information on the duration of these episodes. We can conclude only that, while low intensity expressions (i.e., watery eyes) occur in psychotherapy fairly frequently, intense sobbing episodes are rare.

As clinicians, our bias is to view crying as something that happens secondary to a discussion of intense, personal issues with a client in psychotherapy. A review of the literature demonstrates that this is certainly true, and this approach to crying is the perspective taken below. However, it should also be noted that several reports have been published in which the specific focus of therapy was to treat crying as a behavioral excess which needed to be extinguished, or at least lessened in frequency.

Theories of Crying in Psychotherapy

PSYCHOANALYTIC THEORY

Breuer and Freud (1895/1964), in their treatment of hysteria, noted that a therapist must reawaken traumatic memories in the individual along with their accompanying affect. Cure was achieved through expression of the affect in the therapy situation. They discussed the "cathartic" effect in which the individual would express the emotion overtly. "If such reaction does not result through deeds, words, or in the most elementary case through weeping, the memory of the occurrence retains above all an affective accentuation" (p. 5). If this traumatic material was not released, it could become converted into somatic problems. Once these feelings were released, however, the hysterical symptoms should disappear. Catharsis, the release of pent-up emotion, was a cornerstone of Freud's theory; this idea is found in much of the literature on emotion in psychotherapy today.

Other writers in the psychoanalytic tradition have also theorized about crying. Löfgren (1965) observed that tears are the only human excretion regarded as clean and seem to occur in adults as a response to an object loss or feelings of shame. The purpose of crying, according to Löfgren, is to dissipate aggressive energy in a harmless manner, and can be thought of as regression in the service of the ego. The cleanliness of tears seems to wash away the dirtiness of aggressive energy. He presents several cases of patients crying in psychotherapy, and all involve some aspect of power and/or aggression. Löfgren suggested that women cry more frequently than men in an attempt to keep men from behaving aggressively toward them and/or to enhance sexual attractiveness.

Greenacre (1945) documented several cases of crying in women that appeared to be associated with exhibitionism and a displacement of the need to urinate. Later, she (Greenacre, 1965) noted that crying often occurs in response to a loss, and compared emotional crying to tearing in response to irritants. Tearing is a method by which the irritated eye is protected from foreign objects, but the "disappointed eye" also attempts to defend itself from hurt, using tears. She also suggested that crying may involve an aggressive component, especially because the show of tears is under the control of the crier.

Vitanza (1960) discussed crying in terms of regression, and postulated four stages that correspond to intrauterine or birth-related states to which the individual regresses when frustrated. The first stage is *Screaming*, which corresponds to the reactions of the baby soon after birth, characterized by anger and aggression. The second stage is *Sobbing*, which is analogous to the infant's reaction during the mother's labor. Vitanza's third stage is *Weeping*, representing the need to return to the womb and immerse oneself in wetness. The final stage, *Depression*, is analogous to the embryonic period in which little activity occurs.

Heilbrunn (1955), like Vitanza, also viewed crying as a regressive effort to return to the safety of the intrauterine environment. He presents the case of a 34 year old woman, for whom crying became the most significant symptom in psychotherapy. She cried almost constantly in many sessions, and the crying was not always appropriate to the content of her conversation. The crying, which also increased in frequency outside of sessions, was interpreted as representing the satisfaction of being understood by others, coupled with a fear of being misunderstood (which was her perception of her parents' response to her in childhood). Further exploration of infantile trauma yielded fantasies about her father's death and fears of her mother's aggression. These insights ended the excessive crying. Heilbrunn conceptualized the case as involving an injury to the patient's narcissism which she re-enacted in the transference; gaining confidence in herself also negated the need for her to follow the rules of her parents.

Analytic writers also theorized on the occurrence of crying associated with happy or positive events. Bergler (1952) noted that typically many dreary events marked by a lack of success occur in life, while excessively positive events are rare. During the dreary times, psychic masochism is built up, and must be released. If an event results in a failure experience, depression results. If the outcome is unexpectedly positive, the individual must find an outlet for the built up psychic masochism. Tears result, in which the individual is crying for the unhappiness which preceded the positive event.

Weiss (1952) suggested that grief is delayed until a happy event occurs because it is too threatening to acknowledge when the grief is current. When the happy event occurs, the grief needs to no longer be repressed and its delayed expression is experienced as pleasurable. Feldman (1956) articulated the position that individuals cry "happy" tears when faced with the contrast between the current positive event and the many previous sad ones. The individual cries, knowing the happiness is transitory. These authors, then, would argue that crying in response to "happy" events is not really happy at all.

EXPERIENTIAL PSYCHOTHERAPIES

The experiential or humanistic therapies include a variety of approaches and techniques, e.g., Perls' Gestalt therapy (Perls et al., 1977), the experiential psychotherapy of Mahrer (1996), and the experiential process approach (Greenberg et al., 1993). As a group, the experiential therapies share the belief that individuals are forward-moving, that treatment needs to be present-oriented, and that emotion is an important part of experience, and, as such, is critical to the therapeutic process.

Perls et al. (1977) wrote "It is only in the recognition of your emotions that you can be aware, as a biological organism, either of what you are up against in the environment or of what special opportunities are at the moment presented....it is only if you acknowledge and accept your grief—the sense of despair and not knowing where to turn as you confront the loss of someone or something of great concern to you—that you can weep and say goodbye" (p. 115–116).

Greenberg and Safran (1987) noted that arousing affect is an important component of the change process in psychotherapy; both clients and therapists associate sobbing and other expressions of hurt with significant events in the therapy process. Greenberg (1993) proposes that the elicitation of emotion leads to several important processes: Emotions allow individuals to become aware of their "core" beliefs about themselves, emotions can result in action, and the expression of emotion causes further cognitive processing and emotional release which promote recovery. While the emotional experience and expression are important, work with the individual's belief system is also necessary to promote change (Greenberg & Safran, 1990).

Because of the pivotal role that emotions play in the experiential treatments, therapists have developed specific techniques which are designed to evoke and intensify emotional experiences. For example, in a procedure called *systematic evocative unfolding* (Greenberg et al., 1993), therapists collaborate with clients to re-explore a particular problematic event; one outcome of this process is the individual's greater awareness of his/her emotional reactions to various aspects of the situation. Gendlin (1981) has developed a focusing procedure in which individuals gain skill in accessing, understanding, and using their own internal emotional processes to promote change. Enactment is also often used to facilitate emotionality, for example, to enable individuals to complete unfinished business (e.g., Paivio & Greenberg, 1995; Polster & Polster, 1973).

Gendlin (1991) describes how tears and crying fit into the experiential framework: Tears may be held back in psychotherapy, and all the material associated with them may also be held back. Therefore, the individual must allow or "welcome" the tears. When the tears are allowed, the past material can come with them, and can then be addressed in treatment. Gendlin also notes that tears come in various forms, gentle tears versus uncontrollable sobbing, and that they may be associated with either past or current events. In summary, Gendlin believes the tears are associated with "new life-steps," but that crying is not necessary for these new steps to occur.

Mahrer and Roberge (1993) provide an example of an intense sobbing event from an experiential therapy session. Initially, the client is asked to put all of her attention on a strong feeling, and she describes feelings of panic and craziness. She begins to sob as the therapist encourages her to become more involved with the feelings and images; the client becomes angry as her experience continues. Her new experiences are welcomed and appreciated, i.e., experienced more thoroughly. Next, the client is encouraged to be a new person who can fully experience the earlier events in a way that she had not before. Finally, the client is encouraged to be the new person in her current life. Often this involves trying on new behaviors both in and outside of therapy. This transcript demonstrates specific interventions designed to access and elicit the internal experience of the individual and their connections to experiential theory.

COGNITIVE-BEHAVIORAL PSYCHOTHERAPY

Cognitive-behavioral therapies focus on the reduction of negative emotions (e.g., anxiety, depression) by intervening at the level of cognition. The assumption is that thinking causes feeling; to change feelings one must change cognition. Because of this focus on rationality and cognition, traditionally, cognitive-behavioral therapists have been seen as disinterested in the elicitation of emotion. Cognitive-behavioral therapists today do discuss the importance of emotion in the therapeutic process, but emotions are generally not elicited, enhanced, and made the focus of treatment as in the experiential psychotherapies. For example, Ellis (1995) discusses "vigorous evocative-emotive" techniques that are utilized to dispel irrational thoughts and to alter behavior. These include shame-attacking exercises, rational-emotive imagery, and role-playing, to mention a few.

Several articles specifically address the conceptualization of crying in therapy (Efran & Spangler, 1979; Nichols & Efran, 1985). These authors view crying as the second phase of a two-stage process. Initially, the individual is aroused, often when a goal is blocked (activation stage). Crying occurs when the individual gives up an existing cognitive schema (recovery phase). From this perspective, then, crying is healthy because it occurs following an important therapeutic event. These authors note that crying, itself, does not feel good—recovery does.

Beck et al. (1979) specifically address the issue of crying in the psychotherapy of depression. Their perspective is that, while there is no need for an explicit focus on emotional expression, crying can have various implications for treatment. On the one hand, crying can provide a clue to the client's reactions and attitudes toward another individual; a woman's uncontrollable crying when discussing her "wonderful" relationship with her husband is an example provided by Beck et al. (1979, p. 38). Other examples of the benefits of crying are that it may allow the individual to feel "self-sympathy" after a loss or to experience relief. However, Beck et al. (1979) also describe situations in which crying can interfere with treatment either because it delays therapeutic movement, or because it results in feelings of shame afterwards. In these cases, self-control procedures can be utilized to enable the individual to manage/minimize the emotional expression.

Several case studies have been presented in which behavioral treatment has been used to alter the frequency of crying behavior. Linton (1985) utilized assertiveness training, specifically regarding emotions, combined with systematic shaping of crying behavior, to increase appropriate expressions of emotion in a female client.

The utilization of behavioral treatments to decrease crying frequency is more common. Cases have been reported in which desensitization (Field, 1970) and contingency management (Redd, 1982) have resulted in successful decreases in crying. Two cases have used anger, either

Table 1. The role of emotion in psychotherapy from psychodynamic, experiential, and cognitive-behavioral perspectives.

	Perspective		
	Psychodynamic	*Experiential*	*Cognitive-behavioral*
Relation to cognition	Emotion results from memory of trauma	Emotion interacts with cognition	Emotion results from cognition
Treatment goals	Release the inhibited emotions	Increase emotional experiencing; use information to increase adaptive behavior	Identify and change dysfunctional beliefs which cause maladaptive emotional states
Type of emotion focused on	Unpleasant experiences and states (hurt, anger)	Adaptive emotions (anger, happiness)	Maladaptive emotional reactions (depression, anxiety)
Therapist's stance	Allows expressions to occur	Encourage/evoke emotional experiences	Tolerate or minimize emotion
Role of emotional experience in treatment	Emotion is necessary at some points in treatment	Must be present for productive work	Emotion need not be currently experienced

Note: Adapted from Labott and Elliott (1990).

assertiveness training for anger expression (Rimm, 1967) or imagery of others getting angry (Tasto & Chesney, 1977) to decrease crying frequency in men.

Table 1 presents a summary comparison of the role of emotion in psychodynamic, experiential, and cognitive-behavioral treatments. Here it can be seen that the importance and place of emotion in psychotherapy differs in these three perspectives.

PRIMAL THERAPY

Primal therapy is perhaps the psychotherapy in which crying occurs most frequently and most intensely. Crying is also written about extensively in the literature of primal therapy (Janov, 1991). Crying is conceptualized in primal therapy as a language through which individuals communicate their suffering and also as a curative process. Janov theorizes that the tears which are the most beneficial are those of the child ("infantile wails") rather than those of the adult ("crying about" events). Repression of the expression of sadness should result in mental illness, physical illness, or behavioral problems. While the theory behind primal therapy does address crying specifically, this treatment is less commonly used by therapists and less empirically validated than the others described above.

Therapist Behaviors

Freud's perspective on catharsis has prompted a wealth of writing and theorizing about the role of emotional expression in both physical and psychological health (e.g., Crepeau, 1981; Pennebaker, 1995; Scheff, 1979; see also Chapters 11 & 13, this volume). Especially in the context of bereavement, crying is seen as a part of the adjustment process (e.g., Rando, 1991; Worden, 1991). Guinagh (1987) views crying as a way to weaken the bond with the lost individual; repeated crying episodes gradually allow the individual to make a healthy adjustment to the loss.

From a theoretical perspective, one would assume that the therapist's stance toward a client's emotion in psychotherapy would differ from one orientation to another (see Table 1). Especially in the experiential tradition, specific interventions have been designed to enable therapists to access and intensify the client's emotional experiencing. Nichols and Efran (1985), working from the perspective of the two-stage cognitive theory described above, provide guidelines to help clinicians manage the emotion of their clients. They recommend that therapists be sure that neither they nor the client curtail the recovery phase and its emotional manifestations. They suggest that therapists learn to use catharsis as a clue that the individual is not acting on their emotions in life; allow them to experiment with emotional expression outside of therapy. Lastly, therapists should use catharsis as a first step toward helping individuals to define themselves in the context of their social relationships.

Mills and Wooster (1987) discuss the importance of the therapist's reaction to a crying client. Consistent with Nichols and Efran (1985), they believe that the therapist needs to be sure that there are no implicit suggestions that the client stops crying. Therapists must be aware of their own socially conditioned attitudes toward crying, and be certain to not convey these to the crying client. Mills and Wooster also note the incompatibility between crying and talking about crying, i.e., asking "why" inhibits the expression. Finally, it is important to remember that crying may occur for various reasons, e.g., to express deep feelings versus as a defense to avoid important issues. Lackie (1977) presents a case of a family in which crying was initially used as a resistance, but later in treatment it was used to express appropriate mourning. These authors agree that the therapist needs to be aware of the meaning of the client's communication (and often discuss it with the client) in order to do effective therapy.

In spite of the discussions of ways to handle crying above, little empirical work has evaluated therapist interventions with respect to crying in psychotherapy. Trezza et al. (1988), in their study of crying in psychotherapy, reported that 63% of clinicians (from various orientations) did not try to stop a client's crying, while 34% did. If a client was about to cry, 73% of therapists had in some way encouraged the client to cry, while 24% had never done so. The clinicians in this sample reported that they were not uncomfortable with a client's crying, and that they tended to see the crying as healthy, i.e., that it would help to decrease depression.

Empirical Evidence on the Efficacy of Crying in Psychotherapy

Only a few studies have looked at the role of emotional expression in relation to therapeutic outcome. Two studies were based on Jackins' (1981) re-evaluation counseling; therapists employed specific techniques designed to elicit emotional expressions such as laughter, anger, and crying. (In both studies, the various types of emotional expressions were considered together.) Nichols (1974) compared brief emotive to insight-oriented analytic treatment. Results indicated that the emotive group expressed more emotion and improved more on behavioral goals than the insight-oriented group. Bierenbaum et al. (1976) compared the effects of emotive psychotherapy in three different time frames: (a) one-half hour, once weekly, (b) one hour, once weekly, and (c) two hours on alternate weeks. Results indicated that the one hour group experienced the most catharsis and improved most on behavioral goals and personal satisfaction. Finally, Beutler and Mitchell (1981) compared the efficacy of experiential versus analytic procedures in clinic outpatients. Their results indicated greater improvement in target symptoms and overall change for experiential treatment. These studies provide evidence for the efficacy of emotional expression generally, although not for crying in psychotherapy, specifically.

Only one study has provided an explicit analysis of a crying event in the context of experiential psychotherapy (Labott et al., 1992).

Completed in the context of a study on the psychotherapy of depression, the client picked the crying event as the most helpful event in the third session. A qualitative analysis of this event, using Comprehensive Process Analysis (see Elliott, 1989, for a description of the method) yielded a wealth of information on the unfolding of the event within the session, therapist interventions, and the interpersonal, emotional, and cognitive impacts of the event. Of most interest is the four-factor model of crying in psychotherapy which emerged from these data. The factors that accounted for the crying included the fact that the client had a significant amount of unexpressed and unfinished emotion from earlier events in her life. She also had a great deal of stress and upset in her life currently, e.g., father's death, marital stress. Both of these factors are consistent with the psychodynamic overflow theory. Thirdly, the client cried because she felt safe with the therapist, and felt that the therapist was able to handle the intense sobbing which occurred. This factor clearly speaks to the importance of the relationship and the therapist's behavior with respect to client crying. Lastly, the crying occurred because the client accessed her early memories and feelings, and also the schemata which were associated with them, consistent with both experiential and cognitive perspectives. In terms of consequences, the major outcomes for the client involved a realization about the central issue underlying her current problem, i.e., the role of past experiences. She also reported feeling tired and drained, and complained of a headache. While this event was certainly not easy for the client, either emotionally or physically, she still reported that it was the most helpful therapeutic event in this session.

The studies described above indicate that both clients and therapists perceive crying as helpful in psychotherapy, although their reasons as to why it is helpful may vary. It is important to note, however, that several laboratory studies seem to indicate that crying does not alleviate depressed mood (e.g., Cornelius, 1997; Kraemer & Hastrup, 1988); it may also negatively impact physiology (Labott et al., 1990). The results in the laboratory and psychotherapeutic settings are likely to differ in several ways, however, which may account for the discrepant findings. First, the treatment setting is generally perceived by the client as a safe and supportive atmosphere; the same can not be said for a laboratory setting which is, at best, impersonal and neutral. Secondly, the crying which occurs in psychotherapy is likely to be of greater intensity and duration than that which occurs in laboratory studies. For example, in the Labott et al. (1992) study, the client sobbed at a high level of intensity for approximately five minutes, then had three other brief crying episodes within the next nine minutes. In contrast, most laboratory studies involve only a few moments of tears in the eyes or down the cheeks. Lastly, the contents of the events associated with crying in psychotherapy are likely to be more intensely personal than in laboratory studies. Even while many laboratory studies involve empathic reactions to movies about loss, for example, they are a step removed from the individual's personal losses.

Future Directions

From the material presented above, it is clear that we have a wealth of theoretical ideas about crying in general, and also about its role in psychotherapy specifically. However, while we have the beginnings of a body of empirical data on crying in the laboratory, data on crying in psychotherapy are lacking. It is perhaps the complexity of psychotherapy that makes the study of only one aspect of it difficult. Yet, as noted above, crying in the laboratory and crying in psychotherapy may be very different—therefore our knowledge base would benefit from the study of both.

One might wonder if there exists a typology of crying behavior, e.g., a few quiet tears present a very different stimulus than long and loud sobbing. While the intensity and presentation of the tears may reflect the precipitating event and the social context, individuals seem to have different characteristic styles of crying that predispose them to certain forms of expression. These styles may result from early experiences, and may have very different effects when they occur in psychotherapy, i.e., therapists may respond differently. Cornelius (1997) has argued that the outcome of a crying event is partly determined by its effect upon the social situation, yet the role of the therapist in the outcome of crying in psychotherapy has been largely unexplored.

In summary, crying in some form occurs often in the context of psychotherapy. While the various psychotherapeutic orientations have different theoretical perspectives on its purpose, most work with the emotional expression to promote therapeutic change. Further, a majority of therapists believe that crying is healthy, either psychologically or physically, yet we have very little evidence on which to base this belief. Future study of psychotherapeutic process and outcome would do well to delineate the role of crying. This would move us closer to an understanding of the complicated biopsychosocial processes involved in the phenomenon of crying in psychotherapy.

References

Beck, A.T., Rush, J.J., Shaw, B.F., & Emery, G. (1979). *Cognitive therapy of depression*. New York: Guilford.

Bergler, E. (1952, November). Paradoxical tears—tears of happiness. *Diseases of the Nervous System*, 337–338.

Beutler, L.E., & Mitchell, R. (1981). Differential psychotherapy outcome among depressed and impulsive patients as a function of analytic and experiential treatment procedures. *Psychiatry*, *44*, 297–306.

Bierenbaum, H., Nichols, M.P., & Schwartz, A.J. (1976). Effects of varying session length and frequency in brief emotive psychotherapy. *Journal of Consulting and Clinical Psychology*, *44*, 790–798.

Breuer, J., & Freud, S. (1964). *Studies in hysteria* (A.A. Brill, Trans.). Boston, MA: Beacon Press. (Original work published 1895).

Cornelius, R.R. (1997). Toward a new understanding of weeping and catharsis? In: A.J.J.M. Vingerhoets, F.J. van Bussell, & A.J.W. Boelhouwer (Eds.), *The (non)expression of emotions in health and disease* (pp. 303–321). Tilburg, Netherlands: Tilburg University Press.

Crepeau, M.T. (1981). A comparison of the behavior patterns and meanings of weeping among adult men and women across three health conditions. (Doctoral dissertation, University of Pittsburgh, 1980). *Dissertation Abstracts International*, *42*, 137B–138B.

Efran, J.S., & Spangler, T.J. (1979). Why grown-ups cry. *Motivation and Emotion*, *3*, 63–72.

Elliott, R. (1989). Comprehensive Process Analysis: Understanding the change process in significant therapy events. In: M.J. Packer & R.B. Addison (Eds.), *Entering the circle: Hermeneutic investigations in psychology* (pp. 165–184). New York, NY: State University of New York Press.

Ellis, A. (1995). Fundamentals of Rational Emotive Behavior Therapy for the 1990s. In: W. Dryden (Ed.), *Rational Emotive Behaviour Therapy* (pp. 1–30). London, UK: Sage.

Feldman, S.S. (1956). Crying at the happy ending. *Journal of the American Psychoanalytic Association*, *4*, 477–485.

Field, P.B. (1970). Preventing crying through desensitization. *American Journal of Clinical Hypnosis*, *13*, 134–136.

Gendlin, E.T. (1981). *Focusing*. New York: Bantam.

Gendlin, E.T. (1991). On emotion in therapy. In: J.D. Safran & L.S. Greenberg (Eds.), *Emotion, psychotherapy, and change* (pp. 255–279). New York: Guilford.

Greenacre, P. (1945). Pathological weeping. *Psychoanalytic Quarterly*, *14*, 62–75.

Greenacre, P. (1965). On the development and function of tears. *Psychoanalytic Study of the Child*, *20*, 209–219.

Greenberg, L.S. (1993). Emotion and change processes in psychotherapy. In: M. Lewis & J.M. Haviland (Eds.), *Handbook of emotions* (pp. 499–508). New York: Guilford.

Greenberg, L.S., & Safran, J.D. (1987). *Emotion in psychotherapy*. New York: Guilford.

Greenberg, L.S., & Safran, J.D. (1990). Emotional-change processes in psychotherapy. In: R. Plutchik & H. Kellerman (Eds.), *Emotion: Theory, research, and experience*, vol. 5 (pp. 59–85). San Diego, CA: Academic Press.

Greenberg, L.S., Rice, L.N., & Elliott, R. (1993). *Facilitating emotional change: The moment-by-moment process*. New York: Guilford.

Guinagh, B. (1987). *Catharsis and cognition in psychotherapy*. New York: Springer-Verlag.

Heilbrunn, G. (1955). On weeping. *Psychoanalytic Quarterly*, *24*, 245–255.

Jackins, H. (1981). *The human side of human beings*. Seattle, WA: Rational Island Publishers.

Janov, A. (1991). *The new primal scream*. Wilmington, DE: Enterprise.

Kraemer, D.L., & Hastrup, J.L. (1988). Crying in adults: Self-control and autonomic correlates. *Journal of Social and Clinical Psychology*, *6*, 53–68.

Labott, S.M., & Elliott, R. (1990). *Emotion and change in cognitive therapy of depression*. Unpublished manuscript.

Labott, S.M., Elliott, R., & Eason, P.S. (1992). "If you love someone, you don't hurt them": A Comprehensive Process Analysis of a weeping event in therapy. *Psychiatry*, *55*, 49–62.

Labott, S.M., Ahleman, S., Wolever, M.E., & Martin, R.B. (1990). The physiological and psychological effects of the expression and inhibition of emotion. *Behavioral Medicine*, *16*, 182–189.

Lackie, B. (1977). Nonverbal communication in clinical social work practice. *Clinical Social Work Journal*, *5*, 43–52.

Linton, S.J. (1985). A behavioral treatment for inability to express emotions. *Scandinavian Journal of Behavior Therapy*, *14*, 33–38.

Löfgren, L.B. (1965). On weeping. *British Journal of Psychoanalysis*, *47*, 375–383.

Mahrer, A.R. (1996). *The complete guide to experiential psychotherapy*. New York: John Wiley & Sons.

Mahrer, A.R., & Roberge, M. (1993). Single-session experiential therapy with any person whatsoever. In: R.A. Wells & V.J. Giannetti (Eds). *Casebook of the brief psychotherapies* (pp. 179–196). New York: Plenum.

Marston, A., Hart, J., Hileman, C., & Faunce, W. (1984). Toward the laboratory study of sadness and crying. *American Journal of Psychology*, *97*, 127–131.

Mills, C.K., & Wooster, A.D. (1987). Crying in the counseling situation. *British Journal of Guidance and Counselling*, *15*, 125–130.

Nichols, M.P. (1974). Outcome of brief cathartic psychotherapy. *Journal of Consulting and Clinical Psychology*, *42*, 403–410.

Nichols, M.P., & Efran, J.S. (1985). Catharsis in psychotherapy: A new perspective. *Psychotherapy*, *22*, 46–58.

Paivio, S.C., & Greenberg, L.S. (1995). Resolving "Unfinished Business": Efficacy of experiential therapy using empty-chair dialogue. *Journal of Consulting and Clinical Psychology*, *63*, 419–425.

Pennebaker, J.W. (1995). *Emotion, disclosure, and health*. Washington, DC: American Psychological Association.

Perls, F.S., Hefferline, R.F., & Goodman, P. (1977). *Gestalt therapy*. New York: Bantam.

Polster, I., & Polster, M. (1973). *Gestalt therapy integrated*. New York: Vintage.

Rando, T.A. (1991). *How to go on living when someone you love dies*. New York: Bantam.

Redd, W.H. (1982). Treatment of excessive crying in a terminal cancer patient. *Journal of Behavioral Medicine*, *5*, 225–235.

Rimm, D.C. (1967). Assertive training used in treatment of chronic crying spells. *Behavior Research and Therapy*, *5*, 373–374.

Scheff, T.J. (1979). *Catharsis in healing, ritual, and drama*. Berkeley, CA: University of California Press.

Tasto, D.L., & Chesney, M.A. (1977). The deconditioning of nausea and of crying by emotive imagery. *Journal of Behavior Therapy and Experimental Psychiatry*, *8*, 139–142.

Trezza, G.R., Hastrup, J.L., & Kim, S.E. (1988, April). *Clinicians' attitudes and beliefs about crying behavior*. Eastern Psychological Association, Buffalo, NY.

Vitanza, A.A. (1960). Toward a theory of crying. *Psychoanalysis and Psychoanalytic Practice*, *47*, 65–79.

Weiss, J. (1952). Crying at the happy ending. *Psychoanalytic Review*, *39*, 338.

Worden, J.W. (1991). *Grief counseling and grief therapy*. New York: Springer.

13 CRYING AND HEALTH

Ad J.J.M. Vingerhoets and Jan G.M. Scheirs

In the popular press there is little doubt that crying is healthy. Cornelius (1986) systematically examined the content of popular articles on weeping published mainly in the United States from the Mid-1800's to 1985. This yielded a total of 70 articles. As one of the major themes, the author identified the conviction that failure to cry results in negative (health) consequences. In no less than 94 percent of the articles the advice to the readers was to let their tears flow. In particular, in the articles published after 1950, Cornelius showed, the classic psychosomatic point of view (cf. Groen, 1957) was popularized. Crying was considered to be an important means for releasing physiological tension. If tension was not released by crying, it might find an outlet elsewhere, for instance, by affecting the body and resulting in disease.

Withholding tears has been described as being potentially emotionally or even physically damaging. Headaches, ulcers, hypertension, and insomnia are examples of disorders that were considered to result from the failure to cry. However, one should be aware that these beliefs prevail in our western culture, which does not necessarily mean that other cultures have the same opinions. Wellenkamp (1988), for example, describes the beliefs of the Toraja, living in Indonesia, with respect to crying and catharsis. For the members of this tribe, it is the expression of negative emotions that should be avoided, because it may lead to serious illness, insanity, and even premature death. Crying is only expected and regarded as positive, even preventing illness, following a death and during a funeral. In addition, it is a traditional remedy for an infertile woman to call for the women to cry with her at a rock said to be inhabited by a spirit (Wellenkamp, 1992).

There is an even longer history of the presumed association between crying and health in writings on medicine and in the arts. Cornelius (1997) cites the Dutch physician and philosopher Franciscus Mercurius Van Helmont, who as early as 1694 wrote about the necessity of crying after bereavement in order to prevent the development of distemper or sickness. A further well-known quotation is from the poem "The Princess," written in 1847 by Alfred Lord Tennyson:

Home they brought the warrior dead;
She nor swooned nor uttered cry.
All her maids, watching said,
"She must weep or she will die".

More recently, Frits Zorn, the pseudonym for a Swiss cancer patient, wrote in his 1988 autobiography that all the tears that he had never wept nor had wished to weep during his life had massed in his neck and had created his tumor, because—as he posited—their true function, i.e., being shed, had not been attained.

Darwin (1872/1965) pointed out that "children, when wanting food or suffering in any way, cry out loudly (...) partly as a call for their parents for aid, and partly from any great exertion serving as a relief" (p. 174). He even suggested a dose-response relationship, as evidenced by the following comment: "And by as much as the weeping is more violent and hysterical, by so much will the relief be greater,—on the same principle that the writhing of the whole body, the grinding of the teeth, and the uttering of the piercing shrieks, all give relief under an agony of pain" (p. 175). Menninger et al. (1964) noted that crying may be considered as perhaps the most human and most universal of all relief measures.

Frey (1985) further quotes the famous British physician Sir Henry Maudsley, stating that "Sorrows which find no vent in tears may soon make other organs weep." Other examples of the conviction that crying is healthy and beneficial can be found in Solter (1995), who considers crying "an inborn healing mechanism"(p. 28) and in Mills and Wooster (1987), who describe crying as a "vital part of a healing or growing process that should not be hindered" (p. 125). Nearly thirty years ago, Rees (1972) speculated that the increased mortality observed after bereavement in men may be due to repression of the expression of sadness, including crying. To put it briefly: Cry or die!

The idea that unexpressed emotions, including the inhibition of crying, may result in physical disease has been conceptualized by different models, including what may be referred to as repression theory, first described by Freud (1915/1957) and further elaborated by Pennebaker (e.g., Berry & Pennebaker, 1993; Pennebaker & Susman, 1988). The basic assumption of this theory is that the active inhibition of emotions and behavior requires physiological effort, thus resulting in a more or less chronic strain to the biological system. Early case descriptions by psychosomatically oriented clinicians also provided support for the idea that the chronic inhibition of sadness and crying was specifically related to the development of respiratory disorders such as asthma, as had been hypothesized by psychoanalysts like French and Halliday (see Alexander, 1950, p. 139). In more recent times, the rationale behind the encouragement to cry has been mainly based on Frey's (1985) ideas that crying is important for the excretion of toxic substances released during distress.

There is thus little doubt in the popular media and in scientific literature that crying is important for one's health. An early example of the scientific approach is the study by Borgquist (1906). This author sent out a questionnaire with open-ended questions and found that people rarely indicate that crying is not beneficial. Interestingly, he made a distinction between the immediate physical effects and the delayed effects of crying on well-being and mood. Many respondents in his sample reported being exhausted, sick, or physically tired from crying, but also mentally relieved. Borgquist also paid attention to the frequently reported termination of crying episodes by sleep, which according to the author may be expected after a crying spell characterized by both physical exhaustion and mental relief. On the other hand, he also referred to the ill effects of (prolonged) crying, as evidenced by sickness, loss of appetite, physical weakness, and "unusual activity of the heart" after a crying spell. As more permanent or long-term effects of crying, Borgquist identified symptoms like headache, stupor, sickness, exhaustion, nausea, and sore eyes. He further made mention of theories trying to explain crying and its effects on somatic processes. With regard to the beneficial effects of crying, these theories emphasize the stimulation of circulatory mechanisms, the revival of metabolic processes, and the relief of the overcharged nervous system.

Borgquist (1906), however, did not attempt to measure the effects of crying directly and systematically. Also, some relevant additional studies have appeared since then, which we will summarize below. These will make clear that it is not established whether crying promotes health, nor what the mechanisms are by which the effects, if any, might come about. In this contribution, we want to deal in more detail with these intriguing questions, summarizing the relevant scientific literature, and concluding with some suggestions for future research.

An Overview of Research Findings

Theoretically, the hypothesis that crying has a positive effect on health might be examined from two different viewpoints: (i) What is known about the short-term effects of crying? Does a crying spell positively influence mood and/or psychobiological processes, including autonomic nervous system activity, endocrine activity, and/or immune functions? and (ii) What is known about the relationship between crying frequency as a more or less stable person feature and health status? The last broad question addresses the long term effects of crying.

We will now discuss the available evidence for each of these questions in more detail.

THE IMMEDIATE EFFECTS OF CRYING

With respect to the proposition that crying brings relief and improves mood, the scientific evidence is limited and inconsistent, as made clear by Cornelius (1997, Chapter 11, this volume). In order to avoid unnecessary overlap between chapters, we here refer to Chapter 11 and limit ourselves to its conclusion that there is a remarkable difference in the results of laboratory and retrospective real-life studies. Cornelius suggests that one variable in particular plays an important role in determining the immediate effects of crying, namely, whether or not the negative events that led the individual to cry have been resolved. If the crying person understands that nothing has changed, (s)he will be less likely to report mood improvement.

From the beginning of this century until now, however, an issue that has been fairly neglected in the literature is the physiological mechanism by which the shedding of tears may bring relief and mood improvement. Gross et al. (1994) give an overview of several mechanisms that have been held responsible for the assumed relief. They distinguish between theories fitting in a "recovery view", versus those fitting in an "arousal view." According to the recovery view, the function of crying is to restore homeostasis in one way or another, for instance, by the release of stress-related toxins, or by an increase in parasympathetic activation (as reflected in tears) that follows high levels of sympathetic activation caused by emotions. As an alternative speculation fitting in the recovery view but left unnoticed by Gross et al. (1994), we would like to add the hypothesis that crying induces the release of endorphins which may facilitate restorative processes after having been in distress (e.g., Panksepp, 1986). Crying is here thus seen as part of an arousal reduction response. In contrast, according to the arousal view, crying leads to a state of increased physiological activation that is stressful and aversive to both the crier and the bystander. This state would bring about attempts to lessen the tears and change the situation as well as stimulate emotional support, comforting behavior, and, possibly, inhibition of aggression, resulting in mood improvement. For a more detailed discussion of the effects of crying on others, we would like to refer to Chapter 9.

In the following, we address studies aimed at investigating the relationship between crying and psychobiological processes specifically. A number of laboratory studies, aptly reviewed by Cornelius (1997), focused on the effects of crying on mood as well as psychophysiological functioning (Cornelius et al. (cited in Cornelius, 1997); Gross et al., 1994; Kraemer & Hastrup, 1988; Marston et al., 1984). The results are rather clear and consistent. The data generally fail to support the hypothesis that crying facilitates recovery. Kraemer and Hastrup did not find differences in heart rates of criers and non-criers exposed to a sad film. In contrast, Marston et al. and Gross et al. reported significant increases in heart rate and—in the case of Gross et al.—also in other indicators of autonomic nervous system activity like finger temperature, galvanic skin response, and respiration. Gross et al. explicitly addressed the issue of physiological recovery after crying. Contrary to expectations, it took longer for crying than for non-crying participants to reach baseline levels of arousal after having been exposed to a sad film. These results thus fail to support the physiological recovery hypothesis formulated above.

As far as is known, only one study focused on the effects of crying on levels of stress hormones. Vingerhoets and Kirschbaum (1997) exposed female subjects to emotional movies and measured mood and saliva cortisol. Although moderate but significant correlations between self-reported intensity of crying and reduction in cortisol could be demonstrated, no differences in the hormonal levels of criers and non-criers were found after the film. In this respect, some animal work may also be interesting. Bayart et al. (1990) explored the reactions of monkeys to separation from their mothers. A remarkable finding was that in these monkeys there was a negative correlation between plasma cortisol levels and vocalizations expressing distress. Thus, screaming was associated with reduced cortisol secretion. Assuming that human crying is equivalent to these distress reactions, it may be hypothesized that crying also reduces the cortisol output of the adrenals. These preliminary findings suggest that crying may promote the recovery of the homeostatic balance within the body, a conclusion seemingly in contrast to the findings by Gross et al. (1994) concerning the level of autonomic nervous system activation.

In addition, there are two studies (Labott et al., 1990; Martin et al., 1993) investigating the effects of crying on an immunologic variable that is characterized as a first-line defense against invasion by potential pathogens, namely, secretory immunoglobulin A (S-IgA). The results of both studies yielded evidence of a negative effect of crying on this immunologic parameter and thus of a negative effect of crying on health. When people cried, there were significant decreases of S-IgA levels, representing decreased protection against pathogens. Such decrements in S-IgA were not found when subjects only felt sad. In other words, not the mere feeling of sadness, but the specific act of crying appeared to have a negative influence on the body's defense mechanisms.

A topic that we do not want to leave undiscussed here is the work of Panksepp and coworkers on distress vocalizations in animals (see Panksepp (1998) for an overview). Based on their work, we hypothesize that crying may trigger the release of certain endogeneous opioids, which may have a sedative and pain reducing effect. As mentioned above, Panksepp also suggests as a main function of these substances the facilitation of recovery after having been in distress. These hypothesized effects thus match nicely the supposed functional effects of crying (sedation, pain reduction, restoring the homeostatic balance). Further research is needed to test this intriguing hypothesis in humans.

Finally, there is the already mentioned anecdotal evidence put forth by Borgquist (1906), suggesting that crying results in symptoms like headache, nausea, etc. Moreover, Saul and Bernstein (1941) describe the interesting case of a patient suffering from urticaria. It appeared that crying had a special reciprocal relationship to her symptoms. When she cried, she did not have urticaria and the attacks usually terminated with crying. On the other hand, the suppression of crying implied the onset of the symptoms. These authors comment that they had seen similar cases.

The evidence taken together reveals only weak support for the notion that there are positive short-term effects of crying on bodily functions. Rather, studies that were done in more controlled conditions suggest that crying might have immediate effects opposite to what has generally been believed: it depresses mood, leads to increased autonomic nervous system activation, and negatively affects immunity.

CRYING FREQUENCY AND HEALTH STATUS

Support for the possible positive long term effects of crying on health might be provided by empirical data obtained in studies addressing one of the following four types of research questions:

(a) Do patients, in particular, those with psychosomatic disturbances, cry less frequently than healthy controls? (b) Do people who cry relatively often feel better or are they in a better health than those who cry less? (c) Is it true that people who do not express their emotions and those who inhibit crying, in particular, are more liable to somatic disease? (d) Is it possible that crying acts as a moderating variable in a stress–health relationship, i.e., is crying only relevant with respect to well-being if one has been exposed to life stressors?

Crying in (psycho)somatic patients

With regard to crying in somatic patients, the number of relevant studies is limited. The best known study probably is the thesis by Crepeau (1981), who investigated crying behavior in ulcer patients, colitis patients, and healthy controls. The results corroborated some psychosomatically oriented hypotheses. Her patients reportedly cried less frequently and evaluated crying more negatively than did healthy controls. However, her patient groups were not clearly defined and there were other methodological problems with this study. More recently, Schlosser (1986) failed to find an association between crying frequency and physical disorder. Labott and Martin (1990) also examined the relation between emotional coping (both humor-coping and cry-coping) and physical health. In contrast to Crepeau (1981), they found that both in women and in low-income respondents, there was a *positive* relation between severity of physical disorder and crying. Vingerhoets et al. (1992) failed to establish any differences in crying frequency between hyperprolactineamic patients and healthy controls. Zeifman (personal communication, March 1999) found that infants and newborns who rarely cry do not appear to suffer from more health problems or psychological deficits than "normal" criers.

To summarize, the few studies addressing this issue until now have yielded mixed findings. One should further realize that from the kind of correlational research reported above, it is impossible to infer the direction of causal relationships between variables without ambiguity. As an illustration, we refer to asthma patients who may learn to refrain from crying because this behavior can trigger or exacerbate an asthma attack (Miller, 1987; Miller & Wood, 1997). In conclusion, there is no evidence—mainly due to a lack of adequate studies—of a relationship between crying frequency and somatic health.

Crying in psychiatric patients

A different picture emerges in the case of psychiatric disturbances (see also Chapter 14, this volume). A serious problem here is that it is often not clear whether or not crying should be considered a symptom of a specific disorder, like depression. In the three most recent versions of the Diagnostic and Statistical Manual of Mental Disorders (American Psychiatric Association, 1980; 1987; 1994) the relevance of crying for diagnosing mood disorders underwent some remarkable modification. Whereas in DSM-III (1980) crying was included as an important characteristic of dysthymic disorders, DSM-III-R (1987) and DSM-IV (1994) no longer included this item under this heading. Hastrup et al. (1986) reported a weak positive association between depression and crying, whereas, at the same time, a substantial link was found between depression and withholding tears. Frey et al. (1983) found higher self-reported crying frequencies among women with clinical depression (8.0 ± 1.5 times per month) than among healthy controls (5.3 ± 0.3 per month). The range of the distributions of the two samples, however, overlapped to a large extent (0 to 31 times per month for the depressives and 0 to 19 times for the controls). These observations challenge the use of crying as a specific symptom of depression (however, see Okada, 1991). The precise relationship between crying and affective disorders is thus not clear.

Hamilton (1982) regards continuing spells of crying as becoming less effective for bringing relief. This author describes crying as a feature of milder forms of depression, whereas in severe depressions, patients may be even incapable of crying. Other authors also noticed that the more severely depressed patients seldom cry (Kottler, 1996; Patel, 1993). To complicate things even further, Davis et al. (1969) reported impressive differences in crying frequency in neurotically depressed patients (81.8%) versus psychotically depressed patients (23.8%). These findings once more suggest that the association between crying and depression is weak at best. Additionally, De Jong and Roy (1990) found a strong negative association between levels of corticotropin-releasing hormone in cerebrospinal fluid and crying (as measured by the crying item of the Beck Depression Inventory) in a sample of 17 depressed patients. However, the authors did not provide any hypothesis concerning the possible nature of this relationship.

Mangweth et al. (1999) compared crying behavior of eating disordered females and healthy controls. The patient group did not differ from the controls in general crying proneness and the effects of crying on mood. In contrast, the patients' estimates of crying frequency were significantly higher. Moreover, on some specific crying proneness items, in particular, those addressing features of their psychiatric diagnosis such as feeling humiliated, feeling insulted, having low self-esteem, and traumatic memories, they also obtained higher scores reflecting a greater proneness to cry.

It has been further reported that people with posttraumatic stress disorder show an "emotional numbness," inhibiting the display of any emotional reactions (e.g. Litz et al., 1997). This too is an illustration of the fact that (not) crying should sometimes be considered as a consequence rather than as a determinant of health status.

In addition, one should be aware of the possibility that medication interferes with crying, as has been shown by Oleshansky and Labbate (1996), who report on the inability to cry in patients treated with serotonin reuptake inhibitors.

Finally, there might be a relationship between certain types of psychopathology and/or personality features that may explain remarkable differences in crying behavior. For example, it might be hypothesized that people like psychopaths, who lack any empathic abilities, cry less often than normals (see also Chapter 7, this volume).

In conclusion, there is little reason to assume that crying and mental health are related in some simple way.

Health status of criers and non-criers

As indicated before, another approach is to focus on crying behavior as the independent, rather than the dependent variable. Do those who cry relatively often feel better than those who do not? The results of a study by Vingerhoets et al. (1993) of a group of 131 women yielded a correlation between self-reported health and crying frequency of exactly .00. There was thus no association between amount of crying and subjective health. However, a limitation of this study might have been that the obtained range of the health ratings was too narrow. One may also challenge the simplicity of the hypothesis tested. Given the findings of the previous section, i.e., that among depressive patients there might be distinct subgroups, each associated with its own degree of crying, one might speculate that it makes more sense to expect a *curvilinear* relationship between crying frequency and health status than a linear one. Similar comments have been made by Bronstein et al. (1996), who found a positive association between crying proneness and adjustment in male adolescents and a negative association in female youngsters.

Of further interest is a case-study by Linton (1985), who reports on the treatment of a 26-year old woman seeking help because of her inability to express emotions, in particular, sadness and crying. Ever since she was a child, she had refrained from crying. Applying a comprehensive behavioral treatment including assertiveness training oriented towards emotional expression plus modeling and systematic shaping of crying behavior, this patient learned to express her feelings and to cry. This treatment also had a positive effect on her sleep problems and anxiety, resulting in a significant increase in well-being. Of course, the value of such case studies is limited because not only the crying behavior changed significantly, but probably also her assertiveness and other variables, preventing the possibility of drawing any clear conclusion. Nevertheless, it is tempting to speculate about a causal relationship between the increase in crying behavior and the increased well-being. However, future studies should address this issue in a sound methodological way.

Health and the inhibition of tears

There is suggestive evidence in the literature that the nonexpression of emotions is associated with the development or poor prognosis of somatic disease. Well-known examples are cancer (e.g., Gross, 1989; however, see also Bleiker & Van der Ploeg, 1997), hypertension (see Nyklicek et al., 1998), and myocardial infarction (e.g., DeNollet, 1997). It is remarkable that, in the determination of the level of nonexpression, measures of crying until now have seldom or never been used. Temoshok and Dreher (1993), discussing the Type C behavior pattern, devote some attention to crying. Type C individuals are described as people working hard to keep on a happy face to the world. They are afraid that grief will overtake them and by consequence do not allow themselves to indulge in tears. According to these authors, many of these Type C persons may have lost their capacity for crying in childhood because their parents prohibited, ignored, punished or disapproved of shedding tears.

Accepting the evidence that nonexpression of emotions is associated with ill-health, a crucial question is whether crying occupies a special position. In other words, is not crying in a certain context more negative for your health than not laughing in another context, or not expressing pride or jealousy, etc? Gross and Levenson (1993) examined the effects of suppressing sadness and amusement while watching films on physiological functioning. Their results indicate that in both conditions there was evidence of increased sympathetic activation of the cardiovascular system. In a later study, similar findings were found for disgust (Gross, 1998). To summarize, apparently, inhibition of *all* emotional response tendencies seems to lead to acute increases in physiological activity that in the long term may result in health damage. Since in normal circumstances people will seldom have the urge to cry, it seems very unlikely that inhibition of it may have any significant effect on health status. Only when the inhibition is generalized to all kinds of (negative) emotions and has turned into a more stable personality trait, the chronicity may have serious consequences for bodily processes and health.

In addition to this form of emotion regulation ("response-focused emotion regulation"), Gross and Muñoz (1995) emphasize the existence of a second kind, referred to as "antecedent-focused emotion regulation." This form of emotion regulation concerns behaviors and cognitions that are actively made in order to prevent the occurrence of an emotional response. To mention some examples, we may seek or avoid particular emotional events or situations, pay selective attention to aspects of an environment, or appraise a stimulus in ways that will change its emotional significance. In contrast to the response-focused emotion regulation which seems to require physiological effort, it is not likely that antecedent-focused emotion regulation is accompanied by increased arousal.

The ambiguity of the concept of emotion regulation is that it can be considered both as a means to maintain health and at the same time as an indicator of (mental) health. Poor emotion regulation implies that we fail to regulate which emotions we feel or express, be it in the work setting, the social domain, or in inner life, as is made clear by Gross and Muñoz (1995).

Crying as a moderator variable

In modern stress theory, coping behaviors refer to (behavioral and cognitive) efforts to eliminate stressors or to dampen the emotional distress caused by them. A global distinction can be made between problem-focused coping and emotion-focused coping (Lazarus & Folkman, 1984; Steptoe, 1991). Problem-focused coping refers to efforts to remove the stressors or to reduce their intensity. Emotion-focused coping, in contrast, implies efforts to diminish the intensity of the emotions and to regulate one's emotions adequately. There is thus a strong parallel with the just mentioned emotion regulation strategies.

Vingerhoets et al. (1993, Chapter 7) have put forth the hypothesis that crying may be considered a special kind of coping strategy, serving both of the above mentioned coping functions. On the one hand, they make a comparison with displacement behaviors seen in non-human animals. Ethologists have suggested that organisms in a situation of conflict, being unable to perform the actions for which they are motivated, must find a palliative in order to release the pent-up energy. Dantzer (1991) emphasizes the role of stereotypic rhythmic movements (including leg swinging and sucking a pacifier) as potentially being important in this respect. Note also the everyday observation that crying can be pacifying in young children and the proverbial statement that one can "sob oneself to sleep." One may wonder whether the rhythmic sobbing as can be seen at funerals in non-Western cultures has a similar function. Further note the rhythmic movements often made when praying or exercising religious activities. These speculations also fit nicely with the previously formulated hypothesis, based on Panksepp's (see Panksepp, 1998) work, emphasizing the importance of the release of endogeneous opioids. On the other hand, Vingerhoets et al. emphasize the powerful effects of crying on others, making it a means to manipulate the situation. Lipe (1980) further suggests that crying may alter the physiologic state so as to facilitate problem-solving behaviors. Crying may thus have the potential to serve both coping functions and may be helpful in turning uncontrollable situations in somewhat more controllable ones, in that way significantly reducing the stressfulness of it.

However, only one study known to us (Labott & Martin, 1987) specifically focused on the hypothesis that crying may act as a moderator variable, revealing its positive effects only in adverse and stressful situations. The results of this study did not lend support to the hypothesis that crying facilitates the coping process resulting in a better mood after stressor exposure. In addition, there is one study focusing explicitly on the effects of expressing emotions in a bereavement situation. Znoj (1997) investigated the well-being of women who had lost their spouses. Crying during imaginary talking with their deceased partners proved to be unrelated to subsequent well-being.

INDIRECT EFFECTS OF CRYING

As mentioned above, it should not be overlooked that crying may also have strong effects on the social environment. If it is true that crying has a strong potential to elicit social support and to promote attachment, then it may be expected that also via this route crying promotes health. There is ample evidence that social support may influence psychobiological processes and act as a buffer against the negative health effects of stressful conditions (e.g., Cobb, 1976; Uchino et al., 1996). A distinction is often made between emotional support (offering the shoulder and the arm around you, comforting words), informational support (advice and useful suggestions how to deal with problems) and instrumental support (offering help, lending money or tools).

Cobb (1976), in his classic review, emphasizes the informational and emotional value of social support processes (e.g., that one is cared for and loved) in fostering coping and adaptation. In a recent review of the literature on social support and physiological processes, Uchino et al., 1996) come to conclude that social support is associated with proper functioning of the cardiovascular, endocrine, and immune systems. There is evidence showing that social support lowers cardiovascular reactivity to psychosocial stressors, which is thought to reflect, in part, sympathetic-adrenergic activation. Such observations have led to the formulation of the hypothesis that this may be a mechanism through which social support may be related to positive long-term effects on health. The observations by Seeman et al. (1994) strongly suggest that it is in particular emotional support that has reliable effects on physiological functioning.

Bowlby (1972) strongly advocates the thesis that crying is most important in facilitating attachment to youngsters, together with other behaviors like smiling, suckling, and following. Assuming that this specific function of crying has not been totally lost in adulthood, it is interesting to consider the results of a recent study by Zachariah (1996) among 118 primiparous pregnant women. This author examined predictors of psychological well-being in this special group and found that most important were husband-wife attachment, life stress, and social support. It is tempting to speculate that also in this group crying facilitates attachment, which makes it more functional that at least a minority of pregnant women have a rather low threshold to shed tears, particularly in the first phase of pregnancy (Lutjens, 1998).

Five Causal Models that may Explain the Relationship Between Crying and Health

To summarize the above evidence and to provide an interpretative framework, we would like to propose five different models about how crying and health might be related and which may be helpful to design future studies addressing this issue. The models are presented in Figure 1.

In the first model (a), there exists a direct, possibly biological, influence of crying on health. The data presented in the beginning of this article under the heading "the immediate effects of crying," showed that the evidence for such a relationship is ambiguous and not very compelling.

The second model (b) shows an indirect relationship, brought about by intervening variables which are largely of a social psychological nature. The largely circumstantial evidence presented in the section "indirect effects of crying" showed that such a mechanism cannot be ruled out. It might be interesting to investigate the role of social support in moderating the effects of crying directly. This can be done by designing an experiment

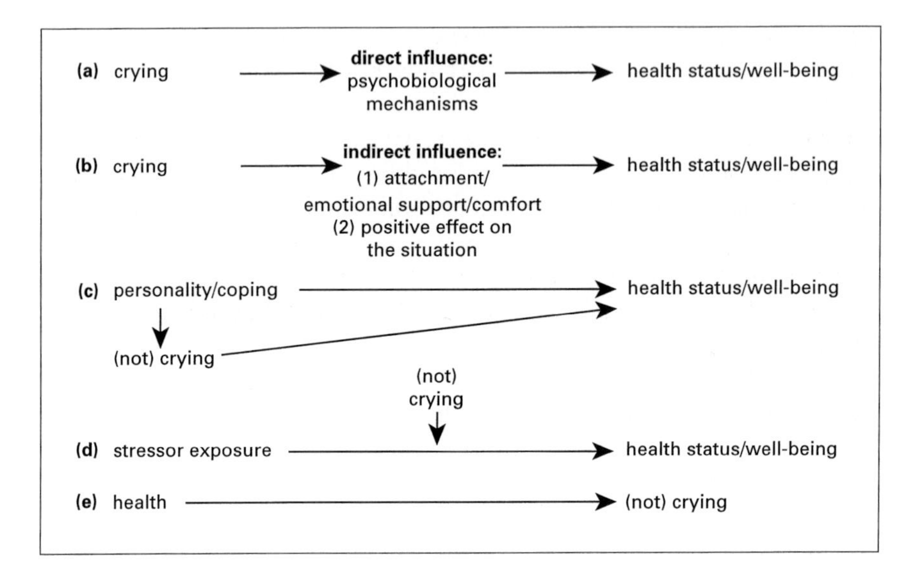

Figure 1. Schematic summary of different possible models for the study of the crying—well-being/health relationship.
From Vingerhoets at al. (2000), Figure 2. Copyright American Psychological Association. Used with permission.

in which people are induced to cry and are assigned to one of several groups, with either bystanders absent or present and in an active versus passive role with regard to comforting behavior.

Model (c) depicts a spurious relationship in which personality factors or coping styles are associated with both crying and health. There is indeed some evidence for a positive relationship between neuroticism, depression, empathy, and repression, on the one hand, and propensity to cry on the other (De Fruyt, 1997; Peter et al., 2000; Vingerhoets et al., 1993; Vingerhoets et al., 1998; Williams, 1982). Neuroticism or "negative affect" is known to correlate negatively with subjective health, but generally not with more objective health indicators (Watson & Pennebaker, 1987). Moreover, women who attain high scores on hardiness (which is related to being resistant to stress) seem to cry less and feel worse after a crying spell than do women who are low on this variable (Schlosser, cited in Goldberg (1987)). Repression in particular is known for its negative association with (at least somatic) health status (e.g., Schwartz, 1990; Schwartz & Kline, 1995).

The fourth model (d) emphasizes the possible buffer functioning of crying. This means that crying only has effects on (somatic and/or mental) well-being when being exposed to stressful conditions and experiencing distress. In normal situations, crying may not have any great effects on the individual.

Finally, model (e) refers to the situation in which crying can be considered a symptom or a consequence of disease, a mood state indirectly induced by the disease or the diagnosis, or a treatment related effect. Examples are the so-called post-partum blues (Beck, 1991), as well as neurological conditions such as CVA's, Parkinson's disease, and multiple sclerosis. Here, sufferers tend to cry more easily than normals and as far as the neurological patients are concerned, they often do this without a discernable cause (Shaibani et al., 1994; see also Chapter 15, this volume). The inhibition or blocking of crying can also be a consequence of disease, as in the example of the asthma patients given earlier, or a side effect of medication. Finally, crying can be a reaction to the diagnosis that the patient is suffering from a serious and/or life threatening disease (Wagner et al., 1997). It is clear that in all these situations the association between crying and (ill) health points to a reversal of cause and effect, where the illness or suffering probably comes first and the increased or decreased crying follows. The plausibility of such a relationship was further supported by the data of a pilot study by Laan (2000, personal communication), which revealed that 16% of the persons who affirmed the question whether their crying proneness had more or less permanently changed after a significant life event mentioned an injury or illness.

Conclusion

It can be concluded that scientific studies until now have yielded little evidence in support of the hypothesis that shedding tears improves mood or health directly, be it in the short or in the long run. This is not to say that crying is a useless behavior. It has strong effects on the environment, promoting comforting and helping behavior, and possibly strengthening attachment.

In that way crying might thus serve both problem-focused and emotion-focused coping functions. However, its effects on health status need further study with adequate attention to the possible pitfalls and drawbacks adhering to the research design that is chosen. One needs to be aware that only experimental designs allow for strong statements about cause and effect relationships. Since most studies concerning crying and health are correlational in character, interpretational difficulties are likely to arise. These could be due to a contaminating "third" variable, or to problems in determining the direction of a causal relationship, as we tried to show in the five explanatory models presented above. It is clear from this discussion, however, that the relationship between crying and health is an important and interesting challenge for investigators. Until enough properly designed studies have been carried out, any statements about the presumed relationship do not surpass the level of mere speculation.

References

Alexander, F. (1950). *Psychosomatic medicine: Its principles and applications.* New York: Norton.

American Psychiatric Association (1980). *Diagnostic and statistical manual of mental disorders (DSM-III)* (3rd. ed.). Washington: Author.

American Psychiatric Association (1987). *Diagnostic and statistical manual of mental disorders (DSM-III-R)* (3rd, revised ed.). Washington: Author.

American Psychiatric Association (1994). *Diagnostic and statistical manual of mental disorders (DSM-IV)* (4th. ed.). Washington: Author.

Bayart, F., Hayashi, K.T., Faull, K.F., Barchas, J.D., & Levine, S. (1990). Influence of maternal proximity on behavioral and physiological responses to separation in infant rhesus monkeys (Macaca mulatta). *Behavioral Neurosciences, 104*, 98–107.

Beck, C.T. (1991). Maternity blues research: A critical review. *Issues in Mental Health Nursing, 12*, 291–300.

Berry, D.S., & Pennebaker, J.W. (1993). Nonverbal and verbal emotional expression and health. *Psychotherapy & Psychosomatics, 59*, 11–19.

Bleiker, E.M.A., & Van der Ploeg, H.M. (1997). The role of (non)expression of emotions in the development of cancer. In: A.J.J.M. Vingerhoets, F.J. van Bussel, & A.J.W. Boelhouwer (Eds.), *The (non)expression of emotions in health and disease* (pp. 221–235). Tilburg: Tilburg University Press.

Borgquist, A. (1906). Crying. *American Journal of Psychology, 17*, 149–205.

Bowlby, J. (1972). *Attachment* (3rd ed.). Harmondsworth UK: Penguin Books.

Bronstein, P., Briones, M., Brooks, T., & Cowan, B. (1996). Gender and family factors as predictors of late adolescent emotional expressiveness and adjustment: A longitudinal study. *Sex Roles, 34*, 739–765.

Cobb, S. (1976). Social support as a moderator of life stress. *Psychosomatic Medicine, 38*, 300–314.

Cornelius, R.R. (1986, April). *Prescience in the pre-scientific study of weeping? A history of weeping in the popular press from the mid-1800's to the present.* Paper presented at the 57th Annual Meeting of the Eastern Psychological Association, New York, NY.

Cornelius, R.R. (1997). Toward a new understanding of weeping and catharsis? In: A.J.J.M. Vingerhoets, F.J. van Bussel, & A.J.W. Boelhouwer (Eds.), *The (non)expression of emotions in health and disease* (pp. 303–321). Tilburg: Tilburg University Press.

Crepeau, M.T. (1981). A comparison of the behavior patterns and meanings of weeping among adult men and women across three health conditions. *Dissertation Abstracts International, 42(01)*, 137B-138B.

Dantzer, R. (1991). Stress, stereotypes, and welfare. *Behavioural Processes, 25*, 95–102.

Darwin, C. (1872/1965). *The expression of emotions in man and animals.* London: John Murray (1965, Chicago: University of Chicago Press).

Davis, D., Lambert, J., & Ajans, A.Z. (1969). Crying in depression. *British Journal of Psychiatry, 115*, 597–598.

De Fruyt, F. (1997). Gender and individual differences in adult crying. *Personality and Individual Differences, 22*, 937–940.

De Jong, J.A., & Roy, A. (1990). Relationship of cognitive factors to CSF Corticotropin-Releasing Hormone in depression. *American Journal of Psychiatry, 147*, 350–352.

DeNollet, J. (1997). Non-expression of emotions as a personality feature in coronary patients. In: A.J.J.M. Vingerhoets, F.J. van Bussel, & A.J.W. Boelhouwer (Eds.), *The (non)expression of emotions in health and disease* (pp. 181–192). Tilburg: Tilburg University Press.

Freud, S. (1915/1957). Repression and the unconscious. In: J. Strachney (Ed. and Transl.) *The standard edition of the complete psychological works of Sigmund Freud* (Vol. 14, pp. 141–195) London: Hogarth Press.

Frey, W.H. (1985). *Crying: The mystery of tears.* Minneapolis, MN: Winston Press.

Frey, W.H., Hoffman-Ahern, C., Johnson, R.A., Lykken, D.T., & Tuason, V.B. (1983). Crying behavior in the human adult. *Integrative Psychiatry, 1*, 94–100.

Goldberg, J.R. (1987). Crying it out. *Health*, February, 64–66.

Groen, J. (1957). Psychosomatic disturbances as a form of substituted behavior. *Journal of Psychosomatic Research, 2*, 85–96.

Gross, J.J. (1989). Emotional expression in cancer onset and progression. *Social Science and Medicine, 12*, 1239–1248.

Gross, J.J. (1998). Antecedent and response-focused emotion regulation: Divergent consequences for experience, expression, and physiology. *Journal of Personality and Social Psychology*, 74, 224–237.

Gross, J.J. & Levenson, R.W. (1993). Emotional suppression: Physiology, self-report, and expressive behavior. *Journal of Personality and Social Psychology, 64*, 970–986.

Gross, J.J, & Muñoz, R.F. (1995). Emotion regulation and mental health. *Clinical Psychology: Science and Practice, 2*, 151–164.

Gross, J.J., Fredrickson, B.L., & Levenson, R.W. (1994). The psychophysiology of crying. *Psychophysiology, 31*, 460–468.

Hamilton, M. (1982). Symptoms and assessment of depression. In: E.S. Paykel (Ed.). *The handbook of affective disorders* (pp. 3–11). Edinburgh UK: Churchill Livingstone.

Hastrup, J.L., Baker, J.G., Kraemer, D.L., & Bornstein, R.F. (1986). Crying and depression among older adults. *Gerontologist, 26*, 91–96.

Kottler, J.A. (1996). *The language of tears*. San Francisco, CA: Jossey-Bass.

Kraemer, D.L., & Hastrup, J.L. (1988). Crying in adults: Self-control and autonomic correlates. *Journal of Social and Clinical Psychology, 6*, 53–68.

Labott, S.M., & Martin, R.B. (1987). The stress-moderating effects of weeping and humor. *Journal of Human Stress, 13*, 159–164.

Labott, S.M., & Martin, R.B. (1990). Emotional coping, age, and physical disorder. *Behavioral Medicine, 16*, 53–61.

Labott, S.M., Ahleman, S., Wolever, M.E., & Martin, R.B. (1990). The physiological and psychological effects of the expression and inhibition of emotion. *Behavioral Medicine, 16*, 182–189.

Lazarus, R. & Folkman, S. (1984). *Stress, appraisal, and coping*. New York: Springer.

Linton, S.J. (1985). A behavioral treatment for inability to express emotions. *Scandinavian Journal of Behaviour Therapy, 14*, 33–38.

Lipe, H.P. (1980). The function of weeping in the adult. *Nursing Forum, 19*, 26–44.

Litz, B.T., Schlenger, W.E., Weathers, F.W., Caddell, J.M., Fairbank, J.A., & LaVange, L.M. (1997). Predictors of emotional numbing in posttraumatic stress disorder. *Journal of Traumatic Stress, 10*, 607–618.

Lutjens, C. (1998). *Mood and crying during pregnancy and postpartum: A pilot study*. Master Thesis, Department of Psychology, Tilburg University, Tilburg, NL.

Mangweth, B., Kemmler, G., Kinzl, J., Ebner, C., De Col, C., Kinzl, J., Biebl, W., & Vingerhoets, A.J.J.M. (1999). The weeping behavior in anorexic and bulimic females. *Psychotherapy & Psychosomatics, 68*, 319–324.

Marston, A., Hart, J., Hileman, C., & Faunce, W. (1984). Toward the laboratory study of crying. *American Journal of Psychology, 97*, 127–131.

Martin, R.B., Guthrie, C.A., & Pitts, C.G. (1993). Emotional crying, depressed mood, and secretory immunoglobulin A. *Behavioral Medicine, 19*, 111–114.

Meerhof, L. (1997). *Alexithymia: Associations with repression, positive and negative affect, crying, and dreaming*. Master Thesis, Department of Psychology, Tilburg University, Tilburg.

Menninger, K., Mayman, M., & Pruyser, P. (1964). *The vital balance*. New York: Viking Press.

Miller, B.D. (1987). Depression and asthma: A potentially lethal mixture. *Journal of Allergy and Clinical Immunology, 80*, 481–486.

Miller, B.D., & Wood, B.L. (1997). The influence of specific emotional states on autonomic reactivity and pulmonary function in asthmatic children. *Journal of the American Academy of Child and Adolescent Psychiatry, 36*, 669–677.

Mills, C.K., & Wooster, A.D. (1987). Crying in the counseling situation. *British Journal of Guidance and Counseling, 15*, 125–130.

Nyklicek, I., Vingerhoets, A.J.J.M., & Van Heck, G.L. (1996). The under-reporting tendency of hypertensives: An analysis of potential psychological and physiological mechanisms. *Psychology and Health, 14*, 1–22.

Okada, F. (1991). Is the tendency to weep one of the most useful indicators for depressed mood? *Journal of Clinical Psychiatry, 52*, 351–352.

Oleshansky, M.A., & Labbate, L.A. (1996). Inability to cry during SRI treatment. *Journal of Clinical Psychiatry, 57*, 593.

Panksepp, J. (1986). The neurochemistry of behavior. *Annual Review of Psychology, 37*, 77–107.

Panksepp, J. (1998). *Affective neuroscience*. New York: Oxford University Press.

Patel, V. (1993). Crying behavior and psychiatric disorder in adults: A review. *Comprehensive Psychiatry, 34*, 206–211.

Pennebaker, J.W., & Susman, J.R. (1988). Disclosure of traumas and psychosomatic processes. *Social Science and Medicine, 26*, 327–332.

Peter, M., Vingerhoets, A.J.J.M., & Van Heck, G.L. (In press). Personality, gender, and crying. *European Journal of Personality*.

Rees, W.D. (1972). Bereavement and illness. *Journal of Thanatology, 2*, 814–819.

Saul, L.J., & Bernstein, C. (1941). The emotional settings of some attacks of urticaria. *Psychosomatic Medicine, 3*, 349–369.

Schlosser, M.B. (1986, August). *Anger, crying, and health among females*. Paper presented at the Annual Meeting of the American Psychological Association, Washington, DC.

Schwartz, G.E. (1990). Psychobiology of repression and health: A systems approach. In: J.L. Singer (Ed.). *Repression and dissociation. Implications for personality theory, psychopathology, and health* (pp. 405–434). Chicago: The University of Chicago Press.

Schwartz, G.E., & Kline, J.P. (1995). Repression, emotional disclosure, and health: Theoretical, empirical, and clinical considerations. In: J.W. Pennebaker (Ed.), *Emotion, disclosure, and health* (pp. 177–194). Washington DC: American Psycho-logical Association.

Seeman, T.E., Berkman, L.F., Blazer, D., & Rowe, J.W. (1994). Social ties and support and neuroendocrine function: The MacArthur studies of successful aging. *Annals of Behavioral Medicine, 16*, 95–106.

Shaibani, A.T., Sabbagh, M.N., & Doody, R. (1994). Laughter and crying in neurologic disorders. *Neuropsychiatry, Neuropsychology, and Behavioral Neurology, 7*, 243–250.

Solter, A. (1995). Why do babies cry? *Pre- and Perinatal Psychology Journal, 10*, 21–43.

Steptoe, A. (1991). Psychological coping, individual differences, and physiological stress responses. In: C.L. Cooper & R. Payne (Eds.). *Personality and stress: Individual differences in the stress process* (pp. 205–234). Chichester UK: Wiley.

Temoshok, L. & Dreher, H. (1993). *The Type C connection. The mind-body link to cancer and your health*. New York: Plume/ Penguin.

Uchino, B.N., Cacioppo, J.T., & Kiecolt-Glaser, J.K. (1996). The relationship between social support and physiological processes: A review with emphasis on underlying mechanisms and implications for health. *Psychological Bulletin, 119*, 488–531.

Vingerhoets, A.J.J.M., & Kirschbaum, C. (1997). *Crying, mood, and cortisol.* Paper presented at the Annual Meeting of the American Psychosomatic Society, Santa Fe, NM (Abstracted in *Psychosomatic Medicine, 59*, 92–93).

Vingerhoets, A.J.J.M., Assies, J., & Poppelaars, K. (1992). Weeping and prolactin. *International Journal of Psychosomatics, 39*, 81–82.

Vingerhoets, A.J.J.M., Cornelius, R.R., Van Heck, G.L. (1998, May). *Crying: To cope or not to cope?* Paper presented at the Second International Meeting on Psychology and Health, Kerkrade, The Netherlands.

Vingerhoets, A.J.J.M., Cornelius, R.R., Van Heck, G.L., Becht, M.C. (2000). Adult Crying: A Model and Review of this Literature. *Review of General Psychology, 4*, 354–377.

Vingerhoets, A.J.J.M., Meerhof, L., & Van Heck, G.L. (1998). Gender differences in the relationship between crying and repression. *Psychosomatic Medicine, 60*, 98. (abstract).

Vingerhoets, A.J.J.M., Van den Berg, M., Kortekaas, R.Th., Van Heck, G.L., & Croon, M. (1993). Weeping: Associations with personality, coping, and subjective health status. *Personality and Individual Differences, 14*, 185–190.

Wagner, R.E., Hexel, M., Bauer, W.W., & Kropiunigg, U. (1997). Crying in hospitals: A survey of doctors', nurses', and medical students' experience and attitudes. *Medical Journal of Australia, 166*, 13–16.

Watson, D., & Pennebaker, J.W. (1989). Health complaints, stress, and distress: Exploring the central role of negative affectivity. *Psychological Review, 96*, 234–254.

Wellenkamp, J.C. (1988). Notions of grief and catharsis among the Toraja. *American Ethnologist, 15*, 486–500.

Wellenkamp, J.C. (1992). Variation in the social and cultural organization of emotions: The meaning of crying and the importance of compassion in Toraja, Indonesia. In: D.D. Frank & V. Gecas (Eds.). *Social perspectives on emotion.* Vol. 1. (pp. 189–216). Greenwich CT: JAI Press.

Williams, D.G. (1982). Weeping by adults: Personality correlates and sex differences. *Journal of Psychology, 110*, 217–226.

Zachariah, R. (1996). Predictors of psychological well-being of women during pregnancy: Replication and extension. *Journal of Social Behavior and Personality, 11*, 127–140.

Znoj, H. (1997). When remembering the lost spouse hurts too much: First results with a newly developed observer measure for tears and crying related coping behavior. In: A.J.J.M. Vingerhoets, F.J. van Bussel, & A.J.W. Boelhouwer (Eds.) *The (non)expression of emotions in health and disease* (pp. 337–352). Tilburg, The Netherlands: Tilburg University Press.

Zorn, F. (1988). *Mars*. Paris: Gallimard.

14 CRYING AND PSYCHIATRIC DISORDER

Vikram Patel

Crying as a form of emotional expressive behavior is thought to be unique to human beings (Frey et al., 1983). It is a dramatic behavior which has traditionally been associated with mood changes, most commonly low mood, but also with happiness, as seen in the contrasting idioms of "crying one's heart out" and "tears of joy." It is not surprising, then, that this behavior has been thought to be a feature of psychiatric disorders, in particular, mood disorders. The supposed relationship between crying and psychiatric disorder was alluded to by Darwin, who noted that "the insane notoriously give way to all their emotions with little or no restraint...nothing is more characteristic of simple melancholia, even in the male sex, than a tendency to weep on the slightest occasions, or from no cause" (Darwin, 1872/1965, pp. 154–155). This chapter will first review the contemporary research literature on the relationship of crying and mood in both nonclinical and clinical samples, and will briefly cover pathological crying (which is more extensively described in chapter 15). Next, it will consider the putative mechanisms which may serve to explain the relationship between mood and crying behavior, and the role of crying as a sign of psychiatric disorder in current psychiatric nosologies. The remainder of the chapter will then be used to examine two questions. First, what is the evidence that crying is associated with mood disorders or any other "functional" psychiatric disorder?; and second, what is the relevance of the syndrome of "emotionalism" or pathological crying to our understanding of the association between crying and mood disorder?

Research on Crying and Mood Disorders

There is little research on the relationship between crying and mood (but see Vingerhoets et al., 1997; chapter 5, this volume), let alone its relationship to mood disorders. Even though most people accept the relationship between crying and low mood, there is a remarkable paucity of well-designed studies examining the association of crying with mood disorder.

STUDIES WITH NONCLINICAL SAMPLES

Frey et al. (1983) studied patterns of crying behavior in participants without any psychiatric disorder. Crying was significantly more common in women, and the stimulus for crying was often interpersonal conflict and in response to unhappy or sad media stories. About half of the women who cried indicated sadness as being the primary emotion related to their crying, but as many as a third reported happiness or anger as primary emotions. Williams (1982) studied crying behavior in 140 adults using a Crying Questionnaire on which participants rated their tendency to cry in response to different situations such as separation, as well as the intensity of episodes of crying, from "just feeling touched and moved" to "a full crying response and sobs." These authors found that women reported crying more frequently than did men, with events such as death, loneliness, separation, and pain being common precipitants. Kraemer and Hastrup (1986) used the Crying Frequency Questionnaire to obtain a global estimate of the frequency of crying in college students; on the questionnaire, participants rated the frequency of their crying episodes for a period of up to one year and rated the degree of their self control over the crying and the stimuli that elicit it. These authors found that there was no relationship between crying and overall depression levels, but did find a significantly greater frequency of crying in women than in men. Hastrup et al. (1986) found a similar sex difference in an elderly population (>65 years) using the same questionnaire. Although crying was also not found to be related to mood in their study, when the sample was categorized into high and low Beck Depression Inventory (BDI) scorers, the former showed a small, but statistically significant, greater frequency of crying behavior. These authors also reported a relationship between low mood and episodes of "feeling like crying" as opposed to actual crying.

Thus, the overall impression from the research on crying and mood in nonclinical samples is that there is no consistent relationship between crying and any specific mood; indeed, crying may be simply one component of any heightened or intense emotional state such as anger, happiness or sadness. There is some evidence to suggest that biological correlates of crying and sad mood also differ. For example, one study examined the relationship between emotional crying and secretory immunoglobulin A (S-IgA) in whole saliva among 42 undergraduates (Martin et al., 1993). Participants completed two separate tasks in a random design. The first involved watching an 18-minute film about a Vietnam veteran's meeting with the family of his slain friend. Participants completed the Depression Adjective Checklist before and after the film, and a 9-point checklist assessing their level of crying during the film. In the second task, participants wrote an imaginative story to each of three Thematic Apperception Test-type pictures. Results indicated that lower S-IgA was associated with crying but not with depressed mood.

Crying behavior is more common in women, and stressful situations or sad events act as triggers to the behavior. The association of crying and gender is more thoroughly described elsewhere in this book (see Chapter 6). However, it is important to note that there are likely to be both social and biological explanations for this gender difference. For example, Okada (1995) explored the neural mechanism underlying the relation between crying and depression and reported that women were more prone to show the behavior. Investigations of blood flow changes demonstrated that the right and left hemispheres have the same activity levels in women, either during rest or during work on a psychological task, while they often have strikingly different levels in men. The neural mechanism for crying that operates when female patients are depressed may be inhibited by a lateralized neural inhibition commonly operating in males. Another biological explanation put forth is that there could be a greater release of lacrimal fluid in women during a crying episode that makes it more difficult for them to suppress tears (Milder, 1970). On the other hand, there is little doubt that gender differences in social roles and stereotypes also play a role not only in the different rates of mental disorder reported among women and men but in common behaviors associated with emotional expression such as crying (Thomas, 1996). For example, in a study by Oliver and Toner (1990), undergraduates of both sexes were classified into "masculine" and "feminine" groups on the basis of responses to a sex role inventory. Gender role typing emerged as an important determinant of the differential reporting of emotional symptoms on the BDI, including crying in women. The authors concluded that while depressed men where more likely to present with somatically oriented symptoms, depressed women were more likely to cry or express cognitive symptoms.

STUDIES WITH CLINICAL SAMPLES

There are very few systematic studies on the association of crying behavior and mood involving samples of patients with psychiatric disorders. Lund (1930), reporting a survey of 19 psychiatric inpatients, found that there was no relationship between crying and depressive psychosis. Instead, crying was more likely to be associated with mood change from depression to elation. As would be expected from a psychiatric population for this period, crying behavior was most commonly associated with neurological disorders such as CNS syphilis and strokes, now referred to as pathological crying or emotionalism. Davis et al. (1969) investigated the view that crying is more typically associated with "neurotic" or "reactive" as opposed to "endogenous" depression. Crying behavior was assessed on the basis of retrospective ward and case note reports on a sample of psychiatric in-patients. These authors reported that more than 80% of neurotically depressed patients exhibited crying, in contrast to less than a quarter of psychotically depressed patients. Frey et al. (1983) reported a marginally higher frequency of crying frequency for currently depressed women as compared to non-depressed women; though this difference just reached statistically significance ($p<0.05$), the considerable overlap in the ranges of crying frequency between the clinical and normal populations suggested a limited usefulness for crying as a diagnostic symptom of depression.

Green et al. (1987) studied crying behavior prospectively in a sample of patients admitted to a general hospital and referred to a liaison psychiatric service. Patients who cried were classified into two groups, viz., *prominent* and *nonprominent* criers. Prominent criers were those in whom the history of crying was dominant in the recent past and crying was consistently present in repeated psychiatric evaluations. Nonprominent criers were those in whom the behavior was minor and inconsistent. These authors reported that crying was more typically associated with neurological disorders as opposed to psychiatric disorders. Thus, the proportion of prominent criers with only a psychiatric disorder was only 20% as compared to 53% of non-prominent criers. Indeed, the presence of a psychiatric disorder, with or without a coexisting neurological disorder, was highest in nonprominent criers (64%). The commonest psychiatric diagnosis was a major depressive disorder. The authors describe clear clinical differences between the neurological and psychiatric groups, with the former exhibiting crying which was more abrupt in onset, stereotyped and disassociated from the prevailing mood. In addition to these two groups of criers, the authors also report a small number of patients whom they refer to as *essential criers*. These were participants who had a history of abrupt episodes of crying occurring several times a day stretching back to several years. While essential criers reported their mood during episodes to be sad, there was no psychiatric or neurological illness. The authors suggest that chronic tearfulness may be a distressing personality trait unrelated to affective disorder.

Crying behavior has often been considered the hallmark of postnatal blues, presenting in over half of all women with the blues, although there is no consensus on why this occurs (Stein, 1982; see also Chapter 10, this volume). The behavior generally occurs in the setting of a change in mood and emotional lability in which depression changes to elation and back again several times a day. While much of the phenomenology resembles that of the premenstrual syndrome, crying is significantly more commonly associated with the former. Crying behavior is also reported to be associated with secondary depressions in the context of endocrine disorders such as Cushing's disease (Starkman et al., 1981). The association of crying with childbirth and endocrine disorders, in particular hypercorticolism, has prompted some authors to suggest that crying may be related to an underlying endocrine or metabolic imbalance (e.g., Stein, 1982).

Crying behavior has also been linked to bereavement, where it is recognized as a culturally widely accepted behavior. Znoj (1997) described a study aiming to determine whether crying is associated with prolonged bereavement symptoms and can predict onset of symptoms up to one year after bereavement. Using a newly developed crying observer measure with a sample of 62 bereaved individuals, this study found that, contrary to expectations, crying was not a helpful behavior in the context of bereavement, and that there was even some evidence that it was an index of prolonged or pathological grief. The author concluded that "weeping sometimes is just the expression of an overwhelming feeling of distress and despair" (p. 351).

Thus, the overall impression from the research on crying in functional psychiatric disorder is that the presence of crying is not a specific indicator of depressive illness. Crying is more often a sign of either reactive low mood (or, in current classifications, adjustment disorders) or neurological disorder. In the unique contexts of childbirth, crying is a feature of the transient labile affective state of postnatal blues. Finally, although one study reported crying as a personality trait, this finding needs replication. There are no data on the relationship between specific personality disorders and crying behavior, although some clinicians consider outbursts of crying with affective instability as a feature of borderline personality disorder and histrionic personality disorder.

THE ASSOCIATION OF CRYING AND PSYCHIATRIC TREATMENTS

Crying behavior is most commonly an indication for psychiatric treatments, particularly in the setting of pathological crying (see below). However, there are also reports of the effect of psychiatric treatment on crying in persons with functional psychiatric problems. For example, selective serotonin-reuptake inhibitors (SSRIs), such as Prozac, have been reported to be associated with complaints of feeling unable to cry (Oleshansky & Labatte, 1996). In contrast, the occurrence of crying behavior during psychotherapy is imbued with important clinical significance. For example, it is suggested that crying during therapy is complexly determined by a matrix of essential processes: specifically, a basic cognitive schema that provides the trigger for the crying, early painful experiences that provide the content, and a safe therapeutic situation that provides the atmosphere. If the event lacks any of these, it is suggested that crying will either not occur at all or will occur in a less intense and less therapeutic form (Labott et al., 1992). A more detailed analysis of the relationship between crying and psychotherapy can be found elsewhere in this book (see Chapter 12).

Research on Crying and Organic Brain Disorders

The topic of pathological crying is dealt with in more detail elsewhere in this book (see Chapter 15). However, a brief overview is necessary in the context of the association of crying with psychiatric disorder, since the divisions between "functional" and "organic" mental disorders are increasingly blurred and arbitrary. Poeck (1969), in his landmark study, described at least five types of pathological crying. Thus, he differentiated between crying that occurred in hysteria and the psychoses, the emotional lability in diffuse organic brain disorders, the "*Witzelsucht*" of basal rostral and brainstem lesions, the pathological crying associated with pseudobulbar palsy, and the *risus sardonicus* of tetanus. Emotional lability is a sudden change in mood in either direction that can be interrupted by environmental stimuli. *Witzelsucht* refers to the excessive fluctuation in emotional behavior accompanied by a similar fluctuation in mood. Pathological crying is characterized by a lack of similarly oriented affective change and a loss of voluntary control of muscles of facial expression. Poeck's classification was thus based not only on the motor characteristics of the behavior but on its association to mood as well. Since this work, there has been a growing body of literature on the occurrence of crying behavior in the setting of brain lesions including pseudobulbar palsy, strokes, dementia, motor neuron disease, multiple sclerosis and epilepsy. In keeping with this wide range of disorder, there is a wide range of anatomical sites of lesions associated with pathological

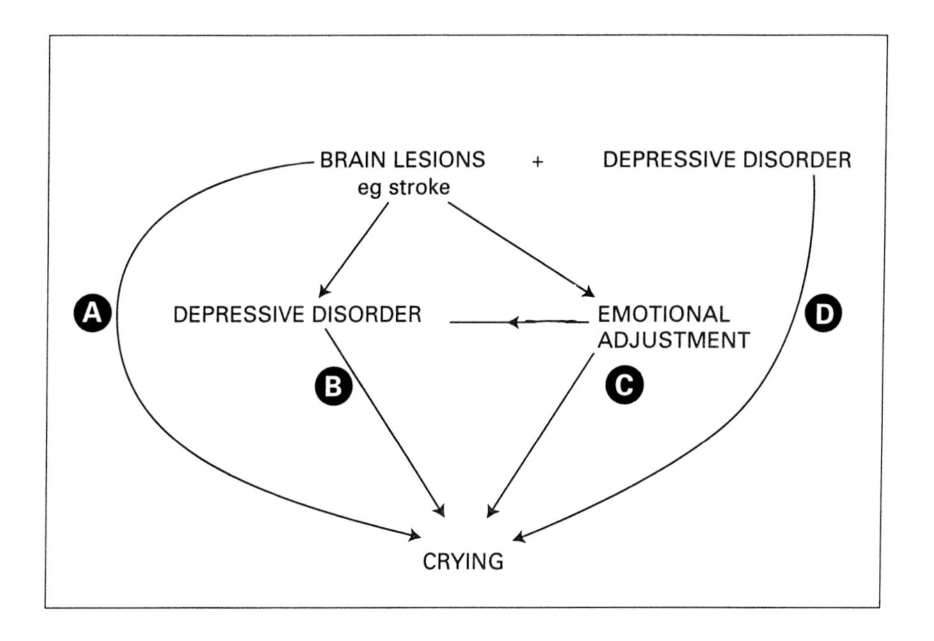

Figure 1. The pathways to crying behavior in organic brain disorders.

crying including the internal capsule, substantia nigra, thalamus and pyramidal tracts (Black, 1994; Shaibani et al., 1994; Chapter 15, this volume). Four mechanisms have been suggested as underlying pathological affect. These are presented in Figure 1.

First, Pathway A depicts crying behavior as a direct result of brain lesions; bilateral lesions in the neocortical upper motor neuron innervation of bulbar motor nuclei (pseudobulbar palsy), basal forebrain, medial temporal lobe, diencephalon, or tegmentum of the brainstem are examples of such lesions. Pathway A, which is described in more detail in the chapter on pathological crying, suggests that the crying behavior is a direct result of brain damage and has no relation to mood. House et al.'s (1989) prospective community-based stroke patient study showed that pathological crying (which is referred to as emotionalism by these authors), though rarely present without concomitant psychopathology, was often associated with focal left anterior hemispheric lesions. They suggested that the behavior was not simply a psychologically understandable reaction to a life threatening and disabling illness but was probably a form of disinhibition, analogous to urinary incontinence in brain-injured patients. Thus, there is loss of control over behavior with low levels of stimulation rather than in response to an irrelevant stimulus or no stimulus at all.

Pathway B depicts crying behavior as a result of post-lesion depressive disorder. However, given the weak evidence of a relationship between depression and crying in "functional" depression, this pathway is unlikely to be a major mediator. Pathway C depicts crying as an emotional response to the life-threatening and disabling nature of many brain lesions such as strokes. On occasion, this adjustment reaction may lead to a major depressive disorder, but again, as with Pathway B, the link with crying is more likely to be due to Pathway C. Some studies have shown a strong association between pathological crying and depression, supporting Pathways B and C as potential mediating mechanisms. For example, Andersen et al. (1995) investigated post-stroke pathological crying in relation to mood score/depression and lesion site in an unselected stroke population the first year following stroke. The frequency of pathological crying in their sample of over 200 participants was 14% at one month, 10% at six months and 11% at one year. Pathological crying correlated significantly with mood score and post-stroke depression, as well as with lesion size, but not with lesion location, sex, age, history of stroke or depression, predisposing disease, or social distress before the stroke incident.

The final Pathway D is postulated by Ross and Stewart (1987) as a result of unilateral lesions of the right frontal opercula when combined with an *existing* major depressive disorder. Using a case series approach, Ross and colleagues (1981, 1987) found that pathological affect, in some brain damaged patients, may be a valuable clinical indicator of an underlying major depression complicating a neurological illness. These authors demonstrate through these unique cases that the pathological affect is the product of an interaction between neurological and psychiatric disorder as shown by some clinical characteristics such as the precipitous onset of the crying behavior (as opposed to the gradual onset in patients who become depressed following brain lesions).

However, there remain considerable difficulties in the assessment of mood in patients with brain lesions (House, 1987). This is especially complicated since the presence of crying cannot, by itself, be considered to be evidence of depressed mood as the earlier discussions in this chapter have shown. Even though there is substantial evidence that antidepressants of both the tricyclic and SSRI groups are effective for the management of pathological crying (e.g., Benedek & Peterson, 1995; Dave, 1996; Lawson & MacLeod, 1969; Peterson et al., 1996), this response cannot be considered to be due to the effect of the medication on an underlying depressive disorder. Indeed, the clinical response of pathological crying to antidepressants is recognized to occur soon after starting treatment (as compared to the lag period for recovery from low mood) and at lower doses than those needed to alleviate depression. Thus, antidepressants may act independently or concurrently on mood-mediating areas and the neural pathways for crying. The anticholinergic effects of some antidepressants would fit with this hypothesis, since crying is a parasympathetic function. This dilemma underscores the great dependence clinical psychiatry has on self-report in the diagnosis of mood disorders and highlights the need for further study of behavioral correlates of mood disorders which may be used as markers of mood, particularly in patients with organic brain lesions whose cognition or speech may be affected.

A Model to Explain the Relationship Between Mood and Crying

Lund (1930) asked the question "why do we weep"? While this topic has generated a considerable volume of theoretical and experimental research, and is covered more extensively in other chapters in this book (see especially Chapters 1 and 5), it is useful to consider this question to explore its relevance to the relationship between mood and crying in psychiatric disorders. The belief that crying may bring about some kind of emotional or physical catharsis is old (Cornelius, 1997; Chapters 11 & 13, this volume). Lund (1930), for example, reported that crying occurred in a variety of emotional states, the key feature being that the behavior appeared when the triggering situation "gained a redeeming feature", or when tension and unpleasant stimulation was followed by a pleasant or alleviating stimulus. Some authors have suggested that crying may have a mood relieving function in mood disorders as well. Thus, crying may be a feature in mild depression and may serve to temporarily relieve unpleasant emotions, but progressive bouts become more frequent and less effective; with increasing depression, patients are rendered "incapable of weeping" (Murphy et al., 1982).

This issue of whether crying plays a role in reducing distressing emotional states has been one of the most popular theories to explain the behavior. Bindra (1972) found that the initial emotional state that triggered an episode of crying often dissipated or changed to a less intense mood after the episode, implying a mood-relieving role for the behavior.

Efran and Spangler (1979) proposed a two-factor theory: in the first stage, the person is aroused and "tension" is created. This emotional state can be positive (such as joy) or negative (such as sadness). In the second stage, there is a shift towards recovery and crying is the behavioral manifestation of this shift. Implicit in this theory is that crying has the function of restoring "organismic equilibrium." However, despite the popularity of this theory, the concept of a relieving function is not consistently supported by other studies (e.g. Kraemer & Hastrup, 1986; Martin & Labott, 1991, Znoj, 1997). For example, in Martin and Labott's (1991) study, film-induced emotional crying was used to examine the effects of five situations on mood in undergraduates in four experiments. Although crying was associated with depressed mood when they were measured closely in time, there was no evidence to suggest that crying facilitated the reduction of subsequent depressed mood. Crying was related to later thought focus on the movie, but depressed mood, independent of crying, was not. A more thorough review of the literature on the relationship between crying and "catharsis" has been presented by Cornelius (1997; Chapter 11, this volume); unfortunately, the issue remains inconclusive at present and it is unlikely that there will be a unidimensional answer to this questions.

The precise mechanism which underlies the relationship between emotional states and crying remains unclear. It is worthwhile considering the emotional regulation model of Gross and Muñoz (1995) in attempting to elaborate a framework which explains the diverse emotional states which may be related to the same behavioral outcome and how the behavior of crying may serve to regulate the emotional state which triggered it. In their model, emotional cues which may represent internal or external events are the starting point of the sequence of events that lead to an emotional response. These cues, in the case of crying, may vary from watching sad films (external cues) or recalling unhappy or nostalgic memories (internal cues). These cues are appraised by the individual, taking into account a variety of environmental and personal factors such as past experiences, personality traits and physical fatigue, triggering a "biologically based emotional program" which begins with the feeling of wanting to cry. Once the program is triggered, the individual will respond in a way to help the person respond adaptively to his or her perceived situation, leading either to an expression of the behavior in a variable intensity (from quiet shedding of tears to full-throated weeping and sobbing) or an inhibition of crying altogether. The emotional response in turn will then modulate and influence the initial emotional cues and appraisal leading to further intensification or inhibition of the behavior. Support for this model is available from empirical studies. For example, Vingerhoets et al. (1997) investigated the contexts and antecedents of crying episodes and reported that the behavior was most commonly triggered by the threat of separation, personal inadequacy and rejection, and that sadness, powerlessness, anger and frustration were the most common associated feelings.

As it can be seen, crying in this model is essentially the result of an emotional experience, as compared to reflex lacrimation as may occur due to stimulation of the cornea. Further, this model would also help explain why any emotional state can trigger crying, the key regulatory factors being the nature of the cue and the way it is appraised by the individual. Further, the model also helps integrate the conflicting findings of the potential role of crying as "relieving tension" and of inhibiting crying as a sign of a more serious mood disorder. Thus, in major depression, it is possible that the inhibitory tendencies are more powerful, whereas in dysthymia, the reverse is the case. However, due to the marked influence of individual and situational variables, it is difficult to find a consistent relationship between these mood states and crying, thus rendering the behavior in an individual of limited value in diagnosis. Instead, the crying behavior would need to be assessed in the context of the individual, his or her coping strategies, methods of appraising emotional cues and so on. This model can also explain the gender differences in crying, using both the biological theories (in terms of the program which may be triggered by an emotional cue) and the gender social role theory (in terms of the way an emotional cue is appraised). Clearly, there is a need for more innovative research designs, especially to examine how crying affects mood (and vice-versa) in clinical samples, since the bulk of existing research has focused on non-clinical populations.

Crying as a Clinical Feature of Psychiatric Disorder

The review of the research literature earlier shows that there has been limited systematic evaluation of the relationship between crying behavior and psychiatric disorder, particularly disorders not associated with obvious neurological lesions. Of the psychiatric disorders which have been examined, mood disorders predominate; indeed a review by this author failed to uncover any specific literature on crying behavior in psychiatric disorders other than mood disorders. It is worthwhile to consider the accepted norms on the clinical significance of crying by scanning standard textbooks and rating scales. It is apparent from such an exercise that there is little mention of crying as a sign of adult psychiatric disorders. Consider the following facts:

(1) Crying has barely no mention in standard psychiatric textbooks, e.g., the *Oxford Textbook of Psychiatry* or the *Comprehensive Textbook of Psychiatry*;
(2) Ethological analyses of depression pay little attention to crying behavior (Ellring, 1989);
(3) "Readiness to tears" is considered a feature of generalized anxiety (Gelder et al., 1989);
(4) It is interesting to see how crying or tearfulness is considered in the three most recent versions of the Diagnostic and Statistical Manual of Mental Disorders (American Psychiatric Association; DSM-III, 1980; DSM III-R, 1987; and DSM IV, 1994). In DSM-III these terms appear as items in a list of symptoms required for a diagnosis of depression in the context of a dysthymic or cyclothymic disorder, but finds no mention in a similar list of items for major depression. The item is altogether dropped in DSM-III-R but reappears in DSM-IV in the category of major depression as sign of depressed mood, i.e., "appearing tearful as observed by others";
(5) ICD-10 mentions crying as a feature of dysthymia as part of a list of 11 symptoms; again, there is no mention of crying in major depression or anxiety disorder;
(6) Rating scales for depression, such as the Hamilton Depression Rating Scale, often include crying or tearfulness as an item (Murphy et al., 1982). However, the item on crying in the BDI has been found to correlate poorly with the overall score, suggesting that this item was not measuring the same dimension of emotional distress as the rest of the questionnaire (Gallagher et al., 1983).

While many of the apparent discrepancies and contradictions above may be accounted for by the repeated changing and revising of psychiatric diagnostic categories and classification systems before a thorough examination of existing ones, it also shows the dubious status of crying as a consistent feature of any psychiatric disorder. Despite this, clinicians would note tearfulness or crying as part of an interview with considerable significance. For example, a patient who appears elated and yet tearful may be considered to display an inappropriate or incongruent affect even though crying is not specific to low mood; indeed, crying is often associated with happiness (Bindra, 1972). On the whole, then, this author is inclined to agree with the advice that clinicians need to exercise caution in interpreting crying behavior as being of diagnostic significance in adult psychiatric disorders (Kraemer & Hastrup, 1986).

Conclusion

Crying is a behavior that is virtually unique to the human species. It is modulated by well-defined cranial nerve pathways that are in turn regulated by higher cortical areas which are also associated with mood regulation. This chapter set out to review the literature on crying behavior and psychiatric disorder with two questions in mind.

The first question asked whether there was any evidence that crying was associated with mood disorders or any other "functional" psychiatric disorder. It is clear that crying behavior is triggered by a wide variety of emotional states, the key feature of which is the intensity of the emotional state rather than the specific type of emotion. There is some evidence that the behavior may play a role in alleviating the intensity of the emotional state, although this is not consistently demonstrated. Similarly, there is inconsistent evidence that the behavior is more typical of dysthymic, neurotic or reactive depressive states as compared to endogenous or psychotic depressive states. If this association is found to be consistent in future research, it may have important clinical and psychopathological implications in understanding the relationship between crying and low mood. With our current state of knowledge, however, clinicians need to be cautious in assuming that the behavior has specific diagnostic significance (Patel, 1993).

The second question concerns the relevance of the syndrome of "emotionalism" or pathological crying to our understanding of the association between crying and mood disorder. It is clear that a variety of mediating mechanisms are possible to explain the phenomenon of pathological crying. What is unclear, though, is whether the expression of the behavior in brain-damaged patients is a sign of an underlying mood disorder or whether it is simply a sign of disorder of the neural substrate of the behavior. In the view of this author, given the fact that crying behavior is not a key feature of depression in patients without neurological lesions, it is highly unlikely that it can be considered a feature of depression in those with neurological lesions. The most likely mechanism to explain pathological crying behavior is that disruption of neural pathways can result in the "release" of the behavior from its higher cortical regulatory influences. It is possible that in some cases, the interaction of depressive disorder and neurological lesions precipitate pathological crying, though this is a rare occurrence. On the whole, then, pathological crying is unlikely to be a sign of a primary mood disorder and may be considered to be a feature of a primarily neurological disorder.

In conclusion, the key point resulting from the research on crying and mood is that there is no specific relationship between the behavior and any functional psychiatric disorder. When crying does occur, it is most often in the context of an intense or heightened emotional state and this may be more likely in those who are reacting to an ongoing stressful situation as compared to those who are "endogenously" depressed. Further research is needed to examine various unresolved issues arising from this review. Potential themes for research include: the diagnostic significance of crying behavior as a sign of minor mood disorders, particularly dysthymia; whether the behavior has any significance in predicting outcome, particularly in terms of responsiveness to antidepressants or psychotherapy, since it may be that those whose mood disorder responds to such treatments are also the same persons who are likely to cry; to investigate further the relationship between crying, brain lesions and mood; and to conduct prospective ethological studies of crying behavior in major depression and dysthymia along with self-report on the effect of crying on mood to uncover both the relationship of crying with mood disorder and its potential role in regulating mood.

References

American Psychiatric Association (1980). *Diagnostic and statistical manual of mental disorders (DSM-III)* (3rd. ed.). Washington: Author.

American Psychiatric Association (1987). *Diagnostic and statistical manual of mental disorders (DSM-III-R)* (3rd, revised ed.). Washington: Author.

American Psychiatric Association (1994). *Diagnostic and statistical manual of mental disorders (DSM-IV)* (4th. ed.). Washington: Author.

Andersen, G., Vestergaard, K., & Ingeman-Nielsen, M. (1995). Post-stroke pathological crying: Frequency and correlation to depression. *European Journal of Neurology*, *2*, 45–50

Benedek, D.M., & Peterson, K.A. (1995). Sertraline for treatment of pathological crying. *American Journal of Psychiatry, 152*, 953–954.

Bindra, D. (1972). Weeping: A problem of many facets. *Bulletin of the British Psychological Society, 25*, 281–284.

Black, K.J. (1994). Pathological laughing and crying. *American Journal of Psychiatry, 151*, 456.

Cornelius, R.R. (1997). Toward a new understanding of weeping and catharsis? In: A.J.J.M. Vingerhoets, F.J. Van Bussel, & A.J.W. Boelhouwer (Eds.), *The (non)expression of emotions in health and disease* (pp. 303–321). Tilburg: Tilburg University Press.

Darwin, C. (1872/1965). *The expression of the emotions in man and animals.* Chicago: University of Chicago Press.

Dave, M. (1996) Paroxetine for pathological crying. *American Journal of Geriatric Psychiatry, 4*, 180–181.

Davis, D., Lambert, J., & Ajans, Z.A. (1969) Crying in depression. *British Journal of Psychiatry, 115*, 597–598.

Efran, J.S., & Spangler, T.J. (1979) Why grown-ups cry: A two-factor theory and evidence from the Miracle Worker. *Motivation & Emotion, 3*, 63–72.

Ellring, H. (1989). *Non-verbal communication in depression.* Cambridge, UK: Cambridge University Press.

Frey, W.H., Hoffman-Ahern, C., Johnson, R.A., Lykken, D.T., & Tuason, V.B. (1983). Crying behavior in the human adult. *Integrative Psychiatry, 1*, 94–98.

Gallagher, D., Breckenridge, J., Steinmetz, J., & Thompson, L. (1983) The Beck Depression Inventory and Research Diagnostic Criteria: Congruence in an older population. *Journal of Consulting & Clinical Psychology, 51*, 945–946.

Gelder, M.G., Gath, D., & Mayou, R. (1989). Neurosis. In: *Oxford textbook of psychiatry* (pp. 175–216). Oxford, UK: Oxford University Press.

Green, R.L., McAllister, T.E., & Bernat, J.L. (1987). A study of crying in medically and surgically hospitalized patients. *American Journal of Psychiatry, 144*, 442–447.

Gross, J. & Muñoz, R. (1995). Emotional regulation and mental health. *Clinical Psychology Science and Practice, 2*, 151–164.

Hastrup, J.L., Baker, J.G., Kraemer, D.L., & Bornstein, R.F. (1986). Crying and depression among older adults. *Gerontologist, 26*, 91–96.

House, A. (1987). Depression after stroke. *British Medical Journal, 294*, 76–78.

House, A., Dennis, M., Molyneux, A., Warlow, C., & Hawton, K. (1989). Emotionalism after stroke. *British Medical Journal, 298*, 991–994.

Kraemer, D.L., & Hastrup, J.L. (1986). Crying in natural settings: Global estimates, self-monitored frequencies, depression and sex differences in an undergraduate population. *Behaviour Research & Therapy, 24*, 371–373.

Labott, S.M., Elliott, R., & Eason, P.S. (1992). "If you love someone, you don't hurt them": A comprehensive process analysis of a weeping event in therapy. *Psychiatry, 55*, 49–62

Lawson, I.R., & MacLeod, R.D.M. (1969). The use of imipramine and other psychotropic drugs in organic emotionalism. *British Journal of Psychiatry, 115*, 281–285.

Lund, F.H. (1930). Why do we weep? *Journal of Social Psychology, 1*, 136–151.

Martin, R.B, & Labott, S.M. (1991). Mood following emotional crying: Effects of the situation. *Journal of Research in Personality*, *25*, 218–244.
Martin, R.B., Guthrie, C.A., & Pitts, C.G. (1993). Emotional crying, depressed mood, and secretory immunoglobulin A. *Behavioral Medicine*, *19*, 111–114.
Milder, B. (1970). The lacrimal apparatus. In: R.A. Moses (Ed.), *Alder's physiology of the eye* (pp. 18–37). St Louis, MO: Mosby.
Murphy, D.L., Pickar, D., & Alterman, I.S. (1982). Methods for quantitative assessment of depressive and manic behavior. In: E. Burdock, A. Sudilovsky, & S. Gershon (Eds.), *The behavior of psychiatric patients* (pp. 355–392). New York: Dekker.
Okada, F. (1995). Weeping and depression: Neural mechanism. *Neuropsychiatry, Neuropsychology, and Behavioral Neurology*, *8*, 293–296
Oleshansky, M.A., & Labbate, L.A. (1996). Inability to cry during SRI treatment. *Journal of Clinical Psychiatry*, *57*, 593.
Oliver, S.J., & Toner, B.B. (1990). The influence of gender role typing on the expression of depressive symptoms. *Sex Roles*, *22*, 775–790.
Patel, V. (1993). Crying behavior and psychiatric disorder in adults: A review. *Comprehensive Psychiatry*, *34*, 206–211.
Peterson, K.A., Armstrong, S., & Moseley, J. (1996). Pathologic crying responsive to treatment with sertraline. *Journal of Clinical Psychopharmacology*, *16*, 333.
Poeck, K. (1969). Pathophysiology of emotional disorders associated with brain damage. In: P.J. Vinken & G.W. Bruyn (Eds.), *Handbook of clinical neurology Vol 3* (pp. 343–367). New York: Wiley Interscience.
Ross, E.D., & Rush, A.J. (1981). Diagnosis and neuroanatomical correlates of depression in brain damaged patients: Implications for a neurology of depression. *Archives of General Psychiatry*, *38*, 1344–1354.
Ross, E.D., & Stewart, R.S. (1987). Pathological display of affect in patients with depression and right frontal brain damage: An alternative mechanism. *Journal of Nervous & Mental Disease*, *175*, 165–172.
Shaibani, A.T., Sabbagh, M.N., & Doody, R. (1994). Laughter and crying in neurologic disorders. *Neuropsychiatry, Neuropsychology, and Behavioral Neurology*, *7*, 243–250.
Starkman, M.N., Schiteingart, D., & Schork, M.A. (1981). Depressed mood and the psychiatric manifestations of Cushings Syndrome: Relationship to hormone levels. *Psychosomatic Medicine*, *43*, 3–17.
Stein, G. (1982). The maternity blues. In: I.F. Brockington & R. Kumar (Eds.), *Motherhood & mental illness* (pp. 119–154). London: Academic Press.
Thomas, P. (1996) Big boys don't cry? Mental health and the politics of gender. *Journal of Mental Health*, *5*, 107–110.
Vingerhoets, A.J.J.M., Van Geleuken, A.J.M.L., Van Tilburg, M.A.L., & Van Heck, G.L. (1997). The psychological context of crying episodes: Toward a model of adult crying. In: A.J.J.M. Vingerhoets, F.J. Van Bussel, & A.J.W.

Boelbouwer (Eds.), *The (non)expression of emotions in health and disease* (pp. 323–336). Tilburg, The Netherlands: Tilburg University press.
Williams, D.G. (1982). Weeping by adults: personality correlates and sex differences. *Journal of Psychology*, *110*, 217–226.
Znoj, H.J. (1997) When remembering the lost spouse hurts too much: First results with a newly developed observer measure for tears and crying related coping behavior. In: A.J.J.M. Vingerhoets, F.J. Van Bussel, & A.J.W. Boelhouwer (Eds.), *The (non)expression of emotions in health and disease* (pp. 337–352). Tilburg, The Netherlands: Tilburg University Press.

15 PATHOLOGICAL HUMAN CRYING

Aziz T. Shaibani, Marwan N. Sabbagh, and Ban N. Khan

Human crying as a medical disorder has largely been neglected by medical investigators with only a few comprehensive reviews in the literature. There is no unifying classification system and the terminology used to describe this condition is confusing. This chapter reviews what is known about the anatomy and pathology of normal and pathological human crying (PC), the clinical syndromes associated with pathological crying, and current treatments of it.

Normal Crying

STRUCTURES INVOLVED IN NORMAL CRYING

Although crying has been noted in utero, newborns normally start crying at birth (initially tearless). Crying is, therefore, likely not a learned behavior. This premise is supported by observations that congenitally blind children cry and shed tears, even though they have had no opportunity to observe others crying. This "reflex" occurs in response to a variety of non-specific stimuli (e.g., tactile, olfactory, auditory, etc.). As the brain matures, socialization of this reflex occurs so that crying can be stimulated by social conditions (Seliger et al., 1992).

The neuroanatomy and neurophysiology subserving crying are not completely understood, although there are clear indications of the involvement of certain brain structures (see Table 1 and also Chapter 2, this volume).

Wilson (1924) has postulated a center linking the facial nerve nucleus in the pons with the tenth motor nucleus in the medulla and with phrenic

Table 1. Proposed anatomic structures involved in normal crying

Facial nerve nucleus (pons)
Vagus motor nucleus (medulla)
Phrenic nuclei (C3,4,5)
Diencephalon (thalamus, hypothalamus, subthalamic nucleus)
Cerebral cortex (frontal lobe)

nuclei in the upper cervical cord (C3,4,5). These connections are needed for the fasciorespiratory coordination in crying. He postulated that such an integrative center could be located in the mesial thalamus, hypothalamus, and/or the subthalamic nucleus. This center's activity is determined by voluntary input through fibers that provide a relay from the cerebral cortex to the brain stem and by involuntary input carried by fibers that descend from the frontal lobes and basal ganglia. The involuntary pathway seems to be inhibited by the voluntary one. Both of these pathways theoretically control the laughter/crying "center" by accelerating and decelerating mechanisms (Davison & Kelman, 1939). Cortical areas that provide the input to the voluntary and involuntary pathways to the crying center are not clearly identified. Cerebral structures that have been implicated to subserve the evolution of crying and laughter include the midline frontolimbic cortex and the adjoining frontal neocortex. There is evidence that the thalamocingulate division of the limbic system may provide reciprocal innervation for crying and laughter (Maclean, 1987).

The presumed crying center sends both tonic and clonic input to the facial muscles. Loss of tonic control leads to a paroxysmal crying disorder which will be discussed later (Shaibani et al., 1994). Diseases that interrupt the involuntary pathway cause impairment of involuntary facial expressions leading to hypomimia ("mask face"), such as that which occurs in Parkinson's disease. Disorders that affect voluntary input lead to loss of volitional control and preservation of emotional facial expressions.

PHYSICAL CONCOMITANTS OF HUMAN CRYING

Crying involves the facial musculature, respiratory tract, and the autonomic nervous system (Wilson, 1924). Expressions involved in crying require continuous contraction of certain groups of facial muscles while the respiratory component involves sudden expiration by forced contraction of intercostal muscles followed by saccadic expiration-inspiration microcycles that occur mainly with inspiration. The vocal cords add a short and broken sound to crying. Autonomic features of crying include activation of both the sympathetic system (arousal, motor, tension) and the parasympathetic system (dilation of the facial blood vessels that causes flushing, dilation of pupils, lacrimation, etc.).

Pathological Crying

Facial expression plays an important role in emotional expression and non-verbal communication. PC is not a primary emotional disturbance but a condition that reflects defective motor expression of emotional affect. It is not appropriate to the situation but spontaneous and is dissociated from emotional stimuli. Accordingly, crying associated with depression and other psychiatric disorders is not included in detail in our discussion (see Chapter 14, this volume). Patients with certain neurological disorders such as stroke, multiple sclerosis, Parkinson's disease, and motor neuron disease frequently experience additional distress from defective emotional control associated with PC.

There is no consensus as to the most appropriate term for this condition and it is not clear if it represents a homogenous syndrome. Terms used to date include pseudo-bulbar affect, pathological crying (Poek, 1985), emotional incontinence (Seliger et al., 1992), emotional lability, emotionalism, and pathological display of affect (Andersen et al., 1995). For the purposes of this discussion, the term pathological crying (PC) will be used.

Clinically speaking, PC is problematic. It is distressing to the patient and caregiver (Allman, 1991). Nevertheless, patients with PC (or their relatives) rarely complain about it, partly because they are often unaware that it is related to their underlying neurologic disorder and that treatment is possible, partly because they are too ashamed to admit to having the condition, and partly because mere discussion of the condition often provokes a crying episode. Additionally, many cases go unrecognized because physicians fail to specifically question their patients about crying episodes. Furthermore, physicians are often unaware of how frequent PC is, how debilitating it can be, and how well it responds to treatment (Andersen, 1997). Features that distinguish normal crying from PC are summarized in Table 2 (Shaibani et al., 1994).

Careful clinical observation suggests that only a few patients fulfill the first criterion. House et al. (1989) described PC triggered by appropriate stimuli in all their stroke patients; the abnormality being one of the magnitude of response. The usual triggers stated in their series were predominantly sad events related to the subject or his relatives. In a few patients, discussing the symptoms of PC was enough to trigger a response (Dark et al., 1996).

Facial expression during PC has been studied using electromyography (EMG) (Tanaka & Sumitsuji, 1991). Increased EMG discharges of the

Table 2. Features of pathological crying

Feature	Additional comments
• Inappropriate to the situation because it follows nonspecific or inappropriate stimuli.	Nonspecific stimuli include contraction of facial muscles, someone approaching the patient, removal of bed covers, feeding the patient, etc.
• Unmotivated.	There is no relation between the affect and the observed expressions. Additionally, there is no relief or mood change after the expression resolves.
• Involuntary.	It has its own automatic pattern and occurs against the patient's will.

upper facial muscles is characteristic of PC while increased EMG discharges of the lower facial muscles have been observed in natural laughing. The patterns of facial expression during PC are similar to those of crying in normal individuals. Since normals show the same expression of crying when experiencing an extreme emotional outburst, it is possible that PC is a maximal manifestation of facial expression caused by a release of inhibition from upper motor control.

NEUROPATHOLOGY

Information about the anatomy of PC comes from a limited number of autopsy reports, studies of congenitally malformed children, and from case reports (see Table 3).

Review of autopsy studies, including Davison and Kelman's (1939) major study of 33 subjects, concluded the following points. First, no single cortical lesion is found to cause PC. For example, PC has been

Table 3. Proposed anatomic structures involved in pathological crying

Descending pyramidal or extrapyramidal tracts to bulbar nuclei
Mesencephalon (mid-brain)
Caudal pons and medulla
Brainstem raphe nuclei
Serotonergic system

described in left middle cerebral artery strokes, tumors in the right frontal lobe, and glioblastoma multiforme of both fronto-parietotemporal regions. This is not surprising, given the high incidence of PC in patients with diffuse disturbances such as vascular dementia. Second, unilateral or bilateral lesions that damage descending tracts (pyramidal or extrapyramidal) to the bulbar nuclei can cause PC.

The relative contribution of right versus left hemisphere structures remains controversial. In addition to pathological laughing, right hemispheric destructive lesions can lead to difficulty in interpreting humor, comedy, and cartoons. A study of six patients with crying seizures revealed non-dominant foci (Luciano et al., 1993). Nevertheless, PC has been associated with predominantly left-sided lesions (Sackeim et al., 1982). Injection of the left carotid with sodium amobarbital has been noted to induce a significant amount of PC (Lee et al., 1990, 1993). However, destructive lesions in one hemisphere can lead to the same abnormalities in emotional expression as irritative lesions (i.e., epilepsy) in the contralateral hemisphere. All these data concerning destructive lesions and the nearly opposite effects of irritative epileptic lesions support the observation that the right hemisphere is dominant for negative emotions and left hemisphere is dominant for positive emotions (Shaibani et al., 1994).

Studies of congenitally malformed babies contribute to localization of PC. Newborns with severe anencephaly, but with preservation of the pons and medulla, do not smile but are able to cry, both spontaneously and after facial stimulation (Poeck, 1985). Congenitally malformed newborns with intact midbrains both cry and smile (Poeck, 1985). One can speculate that crying and associated facial expressions which signify the need for help and defense are more important for survival than laughing. This is consistent with caudal localization of mechanisms required for crying in anencephalic infants.

Clinical case reports further add to our knowledge about localization of crying mechanisms. One clinical case report concerns a patient who presented solely with PC, in whom Magnetic Resonance Imaging (MRI) showed a vascular malformation in the anterior upper part of the brain stem (Andersen et al., 1994). Another patient whose hyponatremia was rapidly corrected subsequently developed PC, gaze paralysis, bilateral face and tongue paresis, and dysarthria. His MRI showed demyelination of the decusating fibers in the pons consistent with central pontine myelinolysis. In a study of amyotrophic lateral sclerosis (ALS) patients, almost all of those with PC had bulbar involvement. Although these studies support a significant role of the brain stem in the generation of crying, it is probable that bilateral cerebral hemispheric disease due to a wide variety of pathological entities is the most common neurologic cause of PC (Shaibani et al., 1994).

The neurochemistry of crying is also not well understood. Recently, it has been proposed that damage to the serotonergic neurotransmission system, particularly the brainstem raphi nuclei, might be directly involved in the pathophysiology of PC (Andersen, 1997; Andersen et al., 1994). The brainstem raphi nuclei give rise to long ascending projections through the basal ganglia to the limbic forebrain of the frontal cortex, from which they run anteriorly to posteriorly through the deep layers of cortex. In the study conducted by Andersen and coworkers, patients with severe post-stroke PC had large bilateral pontine or central hemispheric lesions, while the clinically least affected patients had mainly unilateral large subcortical lesions. Recently, there was a report of a patient who developed severe and long-lasting poststroke PC after a unilateral cerebral infarction (Derex et al., 1997). The authors concluded that a single limited subcortical left-sided lesion with critical topography regarding serotonergic pathways may cause permanent PC. Hence, the greater the damage to the serotonergic system, the more long lasting and easily or non-specifically provoked are the episodes of crying. A role for the frontal cortex was recently suggested for PC in multiple sclerosis (Feinstein et al., 1999) and amyotrophic lateral sclerosis (McCullagh et al., 2000)

Associated Conditions

With the exception of the psychiatric causes of PC where it is a part of the emotional pathology, all other secondary causes of PC affect the motor expression of emotions rather than the emotions themselves. Table 4 lists the conditions associated with PC.

AMYOTROPHIC LATERAL SCLEROSIS (ALS)

In 25% of ALS patients, bulbar involvement develops. Thirty to fifty percent of these patients have PC. Poeck (1985) has observed that PC associated with ALS can occur spontaneously or follow minimal stimulation. This automatic response may eventually be extinguished. ALS patients without bulbar involvement do not develop PC.

Table 4. Conditions associated with pathological crying

- Amyotrophic lateral sclerosis (ALS)
- Multiple sclerosis (MS)
- Cerebrovascular disease
- Brain anoxia
- Extrapyramidal syndromes
- Central pontine myelinolysis
- Olivopontocerebellar atrophy
- Dacrystic seizures
- Structural causes: hypothalamic gliomas and hamartomas; posterior fossa tumors (osteochomdroma, astrocytoma); temporal lobe tumors
- Trauma
- Aneurysms
- Brainstem arteriovenous malformations
- Alzheimer's disease
- Psychiatric disorders: schizophrenia, hysteria, manic-depression, psychosis

MULTIPLE SCLEROSIS (MS)

MS has often been described as one of the classic conditions in which PC occurs. PC develops in 7%–10% of MS patients, sometimes associated with euphoria clinically (Foerster & Gagel, 1993). It may occur in isolation from the other affective disorders frequently associated with MS (e.g., major depression, bipolar disorder).

In a consecutive series of 152 patients with clinically or laboratory definite MS screened for the presence of PC, the demographic and disease profile to emerge was one without gender predilection, with fairly long-standing disease duration (i.e., after patients had had MS for approximately a decade), associated with progressive significant physical disability, and not necessarily of brainstem origin (Feinstein et al., 1997). However, others have suggested PC is associated with pontine, brain stem and periventricular demyelination (Reischies et al., 1988; Robins et al., 1986). MS patients with PC do not experience greater subjective emotional distress, but are more intellectually impaired and, as such, are likely to have more extensive brain damage than patients without PC.

ISCHEMIC VASCULAR DISEASE

Abnormal emotionalism is noted in 40% of patients who have suffered stroke. In addition to PC, depression, mania, bipolar disorder, anxiety and apathy have been observed following stroke, any of which may adversely affect rehabilitation (Ghika-Schmid & Bogousslavsky, 1997; Robinson, 1997). PC affects 20% of post-stroke patients and occurs primarily in the acute stroke period (Andersen, 1997). Left frontal and temporal strokes were more commonly implicated. A recent clinical case study reported a patient who presented with sudden onset of vomiting, left hemisensory loss of the face and body, and progressive loss of verbal responsiveness who was noted to be crying continuously. The computer tomography (CT) scan 12 hours after onset of symptoms showed a hyperdensity in the tip of the basilar artery, suggesting basilar artery thrombosis (Larner, 1998). Cerebral arterial angiography confirmed occlusion of the basilar artery. Hence, PC may be the first symptom of a catastrophic brainstem cerebrovascular event.

DACRYSTIC SEIZURES

Dacrystic seizures or ictal crying is much less common than gelastic seizures (epileptic laughter). One report describes seven patients who cried during or right after partial seizures documented by video-EEG telemetry (Lee et al., 1990). The seizure foci were found in the non-dominant temporal region in five of the patients, in the frontocentral region in one and the non-dominant frontal region in another.

TRAUMATIC BRAIN INJURY (TBI)

PC commonly interferes with both rehabilitation efforts and with patients' quality of life after TBI. Such a disability is especially important to patients with TBI, who are likely to be young adults in the process of trying to establish new personal and professional relationships. In one study, patients displaying pathological laughing and crying had a greater severity of brain injury than patients without the syndrome (Zeilig et al., 1995). These patients also had other associated neurological symptoms compatible with pseudobulbar palsy. Pathological laughter alone, or combined with crying, was more frequent than crying alone. An attempt to correlate clinical features with focal lesions on neuro-imaging studies was inconclusive.

ALZHEIMER DISEASE (AD)

PC may be common in AD. Starkstein et al. (1995) examined 103 patients with AD and found PC in almost 40%. Many of them suffered from mixed laughter and crying. A significant amount of cortical atrophy was reported with this condition. PC has been reported in Alzheimer disease in particular when the latter is associated with generalized anxiety disorder (Chemerinski et al., 1998).

PSYCHIATRIC DISORDERS

These are the most common causes of PC. A syndrome of "essential crying" has been described where no psychiatric or medical/neurological contributing factors were identified. Patients experienced life-long abrupt crying outbursts, usually in a context appropriate to internal and environmental cues. PC is mood-related in affective disorders, and may be a manifestation of the inappropriate affect seen in schizophrenics (Shaibani et al., 1994).

Treatment of Pathological Crying

PC is a socially disabling condition, embarrassing to patients and their families. Fear of developing uncontrollable bouts of PC, secondary phobias and social withdrawal often make symptoms worse (Shaibani et al., 1994).

There are relatively few reports concerning the treatment of PC. Most of the trials were not controlled and included small numbers of patients but a variety of different medications have shown beneficial effects such as tricyclic antidepressants, L-dopa, and fluoxetine (Allman, 1991). Only one study used a objective rating scale for PC, while other studies used clinical judgement alone.

Five open trials and three double-blind trials all reported benefit for treating PC with agents including anti-depressants, L-dopa, and thyrotropin releasing hormone (TRH) (see Table 5). The nature and degree of the response to medication was often dramatic, with significant declines reported in the number and duration of outbursts. In trials that investigated the use of anti-depressants, the amelioration of PC was apparently independent of any anti-depressant effect. Serotonin reuptake inhibitors, such as citalopram, have shown particular efficacy (Andersen, 1995). When scores on depression

Table 5. Therapeutic trials for pathological crying

Drug	Etiology	Number of Responders	Total Dose mg/day	Type of trial	Number of days to response
L-Dopa	CVD/ trauma	10/25	600–1500	Open	2–5
Amantadine	CVD/ trauma	4/8	100	Open	2–5
Imipramine	CVD	5/7	60	Double-blind	7
Amitriptyline	MS	8/12	25–75	Double-blind/ crossover	2
Fluoxetine	MS/CVD	13/13	20	Open	3–14
Fluoxetine	HI/CVD	6/6	20	Open	7
Nortriptyline	CVD	28/28	20 increased to 100	Double-blind/ controlled	14
TRH	CVD/OPCA	2/4	1.0	Open	3–4

CVD, cerebrovascular disease; MS, multiple sclerosis; HI, head injury;
OPCA, olivopontocerebellar atrophy

rating scales were measured during some of these studies, no significant change was noticed despite improvement in the PC. Most of benefits were noted within several day of instituting therapy, a shorter period than the acknowledged 3–4 weeks required for an anti-depressant effect (Shaibani et al., 1994). Unusual change from PC to laughter following drug treatment has been recently described in ALS patients (McCullagh & Feinstein, 2000)

Conclusion

This chapter focused on the features of and neuroanatomic structures involved in PC. In addition, clinical disorders associated with this uninhibited expression of emotions were discussed. Finally, clinical trials testing the effects of different medications on this condition were summarized. It appeared that significant improvement was possible, despite the lack of clear relationship of such amelioration with the antidepressant effects of these agents.

With the rapid new developments in neuro-imaging, our knowledge of the brain mechanisms involved in normal and pathological crying is expected to increase considerably in the coming years. Better insight into the neuroanatomy and neurophysiology of this behavior may help to enhance our understanding of its functions.

References

Allman, P. (1991). Emotionalism following brain damage. *Behavioural Neurology, 4,* 57–62.

Andersen, G. (1995). Treatment of uncontrolled crying after stroke. *Drug and Aging, 6,* 105–111.

Andersen, G. (1997). Post-stroke depression and pathological crying. Clinical aspects and new pharmacological approaches. *Aphasiology, 11,* 651–664.

Andersen, G., Ingeman-Neilsen, M., & Vestergaard, K. (1995). Post-stroke pathological crying, frequency and correlation to depression. *European Journal of Neurology 2,* 45–50.

Andersen, G., Ingeman-Nielsen, M., Vestergaard, K., & Riis, J.O. (1994). Pathoanatomical correlation between post-stroke pathological crying and damage to serotonergic neurotransmission. *Stroke, 25,* 1050–1052.

Chemerinski, E., Petracca, G., Manes, F., Leiguarda, R., & Starkstein, S.E. (1998). Prevalence and correlates of anxiety in Alzheimer's disease. *Depression & Anxiety, 7,* 166–170.

Dark, F., McGrath, J., & Ron, M. (1996). Pathological laughing and crying. *Australian and New Zealand Journal of Psychiatry, 30,* 472–479.

Davison, C., & Kelman, H. (1939). Pathological laughter and crying. *Archives of Neurology and Psychiatry, 42,* 595–643.

Derex, L., Ostrowsky, K., Nighoghossian, N., & Trouillas, P. (1997). Severe pathological crying after left anterior choroidal artery infarct, reversibility with paroxitine treatment. *Stroke, 28,* 1464–1466.

Feinstein, A., Feinstein, K., Gray, T., & O'Connor P. (1997). Prevalence and neurobehavioral correlates of pathological laughing and crying in multiple sclerosis. *Archives of Neurology, 54,* 1116–1121.

Feinstein, A., O'Connor, P., Gray, T., & Feinstein, K., (1999). Pathological laughing and crying in multiple sclerosis: A preliminary report suggesting a role for the prefrontal cortex. *Multiple Sclerosis*, 5, 69–73.

Foerster, O., & Gagel, O. (1993). Ein Fall von Ependymcysts des III Ventrikels. *Zentralblatt für die Gesamte Neurologie und Psychiatrie, 45,* 312–44.

Ghika-Schmid, F., & Bogousslavsky, J. (1997). Affective disorders following stroke. *European Neurology, 38,* 75–81.

House, A., Dennis, M., Molyneux, A., Warlow, C., & Hawton, K. (1989). Emotionalism after stroke. *British Medical Journal, 298,* 991–994.

Larner, A.J. (1998). Basilar artery occlusion associated with pathological crying, "Folles larmes prodromiques"? *Neurology, 51,* 916.

Lee, G., Loring, D., Meader, K., & Brooks, B. (1990). Hemispheric specialization for emotional expression, a reexamination of results from intracarotid administration of sodium amobarbital. *Brain and Cognition, 12,* 267–280.

Lee, G.P., Loring, D.W., & Meador, K.J. (1993). Influence of premorbid personality and location of lesion on emotional expression. *International Journal of Neurosciences, 72,* 157–165

Luciano, D., Devinsky O., & Perrine, K. (1993). Crying seizures. *Neurology, 43,* 2113–2117.

MacLean, P.D. (1987). The midline frontolimbic cortex and the evolution of crying and laughter. In E. Perecman (Ed.), *The frontal lobes revisited* (pp. 121–140). New York: The Irbn Press.

McCullagh, S., Moore, M., Gawel, M., & Feinstein, A. (2000). Pathological laughing and crying in ALS: An association with orefrontal cognitive dysfunction. *Journal of Neurological Science*, 169, 43–48.

McCullagh, S., & Feinstein, A. (2000). Treatment of pathological affect: Variability in response for laughter and crying. *Journal of Neuropsychiatry and Clinical Neuroscience, 12,* 100–102.

Poeck, K. (1985). Pathological laughter and crying. *Handbook of clinical neurology, 1,* 219–224.

Reischies, F.M., Baurn, K., Brau, H., Hedde, J.P., & Schwidt G. (1988). Cerebral resonance magnetic imaging findings in multiple sclerosis. *Archives of Neurology, 45,* 1114–1116.

Robins, P.V., Brooks, B.R., O,Donnell P., Pearlson, G.D., Moberg, P., Jubelt, B., Coyle, P., Dalos, N., & Folstein, M.F. (1986). Structural brain correlates of emotional disorder in multiple sclerosis. *Brain, 109,* 585–597.

Robinson, R.G. (1997). Neuropsychiatric consequences of stroke. *Annual Review of Medicine, 48,* 217–229

Sackeim, H.A., Greenberg, M.S., Weiman A.L, Gur, R.C., Hungerbuhler, J.P., & Geschwindt, N. (1982). Hemispheric asymmetry in the expression of positive and negative emotions. Neurologic evidence. *Archives of Neurology, 39,* 210–18.

Seliger, G., Hornstein, A., & Flax J., Herbert, J., & Schroeder, K. (1992). Fluoxetine improves emotional incontinence. *Brain Injury, 6,* 267–270.

Shaibani, A.T., Sabbagh, M.N., & Doody R. (1994). Laughter and crying in neurologic disorders. *Neuropsychiatry, Neuropsychology & Behavioral Neurology, 7,* 243–250.

Starkstein, S.E., Mignorelli, R., Teson, A., Petracca, G., Chemerinski, E., Manes, F., & Leiguarda, R. (1995). Prevalence and clinical correlates of pathological affective display in Alzheimer's disease. *Journal of Neurology, Neurosurgery & Psychiatry, 59, 55–60.*

Tanaka, M., & Sumitsuji, N. (1991). Electromyographic study of facial expressions during pathological laughing and crying. *Electromyography & Clinical Neurophysiology, 31,* 399–406.

Wilson, S.A.K. (1924). Some problems in neurology, II. Pathological laughing and crying. *Journal of Neurology & Psychopathology, 16,* 299–333.

Zeilig, G., Drubach, D.A., Katz-Zeilig, M., & Karatinos J. (1996). Pathological laughter and crying in patients with closed traumatic brain injury. *Brain Injury 10,* 591–597.

16 THE STUDY OF CRYING: SOME METHODOLOGICAL CONSIDERATIONS AND A COMPARISON OF METHODS FOR ANALYZING QUESTIONNAIRES

Jan G.M. Scheirs and Klaas Sijtsma

"Crying, far more than its opposite, laughing, is a neglected problem of psychology," Borgquist wrote in 1906. Things do not seem to have changed much since the beginning of the 20th century. Several basic questions concerning crying have not been answered yet and uncorroborated theories and speculations still characterize the field (Vingerhoets et al., 2000). Much research is needed for gaining an understanding of both the causes and the functional significance of crying. Examples of important but still barely investigated questions are: What are the typical situations in which crying occurs? What are the typical emotions accompanying or following a crying spell? What are the personality characteristics of those who often cry and those who hardly do? Are there gender or cultural differences in crying? What is the developmental pattern of crying across the life span?

In order to obtain valid information from empirical studies, the design of such studies should be both sound and appropriate. How can that be achieved? Although it is not our aim to discuss this issue fully here, it might be helpful to present some general considerations concerning the choice of methodology, and to briefly discuss some of the stronger and weaker points of different approaches. We will do that first, and we will argue that collecting data by means of questionnaires has been and probably will remain a valuable method for investigating crying. Next, we compare two methods for analyzing ordered rating scale data from questionnaires designed to measure aspects of crying by means of sum scores on sets of items. Finally, we apply these methods to the available Adult Crying Inventory (ACI) questionnaire data (Vingerhoets & Becht, 1996), and we discuss the practical relevance of the results for researchers and practitioners.

Methodological Considerations in Crying Research

CRYING AS A DEPENDENT AND AS AN INDEPENDENT VARIABLE

Crying is the dependent variable when studying typical situations or mood states in which crying occurs, for example, crying at a funeral, at the work place, or at home, or crying in response to sadness, anger, or joy. Research in which crying is the dependent variable is often exploratory and uses correlational methods for studying relations among variables.

Crying is the independent variable when the question is how crying influences other variables, for example, whether crying promotes health or what effects crying has on others (see Figure 1). When crying is an independent variable it is often dichotomous, i.e., crying is distinguished from not crying, and an experimental or quasi-experimental design may be used for determining its effects. The distinction between crying as an independent or a dependent variable thus is relevant, because it often determines the research methodology.

When crying is the dependent variable, several aspects of crying may be studied. First, the focus might be on its specific physiological aspects, for example, the chemical composition of tears, the innervation of the lacrimal system, or changes in activation of the autonomic and somatic

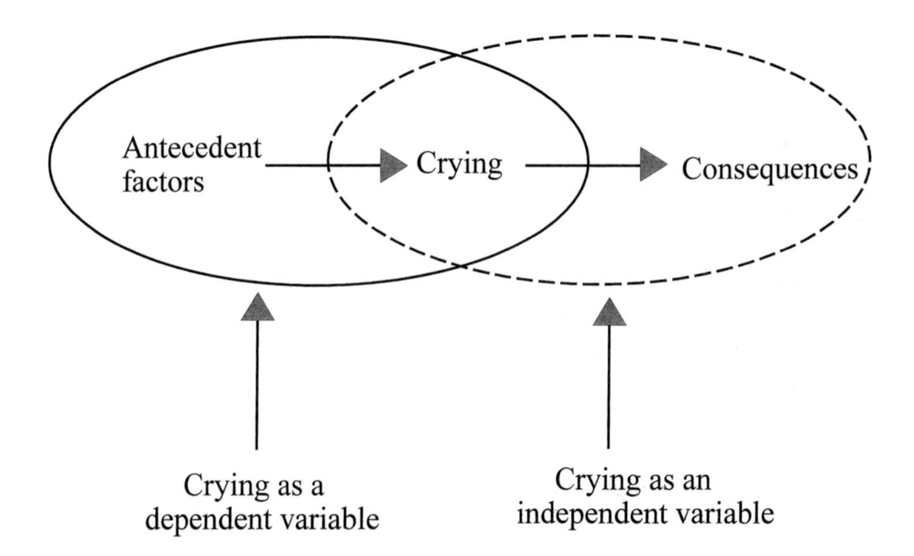

Figure 1. When studying the relationships between crying and other variables, crying can be considered either a dependent or an independent variable.

nervous system that accompany crying (e.g., Gross et al., 1994). Second, there might be interest in the behavioral manifestations of crying, for example, in the frequency, the propensity, the intensity, or the duration of crying (Vingerhoets & Scheirs, 2000). These measures can be studied by observation or by self-report.

When studying the behavioral manifestations of crying, systematic observation is the preferred method theoretically, because the measurement procedure can be made more explicit and objective. In practice, however, the use of behavioral observation is hampered by adult crying being a rare event in most societies, especially in males (e.g., Chapter 6, this volume; Vingerhoets & Scheirs, 2000). Unlike the medical researcher, who can study a rare disease because people suffering from it eventually will come to see him, the researcher investigating crying has hardly any opportunities to study his or her target behavior "in vivo." An exception may be made for crying studied in the laboratory, where crying can be induced by showing a sad film to the subjects or by having subjects recall affectively-laden events (Delp & Sackeim, 1987; Marston et al., 1984). For example, Znoj (1997) investigated a group of recently bereaved women who were asked to imagine that they were talking to their deceased partners. The resulting crying or crying-related behaviors, such as swallowing or pressing lips, were filmed and scored by trained observers.

In the laboratory the amount of crying can thus be assessed in several ways: by obtaining either self ratings or observer ratings, or by measuring the physiological response, i.e., lacrimal flow, as was done by Delp and Sackeim (1987). When the causes of crying are investigated, however, laboratory procedures such as these can hardly be used. This, of course, is due to the unknown chances of "success," i.e., the chances that people will actually cry in response to a particular stimulus. Also, researchers might find it unethical to elicit sadness in study participants. A final argument against observation is that usually there is interest in not only the behavioral manifestations of crying (sobbing at the one end of the scale, or just showing watery eyes at the other end), but also in the emotions and feelings that accompany crying.

Wallbott and Scherer (1989) emphasized that subjective experience is the most important characteristic of emotions. According to these authors, physiological measurements and other objective measures of expressive behavior do not capture this experience. Thus, it is both understandable and justifiable that most investigations of crying have made use of self-reports, i.e., structured questionnaires or open interviews, for collecting data.

Young infants cry more often than adults. For them, and especially for babies for whom crying is the only means of communication, crying can serve many functions (Chapter 3, this volume; Lester, 1985). We do not know yet to what degree the causes and functions of crying in adulthood parallel those in infancy. It is obvious that the study of infant crying is bound to use the method of observation exclusively, either by the mechanical or electronic recording of vocalized expressions or directly by human observers. Infant crying, however, is beyond the scope of this chapter.

Problems Associated with Self-Report

The method of self-report, and the use of questionnaires in particular, has several problems. First, questionnaire studies may suffer from *response bias* or *response set*. For example, respondents may show a tendency to give socially desirable answers, or to use positive answer categories more easily than negative ones. Also, sequence effects resulting in stereotypical response patterns may occur (Stone, 1995). These effects may be due to highly similar questions which are asked closely after each other, or to the assessment of mood on different occasions using the same questionnaire. Bias in the reported frequency of behavior might also be introduced by question wording, choice of response alternatives and length of the reference period (Schwarz, 1990).

Second, details of autobiographical events and the frequency of these events are sometimes badly remembered (Bradburn et al., 1987). Specifically, retrospective reports tend to overestimate the intensity of unpleasant emotional experiences, and intensity and frequency of emotions tend to be confused (Thomas & Diener, 1990). Kraemer and Hastrup (1986) found that self-monitored incidents of crying correlated reasonably well with frequency estimates that were given later, but also that the subjective estimates were generally higher than the recorded incidences.

Whereas the contaminating effects of specific response biases can often be controlled for by careful questionnaire design, the distortions caused by inaccurate recall of life events or associated emotions are more difficult to overcome. When these distortions are random deviations from the true values, they add to measurement error and decrease the power of the statistical tests. Thus, such distortions reduce the likelihood that true relationships are found. When the distortions are systematic instead, the internal validity of the investigation may be questioned, because associations among variables and possible causal relationships may not be established accurately. This is a serious problem because respondents are known to reconstruct their memories when they are unable to retrieve sufficient information (Bradburn et al., 1987). When groups of participants, e.g., men and women, employ different strategies to fill the gaps in their memories, this could lead to distorted conclusions, in this case about the nature of gender differences. However, we are still far from grasping how these mechanisms work precisely and from knowing to what degree the functioning of human memory affects survey outcome.

From the above findings it can be concluded that when frequency or intensity estimates of crying-related behaviors or emotions are reported, the estimates themselves are likely to be inaccurate. Furthermore, one should be careful when comparing the results from questionnaires that were devised to measure the same concept (i.e., crying behavior), but that differed with regard to items included, question wording, or scale values.

An alternative procedure for reducing the problems associated with faulty memory is the use of diaries. It should be noted, however, that the detailed recording of one's behavior might affect this behavior and that, due to the continued attention asked from the participants, recording periods are necessarily limited in length (Lyberg & Kasprzyk, 1991). Thus, diary keeping may not be a method that is generally suited for the study of rare events such as crying. Therefore, questionnaires will probably continue to be used as the preferred research instrument in the field.

CORRELATIONAL RESEARCH VERSUS EXPERIMENTAL RESEARCH

It appears that the predominant use of questionnaires almost predisposes research questions concerning crying to be tackled by correlational methodology. Although this is the state of affairs for much crying research, it does not preclude more rigorous research designs to be used. We already mentioned the research by Marston et al. (1984), Delp and Sackeim (1987), and Znoj (1997), as examples of laboratory research in which crying was induced and aspects of the crying response or its consequences were subsequently measured. As Cornelius (1997, Table 26.2) showed, there are several examples of this type of research that can be traced in the literature.

When the groups compared in an experiment have been composed by random assignment of subjects, the design is called experimental, whereas comparing intact groups is called quasi-experimental. Because only one variable is varied in an experiment and other variables are kept constant, the experiment is well-suited for strong inferences about causal relationships. Internal validity in experiments is usually high, but external or ecological validity may be low because the laboratory environment is different from real life. Thus, the crying observed in an experiment may not resemble the crying outside the laboratory (however, see Cornelius, 1997, p. 307). This may limit the generalization of laboratory results, and because of this some researchers (e.g., Löfgren, 1966) have decided to abandon the experiment completely.

An important question concerning experiments is whether it is possible to obtain two groups of participants, say, criers and non-criers, without further systematic differences between them, and thus whether it is possible to surpass the limitations of the quasi-experiment. Kraemer and Hastrup (1988) showed that this can be done. They randomly assigned subjects to either of two conditions in which the instructions were to either cry or inhibit crying in response to a film. Self ratings and observer ratings indicated that the participants succeeded in acting according to instructions. Experiments like these allow differences in the dependent variable to be ascribed uniquely to the crying response, and thus are promising for obtaining a better understanding of the effects of crying and its suppression.

Crying can also be studied using the ethological method. Behavior is studied by participating in or closely observing the activities of people in their natural environment, especially in non-western societies, and by writing down, recording, photographing or filming what is seen or heard (e.g., Eibl-Eibesfeldt, 1984). The ethological method has been criticized because of its subjectivity of measurement and difficulty of interpretation, and praised because of its high external validity.

Except for the ethological method, the use of questionnaires for the measurement of adult crying, emotions, or crying-related behaviors has proven to be important in most of the methodologies discussed here. That is why in the remainder of this chapter, we will concentrate on the questionnaire as a widely applicable device for collecting data on crying. More specifically, we will explain and illustrate two methods for analyzing data obtained from questionnaires, and discuss the results of a data analysis and the consequences for further research.

Two Methods for Analyzing Questionnaire Data

We will first briefly describe the questionnaire that was used by Vingerhoets and Becht (1996) in a cross-cultural study on situations and mood states eliciting crying. Then we will explain both a classic and a modern psychometric method for analyzing questionnaires. The questionnaire—the Adult Crying Inventory (ACI)—is reprinted in the appendix of this book. The cross-cultural aspects are discussed by Becht et al. (Chapter 8, this volume). We will here treat the data obtained with the ACI as if they were collected in a homogeneous population, thus ignoring the subgroup structure by countries.

THE ADULT CRYING INVENTORY (ACI)

The ACI consists of 54 statements about situations and mood states which may elicit adult crying. Vingerhoets and Becht (1996) derived the items partly from existing questionnaires, while others were newly formulated to describe theoretically relevant situations not covered by existing items. Some of the statements have a negative connotation and others have a positive connotation. Responses were given on an ordered 7-point rating scale, which runs from "never" (lowest score: 0) to "always" (highest score: 6). The five intermediate response categories are unlabeled. Examples of statements are: "I cry when I experience opposition from someone else" (Item 28), "I cry when I remember sad things that have happened to me" (Item 16), and "I cry when I hear the national anthem and/or see the national flag rise" (Item 47).

The examples already suggest that the situations and mood states may refer to different causes of crying, thereby perhaps inducing different kinds of crying. For example, Item 47 may bring back memories of a highly appreciated sports event which then caused the respondent to cry from pride or joy, whereas Item 16 for that same respondent may strongly refer to the loss of a beloved friend which caused crying from sadness. Moreover, Item 28 seems to refer to another cause which may have to do with distress experienced as a reaction to external threats, for example, in a work situation. Thus, the ACI does not seem to measure a unitary trait of crying proneness, but rather a mixture of related traits.

Obviously, if the researcher *suspects* his/her questionnaire to be multidimensional, or if the researcher *expects* multidimensionality due to the deliberate choice of items measuring different aspects of a particular trait, this hypothesis has to be investigated explicitly. The classical method for identifying multidimensionality in questionnaires is factor analysis followed by classical reliability analysis intended to estimate for each factor the reliability of the total score on the items having high loadings on that factor. Besides these well-known and successful methods, during the past decades new psychometric methods have been proposed for analyzing multidimensional rating scale data. Sometimes these newer methods may replace classical methods, and sometimes they may be used in addition to these classical methods.

The main purpose here is to report on the analysis of the ACI data using factor analysis and classical reliability analysis, and critically compare the results with results obtained from a method from modern psychometrics known as Mokken scale analysis (Hemker et al., 1995; Mokken, 1997; Mokken & Lewis, 1982; Molenaar, 1997; Sijtsma, 1998). We assume that the reader has knowledge of the general ideas behind factor analysis, but that Mokken scale analysis may be less well known. Thus, before discussing the results from the ACI data analysis, we discuss Mokken scale analysis in some detail, and compare the method with factor analysis in order to clarify differences and similarities.

MOKKEN SCALE ANALYSIS

Mokken scale analysis (MSA) selects items from a larger item set so as to satisfy a measurement model, which is the monotone homogeneity model (Mokken & Lewis, 1982; Sijtsma, 1998). A property of the monotone homogeneity model is that a higher item score (here, ranging from 0 to 6) indicates a higher level of the latent trait being measured by the questionnaire; here, crying proneness. Thus, a person responding in the lowest category of Item 16 ("I cry when I remember sad things that have happened to me") has a relatively weak crying proneness, and persons giving responses in higher categories are progressively more prone to cry. In addition, adding the scores on different items each measuring crying proneness yields a sum score of which higher values indicate stronger proneness to cry. Moreover, this sum score more reliably indicates the position of the respondent on the latent trait than do the individual item scores. If the monotone homogeneity model can be fitted to the data, these scoring and ordering properties are valid for those data; in practice, the questionnaire then can be considered a proper measurement device for ordering respondents on crying proneness.

The monotone homogeneity model is based on the following assumptions. First, it assumes that the items all measure the same latent trait (*unidimensionality* assumption); in this case crying proneness.

Second, the model assumes that respondents with a higher crying proneness are more likely to obtain higher item scores than respondents with a lower crying proneness (*monotonicity* assumption). These are intuitively appealing properties of psychological measurement that need to be checked on empirical data, however, because data can be multidimensional and the relation between item score and latent trait may not be monotonically nondecreasing. MSA performs these checks by investigating whether the observable consequences from unidimensionality and monotonicity hold for the data.

These observable consequences are that (1) all covariances or correlations between items must be nonnegative (a check for unidimensionality), and (2) the item scalability coefficients H_i and the scalability coefficient H for the questionnaire as a whole are positive (a check for monotonicity). Hemker et al. (1995) used the following guidelines for interpreting H:

$H < 0.3$:	itemset is unscalable;
$0.3 \leq H < 0.4$:	itemset is weakly scalable;
$0.4 \leq H < 0.5$:	itemset has moderate scalability;
$0.5 \leq H \leq 1.0$ (maximum):	itemset is strongly scalable.

Further support for the monotonicity investigation comes from inspection of the nonparametric (and also nonlinear) regressions of each of the item scores on the total score of the other items of the questionnaire, called the item-restscore regressions. Nondecreasingness of these regressions supports the monotonicity assumption.

The fit investigation of the monotone homogeneity model can be quite troublesome if the data are multidimensional, which is probably true for the ACI data. Thus, in a first item analysis step, we try to identify clusters of items that satisfy the monotone homogeneity model to at least some degree, and in a second step we investigate within each cluster in more detail whether the monotone homogeneity model fits the data. In the first step, we use an automated item selection procedure that starts with a pair of items and then extends the scale by adding items one-by-one. The automated item selection procedure is part of the software package MSP (Mokken Scale analysis for Polytomous items; Molenaar and Sÿtsma, 2000), and selects items into clusters under the following definition of a scale:

(i) Covariance between all item pairs (i,j) is positive;
(ii) $H_i \geq c > 0$; where c is a constant, for all items i; and $H \geq c$.

Condition (i) follows from the unidimensionality and monotonicity assumptions, and condition (ii) selects items to form a scale on which respondents can be ordered with an accuracy determined by the choice of c. Usually, c is 0.3, which guarantees at least weak scalability, but the researcher may wish to select other values. The evaluation of item-restscore regressions is not part of the automated item selection procedure, and has to be done in the second item analysis step after item clusters satisfying conditions (i) and (ii) have been selected.

Item selection into the first cluster stops if all items have been selected, or if there are no more items left that satisfy both conditions (i) and (ii). In the latter case, the automated item selection procedure tries to select a second cluster from the items left; a third cluster; and so on. After an item has been selected into a scale, it is no longer a candidate for later scales. The MSP software has facilities for letting the second scale be the start set for a secondary item selection round in which all other items are candidates for selection; and so on. Moreover, the researcher may also wish to try different values of the lower bound c of H, thereby manipulating the threshold for admission of items into a cluster. Finally, the researcher may define on a priori grounds (for example, based on item contents or knowledge from previous research) a set of items to be a scale and test this hypothesis without any item selection taking place.

COMPARING MOKKEN SCALE ANALYSIS AND FACTOR ANALYSIS

Because we had no clear conception of the composition of the ACI in terms of underlying dimensions, an exploratory factor analysis seemed to be most appropriate here. Thus, we analyzed the 54 items of the ACI using principal components analysis (PCA), followed by both orthogonal (varimax) and oblique rotations of the loadings matrix (e.g., Nunnally, 1978). Because MSA and PCA pursue the same purposes to a high degree, in this section we will briefly compare these techniques. After that, we discuss the results of analyzing the ACI data using both MSA and PCA.

(1) PCA first constructs a weighted sum (the first principal component) of all 54 item score variables ($X_1, \ldots, X_{54}$) that has the highest variance given all possible choices of item weights. This maximum variance equals the first eigenvalue of the correlation matrix. Next, PCA constructs a second weighted sum (the second principal component) of residual variables ($X'_1, \ldots, X'_{54}$), which are uncorrelated to the first principal component, that has the highest possible variance given all possible choices of items weights; this is the second

eigenvalue of the correlation matrix; and so on for the other principal components. In constructing each of the principal components, PCA considers all items *simultaneously*, and provides loadings for each item on each principal component. Automated item selection using MSA is *sequential*, meaning that once an item has been selected into a certain cluster, the item is no longer a candidate for selection into the next clusters. This restriction is surpassed by letting the second scale be the start set for item selection in a next selection round, in which all other items including those from the first scale are candidates for selection into the second scale. Similarly, the items from the third scale selected in the first selection round can be start set, and so on.

(2) The automated item selection procedure in MSA uses a particular measurement model, which is the monotone homogeneity model. This combination of item clustering and simultaneous measurement modeling is unique to MSA. PCA clusters items based on relatively high mutual correlations, but not so as to satisfy a particular measurement model. After the researcher has decided, based on factor loadings, which items "belong together", usually an additional reliability analysis is carried out to further study the measurement properties of the selected item set. The combination of PCA or other methods of factor extraction and classical reliability analysis often is encountered in practice.

Results

RESULTS OF FACTOR ANALYSIS AND RELIABILITY ANALYSIS

The data set consisted of responses of 3906 subjects from 30 countries. There were 1630 males (41.7%) and 2276 females. Because of the large sample size and because there were relatively few missing values, we decided to delete subjects with missing values listwise. This resulted in 3547 complete data records to be used for data analysis.

A PCA was carried out to investigate the dimensionality structure of the ACI. The first principal component explained 31.9% of the variance, the second 6% and the third 3.4%. All other principal components with eigenvalues higher than 1 together explained 14% of the total variance. Based on these results and on the degree of interpretability of different factor solutions that resulted from orthogonal (varimax) or oblique (oblimin) rotation, it was considered best to maintain both the three- and two-factor solutions.

The factor pattern matrix obtained after orthogonal rotation of three principal components is shown in Table 1. This procedure led to a solution in which items with loadings of at least 0.3 on the first factor and relatively low loadings on the other two factors could be interpreted as measuring a person's integrity being afflicted. For example, people tend to cry when they cannot reach their goal, loose grip, experience guilt or feel threatened. There is a clear aspect of frustation and agitation in several items. The items with the highest loadings were: "I cry when things don't go as I want them to go," "I cry when having been humiliated or insulted" and "I cry when I experience opposition from someone else." We labeled this factor "*Distress*." The 22 items constituting this factor are (ordered by decreasing loadings): 35, 24, 28, 32, 23, 9, 36, 12, 29, 41, 40, 30, 34, 39, 19, 5, 10, 44, 49, 38, 6, 43.

Based on the interpretation of loadings patterns, we labeled the second factor "*Sadness*." As opposed to the first factor, several items are characterized by a passive rather than by an active attitude. The items with the highest loadings were: "I cry at funerals," "I cry while I watch sad movies or television programs", and "I cry when a tragic event happens to me." The 14 items that constitute this factor are: 26, 45, 31, 50, 8, 1, 2, 48, 16, 53, 21, 17, 37, 25.

The third factor was labeled "*Joy*". Items loading at least 0.3 on this factor refer to the experience of positive emotions. The items with the highest loadings were: "I cry when watching or hearing an admired person," "I cry when I hear a happy song," and "I cry when watching the awards ceremony at sporting events such as the Olympics." The 18 items comprising this factor are: 54, 14, 33, 18, 55, 27, 47, 15, 3, 11, 22, 52, 20, 46, 13, 51, 7, 4.

Next, we considered the three groups of items as constituting scales and calculated Cronbach's alpha for each total score. This resulted in 0.94 for "Distress," 0.90 for "Sadness," and 0.87 for "Joy."

Oblique rotation of three principal components led to a first factor which consisted of *negative (unpleasant) emotions* (marker items are: 35, 24, 12 and 32), and a second factor which consisted mainly of *positive (pleasant) emotions* (marker items are: 33, 54, 47, 27). The third factor consisted of only three items (items 26, 45 and 2) that conceptually belong to either of the former two categories and that can hardly be interpreted as a separate factor. Based on these results and on the low proportion of variance explained by the third principal component, it might be hypothesized that a two-factor structure underlies the data[1]. The loadings structures resulting from orthogonal and oblique rotations

---

1 See also Chapter 8 in which Becht et al. identify two ACI factors, which they label *Distress* and *Eustress*.

Table 1. Factor pattern matrix of the ACI after orthogonal rotation of three principal components

Item no.	Distress	Sadness	Joy
35	**.72**	.29	.12
24	**.65**	.36	.13
28	**.64**	.14	.22
32	**.63**	.30	.13
23	**.62**	.30	.14
9	**.62**	.32	.13
36	**.62**	.25	.20
12	**.62**	.36	.11
29	**.61**	.32	.18
41	**.59**	.40	.09
40	**.57**	.47	.10
30	**.56**	.31	.13
34	**.56**	.34	.09
39	**.55**	.53	.04
19	**.54**	.37	.11
5	**.53**	.07	.23
10	**.51**	–.05	.27
44	**.49**	.19	.25
49	**.47**	.23	.31
38	**.47**	.40	.12
6	**.40**	–.04	.21
43	**.37**	.02	.32
26	.12	**.65**	.15
45	.11	**.64**	.32
31	.39	**.63**	.02
50	.33	**.56**	.31
8	.25	**.55**	.02
1	.39	**.54**	.12
2	.19	**.53**	.23
48	.41	**.52**	.20
16	.39	**.50**	.23
53	.29	**.49**	.33
21	.23	**.48**	.41
17	.29	**.46**	.34
37	.33	**.45**	.33
25	.23	**.44**	.37
54	.20	.10	**.61**
14	.13	.00	**.58**
33	–.03	.13	**.58**
18	.24	.18	**.56**

Table 1. (*Continued*)

Item no.	Distress	Sadness	Joy
55	.28	.07	.55
27	.07	.24	.55
47	.01	.06	.55
15	.20	.35	.54
3	.06	.16	.52
11	.14	.27	.50
22	.19	.18	.50
52	.26	.24	.50
20	.12	.28	.50
46	.24	–.05	.45
13	.06	.40	.44
51	.13	.41	.44
7	.29	.15	.37
4	.30	.01	.35

of the first two principal components were very similar, because for both solutions negative emotions represented the first factor and positive emotions represented the second factor. The correlation between the oblique factors was 0.53.

To summarize, the data can be represented reasonably well by both a three-factor and a two-factor solution. The three-factor solution is characterized by factors labeled "Distress," "Sadness" and "Joy," and the two-factor solution by factors labeled "Negative emotions" and "Positive emotions." Our three-factor solution corresponds nicely to the conclusions arrived at by Borgquist (1906) and Bindra (1972) who, although employing somewhat different labels, distinguished the same three types of tear eliciting situations (see also Williams & Morris, 1996).

RESULTS OF MOKKEN SCALE ANALYSIS

Automated item selection. Following Hemker et al. (1995), we ran the automated item selection procedure on all 54 items using consecutive lower bounds of c = 0.0, 0.2, 0.3, 0.4, and 0.5, respectively. Table 2 shows that with the borderline c = 0.0 all 54 items were selected into the same weak scale (H = 0.33), and with c = 0.2, 53 items were selected into the same weak scale (H = 0.34; item 47 was excluded due to too low a H_i value).

Table 2. ACI-item clustering for several values of lowerbound H = c. For each factor, item numbers reflect rank number as items appear in ACI. Lower case letters in columns reflect clustering of items for specific values of c.

	Lowerbound c				
Item no.	0.0	0.2	0.3	0.4	0.5
	Factor: Distress				
5	a	a	a		
6	a	a			
9	a	a	a	a	a
10	a	a			
12	a	a	a	a	a
19	a	a	a	a	g
23	a	a	a	a	a
24	a	a	a	a	a
28	a	a	a	a	g
29	a	a	a	a	a
30	a	a	a	a	g
32	a	a	a	a	a
34	a	a	a	a	
35	a	a	a	a	a
36	a	a	a	a	
38	a	a	a	a	d
39	a	a	a	a	a
40	a	a	a	a	a
41	a	a	a	a	a
43	a	a			
44	a	a	a		d
49	a	a	a		
	Factor: Sadness				
1	a	a	a	a	
2	a	a	a	g	
8	a	a	a		
16	a	a	a	a	b
17	a	a	a	a	c
21	a	a	a	b	h
25	a	a	a	b	
26	a	a	a	c	f
31	a	a	a	a	b
37	a	a	a	a	c
45	a	a	a	b	e
48	a	a	a	a	b

Table 2. (*continued*)

	Factor: Sadness				
50	a	a	a	a	c
53	a	a	a	a	
	Factor: Joy				
3	a	a	b	d	
4	a	a			
7	a	a			
11	a	a	b	f	
13	a	a	b	b	e
14	a	a	b		
15	a	a	a	f	
18	a	a	a	f	
20	a	a	a	c	
22	a	a	b	d	h
27	a	a	b	d	
33	a	a	c	e	i
46	a	a			
47	a	(a)*	c	e	i
51	a	a	a	c	f
52	a	a	a	g	
54	a	a	a		
55	a	a	b	f	

* This item was excluded due to rounding error in H_{47}.

For c = 0.3, 39 items, marked in Table 2 by a lower case "a", were selected into a weak scale. Furthermore, a small weak scale and a small strong scale were formed. The large scale included 19 of the 22 items that had their highest loadings on the "Distress" factor; all 14 items that had their highest loadings on the "Sadness" factor; and 6 of the 18 items that had their highest loadings on the "Joy" factor. Thus, the first scale included most of the items that measured crying proneness due to unpleasant causes. All 7 items (Table 2, lower case "b") that were selected into the second scale had their highest loadings on the "Joy" factor. The third two-item scale consisted of items (Table 2, lower case "c") which also had their highest loadings on the "Joy" factor.

For c = 0.4, the automated item selection procedure selected 6 item clusters of moderate scalability and one item cluster of strong scalability. The first scale was a subset (24 items; Table 2, lower case "a") of the first scale from the analysis for c = 0.3. Now, 16 "Distress" items, 8 "Sadness", and no "Joy" items were selected. The other six scales had

2, 3, or 4 items (Table 2, lower cases "b", "c", "d", "e", "f", "g"), and were mostly composed of "Joy" items, and sometimes of "Joy" and "Sadness" items, but never of only "Sadness" items. None of these small scales contained "Distress" items.

Lower bound c = 0.5 proved to be too high for 25 items; thus, only 29 items could be selected, which then were subdivided across 9 high-scalability scales. The first scale contained 10 "Distress" items (Table 2, lower case "a"). The other 8 scales each contained only 2 or 3 items (Table 2, lower cases "b", "c", "d", "e", "f", "g", "h", "i").

On the basis of the results for several lower bound values c, it can be concluded that, on the lowest c level (0) and also on a somewhat higher c level (0.2), all items were selected into the same scale. This suggests that they together measure a general crying proneness. Raising c to 0.3 led to one scale for crying proneness due to unpleasant causes, which might be associated with either distress or sadness, and one shorter scale for crying proneness in pleasant situations. Further raising c to 0.4 created a shorter "Unpleasantness" scale, and several very short scales; and for c = 0.5 the item set further crumbled into many short scales, whereas almost one half of the items were unscalable.

Hemker et al. (1995) found this pattern of results across increasingly higher c values to be typical of unidimensionality. Thus, according to these authors the ACI would measure one trait: general crying proneness. The results for c = 0.3, in particular, suggest that it also may be useful to distinguish crying proneness in unpleasant situations from crying proneness in pleasant situations.

"Sadness" and "Joy" as start set. Using the 14 items comprising the second factor and interpreted as measures of "Sadness" as the start set for the automated item selection procedure, for lower bound c = 0.3 we found that 25 additional items were selected. Nineteen of these 25 items were "Distress" items, and the last six items selected were "Joy" items. This result corroborates the interpretation of "Distress" and "Sadness" as common measures of "Unpleasantness." Taking the 18 "Joy" items (third factor) as the start set (c = 0.3), led to the selection of another 32 items; thus, only four items were not selected. This result demonstrates that "Joy" is not a strong factor.

Testing a priori scales. We considered each of the three factors found using PCA as an a priori defined scale, and tested this assumption separately for "Distress", "Sadness", and "Joy". Table 3 shows that "Distress" and "Sadness" were moderate scales. "Joy" had H = 0.30 and could thus be characterized as a borderline weak scale. This result corroborated the item selection results where no cluster could be selected

Table 3. Item scalability coefficients (H_i) and number of violations (#vi) of item-restscore regressions for each of the three ACI factors

Factor								
Distress (H = .43)			Sadness (H = .41)			Joy (H = .30)		
No.	H_i	#vi	No.	H_i	#vi	No.	H_i	#vi
5	.35	0	1	.42	0	3	.28	1
6	.26	2	2	.37	0	4	.22	2
9	.45	0	8	.34	1	7	.26	1
10	.31	4	16	.44	0	11	.34	0
12	.47	0	17	.43	0	13	.30	0
19	.43	0	21	.41	0	14	.30	1
23	.46	0	25	.37	0	15	.37	0
24	.49	0	26	.39	0	18	.34	0
28	.45	0	31	.44	0	20	.31	0
29	.46	0	37	.42	0	22	.29	1
30	.42	0	45	.42	0	27	.31	1
32	.46	0	48	.45	0	33	.29	0
34	.42	0	50	.47	0	46	.22	0
35	.50	0	53	.42	0	47	.26	0
36	.44	0				51	.29	0
38	.41	1				52	.32	0
39	.49	0				54	.34	0
40	.49	0				55	.32	0
41	.47	0						
43	.30	0						
44	.37	0						
49	.37	0						

that contained at least half of the "Joy" items. For all three scales, the number of violations from the expected orderings of each of the item-restscore regressions was small. This result supports the monotonicity assumption which in turn, together with the weak to moderate scale results, supports the use of the total score on each of the three scales for ordering respondents.

Item selection from a priori scales. Selecting items from the three a priori scales using c = 0.3 led to the following results. Except for item 6 (too low H_i value), all other "Distress" items were selected into one moderate scale (H = 0.44). Without exception, all "Sadness" items were selected into the same moderate scale (H = 0.41; see Table 3). Twelve "Joy" items were selected into one weak scale (H = 0.34), 3 items were selected into another moderate scale (H = 0.40), and 3 items were excluded from these scales because they had too low H_i values with respect to the items in each scale. This result again corroborates the somewhat problematic nature of the "Joy" items taken as measuring one unitary trait.

Conclusions

This chapter started with a review of research methods which are potentially useful for studying antecedents and consequences of crying. The usefulness of several of these methods was acknowledged while other methods were discussed more critically. We concluded that data collection, often concerning trait measurement, by means of questionnaires applied in large sample surveys is a predominant method. This led us to further consider the questionnaire, and discuss two methods for analyzing scores obtained from ordered rating scales. The methods were principal components analysis and Mokken scale analysis. As an illustration, both methods were used for analyzing rating scale scores from the ACI (Vingerhoets & Becht, 1996), which is used for measuring crying proneness. We used PCA and MSA for investigating whether its items measured crying proneness irrespective of situation or mood state, or whether subsets of items could be distinguished each measuring crying proneness in response to a particular kind of situation or mood state.

The PCA led to one solution with three factors representing crying from Distress, Sadness or Joy, but also to another solution with two factors representing crying from either Unpleasant or Pleasant Emotions. The MSA led to a unidimensional solution suggesting that the ACI measured General Crying Proneness, but indications were also found that Unpleasant and Pleasant Emotions may be distinguished. The three-factor solution was corroborated by the literature, but not by the MSA. Thus, it appears that our methods have not completely resolved the problem of determining the underlying dimensionality of the ACI.

The results from an MSA and a PCA are determined in the first place by the correlation structure among the items. Moreover, the results from MSA are also influenced by the choice of a lower bound c for the selection of items into a scale and by the order in which scales are constructed, and the results from PCA are influenced by the choice of the number of principal components to retain for rotation and, next, the choice of a lower bound for the loadings on the same factor. A higher threshold for including items into a cluster, be it a higher lower bound c or a higher minimum factor loading, leads to the selection of smaller item clusters. These item clusters usually are more homogeneous with respect to their contents and, thus tend to have a more restricted and often clearer meaning.

The 54 items comprising the ACI were collected so as to cover a wide range of situations and mood states each of which was capable of eliciting crying, but without a clear taxonomy guiding item selection. Thus it may come as no surprise that the responses to the items of the ACI seem to be guided by several underlying traits but, as is typical of exploratory research, that the trait structure is not entirely clear from the correlations among the items. This outcome calls for more research. For example, a revision of the ACI may be aimed more systematically at measuring particular aspects of crying. Also, the complex composition of the present data with respect to covariates such as country, culture, and gender may partly explain the indeterminacy of the results, and future research should address this covariate structure.

What can be concluded about the practical usefulness of the ACI? This depends on the application envisaged. For simple classification of respondents, one could imagine that a measurement practitioner prefers one summary score for the complete questionnaire which reflects general crying proneness. A researcher more interested in the structure of crying proneness or in situation-specific crying, however, may want to use sum scores on either two or three subscales as were identified here. The alpha coefficients found for the subscales Distress, Sadness, and Joy were each above .8s, which seems to be sufficiently high to justify such applications.

References

Bindra, D. (1972). Weeping, a problem of many facets. *Bulletin of the British Psychological Society*, *25*, 281–284.

Bradburn, N.M., Rips, L.J., & Shevell, S.K. (1987). Answering autobiographical questions: The impact of memory and inference on surveys. *Science*, *236*, 157–161.

Borgquist, A. (1906). Crying. *American Journal of Psychology*, *17*, 149–205.

Cornelius, R.R. (1997). Toward a new understanding of weeping and catharsis? In: A.J.J.M. Vingerhoets, F.J. Van Bussel, & A.J.W. Boelhouwer (Eds.), *The (non)expression of emotions in health and disease* (pp. 303–321). Tilburg, The Netherlands: Tilburg University Press.

Delp, M.J., & Sackeim, H.A. (1987). Effects of mood on lacrimal flow: Sex differences and asymmetry. *Psychophysiology, 24,* 550–556.

Eibl-Eibesfeldt, I. (1984). *Die Biologie des menschlichen Verhaltens: Grundriss der Humanethologie.* München: Piper.

Gross, J.J., Fredrikson, B.L., & Levinson, R.W. (1994). The psychophysiology of crying. *Psychophysiology, 2,* 460–468.

Hemker, B.T., Sijtsma, K., & Molenaar, I.W. (1995). Selection of unidimensional scales from a multidimensional itembank in the polytomous Mokken IRT model. *Applied Psychological Measurement, 19,* 337–352.

Kraemer, D.L., & Hastrup, J.L. (1986). Crying in natural settings: Global estimates, self-monitored frequencies, depression and sex differences in an undergraduate population. *Behaviour Research and Therapy, 24,* 371–373.

Kraemer, D.L., & Hastrup, J.L. (1988). Crying in adults: Self-control and autonomic correlates. *Journal of Consulting and Clinical Psychology, 60,* 53–68.

Lester, B.M. (1985). Introduction: there's more to crying than meets the ear. In: B.M. Lester & C.F.Z. Boukydis (Eds.), *Infant crying: theoretical and research perspectives* (pp. 1–27). New York: Plenum Press.

Löfgren, L.B. (1966). On weeping. *International Journal of Psychoanalysis, 47,* 375–381.

Lyberg, L., & Kasprzyk, D. (1991). Data collection methods and measurement error: an overview. In: P.P. Biemer, R.M. Groves, L.E. Lyberg, N.A. Mathiowitz, & S. Sudman (Eds.), *Measurement errors in surveys* (pp. 237–257) . New York: John Wiley & Sons.

Marston, A., Hart, J., Hileman, C., & Faunce, W. (1984). Toward the laboratory study of sadness and crying. *American Journal of Psychology, 97,* 127–131.

Mokken, R.J. (1997). Nonparametric models for dichotomous responses. In: W.J. van der Linden & R. K. Hambleton (Eds.), *Handbook of modern item response theory* (pp. 351–367). New York: Springer.

Mokken, R.J., & Lewis, C. (1982). A nonparametric approach to the analysis of dichotomous item responses. *Applied Psychological Measurement, 6,* 417–430.

Molenaar, I.W. (1997). Nonparametric models for polytomous responses. In: W.J. van der Linden & R.K. Hambleton (Eds.), *Handbook of modern item response theory* (pp. 369–380). New York: Springer.

Molenaar, I.W., & Sijtsma, K. (2000). *User's Manual MSP-5 for Windows.* Groningen, The Netherlands: iecProGAMMA.

Nunnally, J.C. (1978). *Psychometric theory.* New York: McGraw-Hill.

Schwarz, N. (1990). Assessing frequency reports of mundane behaviors: Contributions of cognitive psychology to questionnaire construction. In: C. Hendrick & M.S. Clark (Eds.), *Review of personality and social psychology: Vol. 11. Research methods in personality and social psychology* (pp. 98–119). Newbury Park: Sage

Sijtsma, K. (1998). Methodology review: Nonparametric IRT approaches to the analysis of dichotomous item scores. *Applied Psychological Measurement, 22,* 3–31.

Stone, A.A. (1995). Measurement of affective response. In: S. Cohen, R.C. Kessler, & L. Underwood Gordon (Eds.), *Measuring stress. A guide for health and social scientists* (pp. 148–171). New York: Oxford University Press.

Thomas, D.L., & Diener, E. (1990). Memory accuracy in the recall of emotions. *Journal of Personality and Social Psychology*, *59*, 291–297.

Vingerhoets, A.J.J.M., & Becht, M.C. (1996, August). *The ISAC study. Some preliminary results.* Paper presented at the International Study on Adult Crying (ISAC) symposium. Tilburg, The Netherlands, The Netherlands.

Vingerhoets, A.J.J.M., & Scheirs, J. (2000). Gender differences in crying: Empirical findings and possible explanations. In: A. Fischer (Ed.), *Gender and emotion. Social psychological perspectives* (pp. 143–165). Cambridge: Cambridge University Press.

Vingerhoets, A.J.J.M., Cornelius, R.R., Van Heck, G.L., & Becht, M.C. (2000). Adult crying: A model and review of the literature. *Review of General Psychology, 4,* 354–377.

Wallbott, H.G., & Scherer, K.R. (1989). Assessing emotion by questionnaire. In: R. Plutchik & H. Kellerman (Eds.), *Emotion. Theory, research, and experience: Vol. 4. The measurement of emotions* (pp. 55–82). San Diego: Academic Press.

Williams, D.G., & Morris, G.H. (1996). Crying, weeping or tearfulness in British and Israeli adults. *British Journal of Psychology*, *87*, 479–505.

Znoj, H.J. (1997). When remembering the lost spouse hurts too much: First results with a newly developed observer measure for tears and crying related coping behavior. In: A.J.J.M. Vingerhoets, F.J. Van Bussel, & A.J.W. Boelhouwer (Eds.), *The (non)expression of emotions in health and disease* (pp. 337–352). Tilburg, The Netherlands: Tilburg University Press.

Epilogue

The preceding chapters have presented data on a wide variety of phenomena associated with crying and many have contained speculation about the mechanisms underlying various aspects of crying. In this chapter we offer a preliminary model of adult crying that we hope may provide a framework for synthesizing what we know and what we think we know about crying and that may serve to stimulate and guide future research on crying.

Let us begin with a summary of what we can conclude about crying given the present state of our knowledge.

(1) Adult crying is a response to either an emotional event or the activation of memories about emotional experiences.
(2) Such situations and memories most often represent social situations (e.g., loss, reunions, conflicts) or evaluations of one's personal functioning (e.g., perceived inadequacy). The emotions felt are generally negative, and are often accompanied by feelings of powerlessness or at least the conviction that one has no proper behavioral response available in the situation.
(3) There are large individual differences in crying proneness. These differences manifest themselves in both strong emotional situations (some people are better at inhibiting their tears than others) and weak emotional situations (some people are more prone to cry than others). These differences are associated with culture, gender, age, socialization, and personality. In addition, within an individual, the threshold of crying may vary as a function of physical (e.g., fatigue, pregnancy, menstrual cycle) or mental state (e.g., level of depression, frustration).
(4) Actual crying is not only a function of crying proneness but also of the capacity and willingness to avoid or seek out emotional situations, both in one's professional and private life. To put it differently, active mood management by antecedent-focused emotion regulation appears to be an important aspect of crying.
(5) Situational context plays an important role in moderating crying. In particular, the presence of others may either inhibit (because one feels ashamed or because one does not want to burden others with one's own sadness or distress) or stimulate (through emotional contagion or because one want to manipulate others). In addition, cultural norms and display rules govern patterns of crying behavior.

(6) Crying has a strong effect on the social environment. It may have dramatic effects on others and may influence their behavioral tendencies toward and evaluations of the crying individual. Therefore, crying is subject to operant conditioning. It can be reinforced or punished, with well-known effects on the frequency of this behavior.
(7) The act of crying might also bring emotional relief or reduce tension by psychobiological and/or social mechanisms.

As one way of encompassing these diverse findings into a single framework that may serve as a guide for future research, we offer the model of adult crying presented in Figure 1. The model was derived initially from a comprehensive review of the literature on crying and a consideration of the role that crying plays in both emotion regulation and social interaction (see Vingerhoets, Cornelius, Van Heck, & Becht, 2000).

Within the model we propose, crying is considered a physiological/expressive response that may be elicited by a number of different appraisals (e.g., that one is powerless to do anything, that a situation is hopeless) and

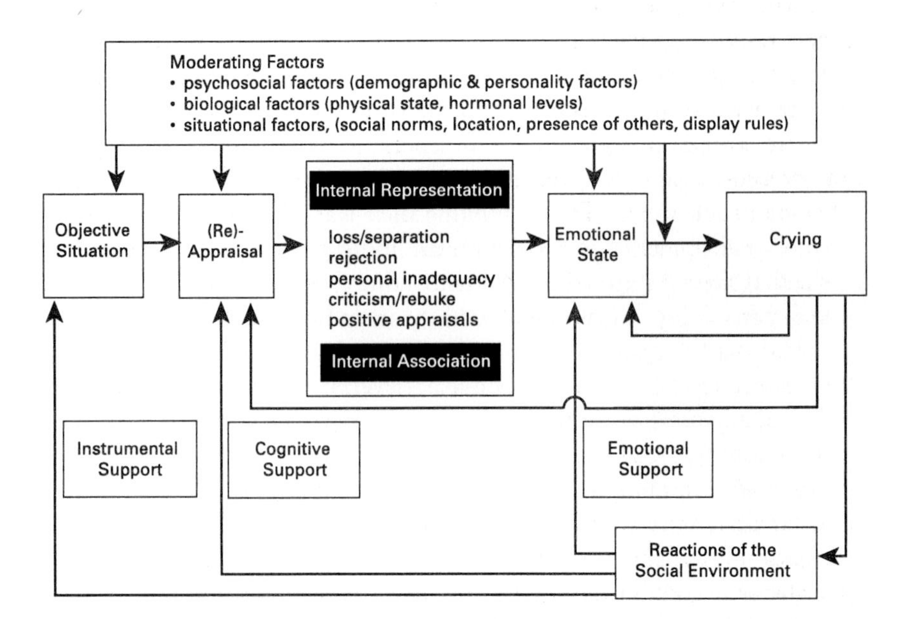

Figure 1. Model of adult crying demonstrating the role of antecedents and moderating factors as well as the influence of crying on the crying person's emotional state and the social environment.
From Vingerhoets et al. (2000), Figure 1. Copyright American Psychological Association. Used with permission

that is associated with a number of different situations (e.g., loss/separation, rejection, reunion) and emotional states (sadness, anger, happiness). Physiologically, the shedding of emotional tears is under parasympathetic nervous system control, although crying as a behavioral event appears to be associated with a combination of sympathetic and parasympathetic arousal. In addition to several different but related emotional states, crying is associated with physical injury, pain, and fatigue. Whether or not a person will cry in a particular situation is moderated by a number of personality (e.g., alexithymia, self-monitoring status) and sociodemographic factors (e.g., gender, age) as well as a variety of situational factors (e.g., the presence of others who are crying, salience of social norms concerning crying). Crying, as both an expressive reaction and instrumental response, may elicit both positive and negative reactions from the social environment, which may have powerful effects on the experience of crying. Emotional support or disapproval from others may feedback directly onto one's emotional state and may also lead one to reappraise the situation that led one to cry in the first place. Crying, by eliciting particular reactions from others, may also more or less directly alter the objective characteristics of the situation (e.g., by eliciting sympathy from an antagonist during an argument). Crying may also have direct effects on both primary and secondary appraisal. Although there is yet little evidence for it, crying may have direct physiological effects on the level of one's arousal.

We are aware that some basic questions remain unanswered by the data presented in the preceding chapters and are not adequately addressed in our model. First, there is the important question of whether there are several qualitatively different kinds of crying. Frijda (1986) has argued that we should make a distinction between silent crying as often occurs when one feels emotionally touched and the kind of overt crying associated with grief or other strong emotions. Similarly, Cornelius (1988) asked whether solitary crying, getting watery eyes while watching a sad film, and sobbing in the arms of another person represent the same phenomenon. Williams and Morris (1996), after factor analyzing the responses of a sample of English and Israeli adults to a detailed crying questionnaire, suggested that at least two types of crying may be distinguished. The first is intense and long-lasting, is difficult to stop once it has started, has a strong impact on the person, and interferes with his or her ongoing activities. This kind of crying appears to be elicited by strongly negative situations such as losing a loved one. The second kind of crying is less intense, more controllable, and is more diverse in terms of its form and eliciting conditions. This is the kind of crying associated with watching a sad film or being touched by the plight of others.

Although each of these distinctions has some *prima facie* validity, none has yet been tested empirically. It may be the case that the model we propose—or at least particular aspects of it—is more appropriate for some forms of crying than others. Our suspicion is that the model will turn out to best fit more intense forms of crying associated with strong negative emotions.

A second unanswered question is in some ways the most basic and important of them all and concerns why crying accompanies some episodes of a particular emotion but not others. What determines whether one is very sad and sheds tears or simply very sad? Is this related to the cognitive evaluation of the situation, the context, the function of crying, or are there other relevant factors yet to be identified? One way to begin thinking about this question is in terms of the signal value of tears. Data suggest that without tears a sad face is far less easy to recognize as a sad face and that the appearance of tears unequivocally communicates the presence of strong emotion (Cornelius, 1999). If this is indeed the case, crying may serve the same function as laughing, which communicates joy, and swearing, which emphasizes one condition of being angry. Further analysis of the communicative aspects of tears is clearly called for.

We are fully aware that this book leaves many questions unanswered. However, we feel that an adequate inventory of the gaps in our knowledge of crying and of the basic questions that remain to be answered is a very important first step towards a better understanding of this most intriguing human emotional responses.

References

Cornelius, R.R. (1988, April). *Toward an ecological theory of crying and weeping*. Ninth Annual Meeting of the Eastern Psychological Association, Buffalo, NY.

Cornelius, R.R. (1999, June). *The evolution and behavioral ecology of crying: Implications for understanding the health benefits of crying*. Second International Conference on the (Non)Expression of Emotions in Health and Disease, Tilburg, The Netherlands.

Frijda, N.H. (1986). *The emotions*. Cambridge: Cambridge University Press.

Vingerhoets, A.J.J.M., Cornelius, R.R., Van Heck, G.L., & Becht M.C. (2000). Adult Crying: A model and review of the literature. *Review of General Psychology, 4*, 354–377.

Williams, D.G., & Morris, G.H. (1996). Crying, weeping or tearfulness in British and Israeli adults. *British Journal of Psychology, 87*, 479–505.

Appendix

ADULT CRYING INVENTORY (ACI)[1]

This questionnaire focuses on feelings and emotional experiences. When the word "crying" is used, it refers to **tears in one's eyes** due to emotional reasons (sobbing and sniffling is not a necessary condition to meet our definition of crying), not because of irritation to the eye.

Please, read every item carefully; answer it honestly, but do not spend too much time on any one item. The first thought that comes to mind is most often the best answer. There are no right or wrong answers; the only thing that is important is that you are honest.

Questionnaires with missing data are unsuited for analysis, so please do not skip any item, even if you feel it is difficult to answer. Unless explicitly stated otherwise, **always check only that response alternative that applies the most.**

The questionnaire consists of five parts. Part A concentrates on the situations and mood states, that make you cry, while Part B focuses on the functions of crying, i.e., the purposes of crying. In Part C, we ask you to indicate which factors are important as determinants of your shedding tears. Part D focuses on your last crying episode, in particular, its context. The last part, Part E, contains questions that only pertain to females. These questions concern the relationship between crying and phase of the menstrual cycle and pregnancy.

---

1 This questionnaire has been developed by Ad Vingerhoets, Department of Psychology, Tilburg University, P. O. Box 90.153 5000 LE Tilburg, the Netherlands. the questionnaire is available in the following languages: Arabic, Bulgarian, Chinese, Dutch, English, Finnish, French, German, Greek, Hungarian, Hindi, Icelandic, Indonesian, Italian, Lithuanian, Malay, Nepali, Polish, Portuguese, romanian, Spanish, Swedish, and Turkish. A revised version of the questionnaire (ACI-R) is available from the author. For some languages different versions are available created in different parts of the world. Copies of the questionnaire are free available on the condition that the author also receives a copy of the data set: Ad Vingerhoets, Department of Psychology, Tilburg University, P. O. Box 90.153, 5000 Le Tilburg, The Netherlands, Tel: +31-13-466.2087; Fax +31-13-466.2370; e-mail: Vingerhoets@kub.nl

Please, do not write your name or put any other identifying remarks on the questionnaire. We would like your responses to be completely anonymous.

Sex: ○ Male ○ Female (Check alternative that applies)

Age: ________________

Education: _______________________

If you have never cried since you were 17,
put a check in this box ....☐

In that case you are not supposed to fill in the rest of the questionnaire.

We nevertheless expect that you return the questionnaire.

!! PLEASE DO NOT OMIT ANY QUESTIONS !!

Part A

EACH OF THE FOLLOWING ITEMS DESCRIBE A SITUATION IN WHICH ONE MIGHT CRY, OR AN EMOTION THAT MAKES ONE CRY. PLEASE, INDICATE AFTER EACH ITEM HOW FREQUENTLY YOU CRY IN SUCH CONDITIONS.

	never=	1	2	3	4	5	6	7	=always
1. I cry when I feel sad		1	2	3	4	5	6	7	
2. I cry when I have to say goodbye to beloved persons		1	2	3	4	5	6	7	
3. I can be moved to tears by the beauty of natural scenes		1	2	3	4	5	6	7	
4. I cry when making love		1	2	3	4	5	6	7	
5. I cry when I feel ashamed		1	2	3	4	5	6	7	
6. I deliberately cry in order to make someone else feel sorry for me		1	2	3	4	5	6	7	
7. I cry when I feel relief		1	2	3	4	5	6	7	
8. I cry over the loss of a love relationship		1	2	3	4	5	6	7	
9. I cry when I do not succeed in getting things together		1	2	3	4	5	6	7	
10. I cry when I experience disgust or contempt for something or somebody		1	2	3	4	5	6	7	
11. I cry when I feel very happy		1	2	3	4	5	6	7	
12. I cry when things do not go well at work or at school		1	2	3	4	5	6	7	
13. I cry when a movie or television program has a happy ending		1	2	3	4	5	6	7	
14. I cry when I hear a happy song		1	2	3	4	5	6	7	
15. I cry when someone does something very special for me or someone else		1	2	3	4	5	6	7	
16. I cry if I remember sad things that have happened to me		1	2	3	4	5	6	7	
17. I cry because of the problems of someone else		1	2	3	4	5	6	7	
18. I cry at happy memories of the past		1	2	3	4	5	6	7	

	never=	1	2	3	4	5	6	7	=always
19. I cry when involved in quarrels and conflicts		1	2	3	4	5	6	7	
20. I cry at weddings		1	2	3	4	5	6	7	
21. I cry when I hear a sad song		1	2	3	4	5	6	7	
22. I cry while reading poetry		1	2	3	4	5	6	7	
23. I cry when I feel powerless		1	2	3	4	5	6	7	
24. I cry when having been humiliated or insulted		1	2	3	4	5	6	7	
25. I cry when reading certain books		1	2	3	4	5	6	7	
26. I cry at funerals		1	2	3	4	5	6	7	
27. I cry in response to the beauty of arts (music, literature, visual arts)		1	2	3	4	5	6	7	
28. I cry when I experience opposition from someone else		1	2	3	4	5	6	7	
29. I cry when I feel frightened		1	2	3	4	5	6	7	
30. I cry when I feel angry		1	2	3	4	5	6	7	
31. I cry when a tragic event happens to me		1	2	3	4	5	6	7	
32. I cry when someone criticizes or lectures me		1	2	3	4	5	6	7	
33. I cry when watching the awards ceremony at sporting events such as the Olympics		1	2	3	4	5	6	7	
34. I cry when feeling self-pity		1	2	3	4	5	6	7	
35. I cry when things don't go as I want them to go		1	2	3	4	5	6	7	
36. I cry when I feel guilty		1	2	3	4	5	6	7	
37. I cry out of pity for others		1	2	3	4	5	6	7	
38. I cry when I experience (physical) pain		1	2	3	4	5	6	7	
39. I cry when I am in despair		1	2	3	4	5	6	7	
40. I cry when I feel rejected by others		1	2	3	4	5	6	7	
41. I cry when I feel that I am in a blind-alley situation		1	2	3	4	5	6	7	

42.	I sometimes laugh so hard that I start to cry	never=	1	2	3	4	5	6	7	=always
43.	I cry when talking with my therapist/doctor		1	2	3	4	5	6	7	
44.	I cry when I am ill		1	2	3	4	5	6	7	
45.	I cry while I watch sad movies or television programs		1	2	3	4	5	6	7	
46.	I cry when practicing religious activities such as praying, listening to preachers, reading holy books		1	2	3	4	5	6	7	
47.	I cry when I hear the national anthem and/or see the national flag rise		1	2	3	4	5	6	7	
48.	I cry when I experience painful memories		1	2	3	4	5	6	7	
49.	I cry when I realize my own vulnerability and mortality		1	2	3	4	5	6	7	
50.	I cry when I see or witness other people suffering		1	2	3	4	5	6	7	
51.	I cry when I attend or witness memorial meetings		1	2	3	4	5	6	7	
52.	I cry when I am reuniting with friends or family members		1	2	3	4	5	6	7	
53.	I cry when I watch other people crying		1	2	3	4	5	6	7	
54.	I cry when watching or hearing an admired person		1	2	3	4	5	6	7	
55.	I cry when I have achieved success		1	2	3	4	5	6	7	

56. Has your tendency to cry *permanently* significantly changed (either increased or decreased) after having experienced a dramatic life event as compared with before that event? yes no

If yes, could you specify the kind of event? ....

57. How do you generally feel after a crying episode as compared with just before?
(**check for *each* emotion/mood the most appropriate alternative!**)

	1	2	3	
(a) relaxed:	less	same	more	as before
(b) tense:	less	same	more	as before
(c) in control:	less	same	more	as before
(d) depressed:	less	same	more	as before
(e) sad:	less	same	more	as before
(f) happy:	less	same	more	as before
(g) relieved:	less	same	more	as before

58. Can you estimate how often you have cried in the *last four weeks*? ____________________ times
59. How would you rate your general tendency to cry? I hardly ever cry 1 2 3 4 5 6 7 8 9 10 I can very easily cry

Part B

THE FOLLOWING ITEMS DESCRIBE SOME OF THE FUNCTIONS OF CRYING AS WELL AS SOME EMOTIONS THAT ARE ASSOCIATED WITH CRYING. PLEASE, INDICATE AFTER EACH ITEM TO WHAT EXTENT YOU AGREE.

	do not agree =	1	2	3	4	5	6	7	= agree
1.	Crying helps me to deal with my problems	1	2	3	4	5	6	7	
2.	I believe that it is useful to cry when life becomes stressful	1	2	3	4	5	6	7	
3.	My life would be better if I were able to really have a good cry	1	2	3	4	5	6	7	
4.	I use crying to help me feel better, when I have problems	1	2	3	4	5	6	7	
5.	Crying is an important and effective way of dealing with life's difficulties	1	2	3	4	5	6	7	
6.	I would rather cry about a problem than keep all my sadness inside	1	2	3	4	5	6	7	
7.	Under certain conditions, when things are bad enough, I have cried almost uncontrollably	1	2	3	4	5	6	7	
8.	If I let myself cry deeply I sleep better	1	2	3	4	5	6	7	
9.	I find that I feel better after a good cry	1	2	3	4	5	6	7	
10.	I feel relaxed after a good cry	1	2	3	4	5	6	7	
11.	After a good crying spell I am better able to cope with my problems	1	2	3	4	5	6	7	
12.	After a good cry I am more optimistic about the future	1	2	3	4	5	6	7	
13.	I try not to cry when I am upset	1	2	3	4	5	6	7	
14.	After crying I feel warm all over	1	2	3	4	5	6	7	
15.	I feel peaceful after a good cry	1	2	3	4	5	6	7	
16.	Crying is the healthiest thing you can do when you are feeling sad	1	2	3	4	5	6	7	
17.	When I am not able to cry in a stress situation I stay feeling tense	1	2	3	4	5	6	7	

18. Mostly I can control my tears	do not agree =	1	2	3	4	5	6	7	= agree
19. I feel ashamed when I am crying		1	2	3	4	5	6	7	
20. After crying I feel often more miserable than before		1	2	3	4	5	6	7	
21. I like to cry		1	2	3	4	5	6	7	
22. Other people generally become gentler when I cry		1	2	3	4	5	6	7	
23. I hate to cry		1	2	3	4	5	6	7	
24. I can manipulate others with tears		1	2	3	4	5	6	7	

Part C

WHICH FACTORS DETERMINE WHETHER OR NOT YOU BREAK OUT IN TEARS AT A CERTAIN MOMENT? INDICATE FOR EACH OF THE FACTORS LISTED BELOW HOW IMPORTANT THEY ARE FOR YOU.

	not at all	0	1	2	3	4	5	very much
(a) the actual occurrence or actual memory of an emotional event		0	1	2	3	4	5	
(b) my mood *before* the event takes place		0	1	2	3	4	5	
(c) hormones		0	1	2	3	4	5	
(d) whether I am willing to give in to shedding tears		0	1	2	3	4	5	
(e) my physical condition (fatigue, hunger, etc.)		0	1	2	3	4	5	
(f) my location (at home, outside, at the workplace)		0	1	2	3	4	5	
(g) the nature of the emotion evoked by the event		0	1	2	3	4	5	
(h) the presence of other people with whom I am intimate		0	1	2	3	4	5	
(i) preceding drugs or alcohol use		0	1	2	3	4	5	

Item		0	1	2	3	4	5	
(j) my personality	not at all	0	1	2	3	4	5	very much
(k) the way I was brought up		0	1	2	3	4	5	
(l) how happy I generally feel in my life		0	1	2	3	4	5	
(m) my education		0	1	2	3	4	5	
(n) my age (as an adult)		0	1	2	3	4	5	
(o) my social status		0	1	2	3	4	5	
(p) the season		0	1	2	3	4	5	
(q) specific time of the day		0	1	2	3	4	5	
(r) my level of self-esteem		0	1	2	3	4	5	
(s) genetic factors (I was born this way)		0	1	2	3	4	5	
(t) whether or not I have an adequate coping response available		0	1	2	3	4	5	
(u) the presence of strangers		0	1	2	3	4	5	
(v) the food and drinks I had just before crying		0	1	2	3	4	5	
(w) whether the situation or event that led me to cry will be resolved in a positive manner		0	1	2	3	4	5	
(x) ever having experienced a dramatic event *in the past*		0	1	2	3	4	5	
(y) any other factors, namely..		0	1	2	3	4	5	
....		0	1	2	3	4	5	

Part D

In this part of the questionnaire we want to learn about your **last** (most recent) crying episode. First, we ask you to give a short description of the situation that elicited your crying response. In addition, there are some questions concerning several aspects of your crying and of the context of your crying.

1. Please, give a short description of the most recent situation or event that made you feel tears in your eyes. Try to be as accurate as possible.

____

____

____

____

____

____

2. How long ago did this most recent crying episode occur?
   1 less than 1 day
   2 2–5 days ago
   3 6–10 days ago
   4 11–30 days ago
   5 1–6 months ago
   6 7–12 months ago
   7 more than 1 year ago

3. How long did the crying episode last?
   1 less than 5 minutes
   2 5–15 minutes
   3 16–30 minutes
   4 31–60 minutes
   5 more than 60 minutes
   6 it were repeatedly recurring spells

4. How intense was your crying?
   1 just wet eyes
   2 wet eyes and silent sobbing
   3 wet eyes, sobbing and howling
   4 wet eyes, sobbing, howling, body movements and vocalizations

5. How much time passed between the confrontation with the situation described above and your crying response?
   0 it is/was an ongoing situation
   1 less than 5 minutes
   2 5–15 minutes
   6 1–7 days
   7 1–4 weeks

3	15–60 minutes	8	1–6 months
4	1–8 hours	9	7–12 months
5	8–24 hours	10	more than 1 year

6. Where were you when you cried (please, answer in as much detail as possible; e.g., 'in the bedroom' or 'in the kitchen' instead of 'at home')?

___

7. What time was it when you cried (indicate the exact time e.g. 5 PM, 9.20 AM)? ___

8a. How many other people were present? ___

8b. If less than 6 persons were present, please specify who (boyfriend, mother, colleague, sister, etc). Otherwise, give a short description of the composition of the group.

___
___
___
___

9. Who or what was responsible for the situation that made you cry? (Check all that apply)

0	does not apply	6	superiors, authorities
1	myself	7	strangers
2	my partner or boy/girl friend	8	fate, (supra)natural forces
3	family members or relatives	9	otherwise, please specify
4	close friends		
5	colleagues, acquaintances		___

10. Which emotions or feelings did you experience? (Check all that apply)

0	don't know	10	anger
1	relief	11	disgust
2	joy	12	guilt
3	contempt	13	elation
4	sadness	14	frustration
5	fear	15	dismay
6	humiliation	16	being touched
7	(self)pity	17	rapture
8	powerlessness	18	any other emotion or feeling
9	satisfaction		(please specify) ___

11. Did your crying change the relationship with somebody who was present while you cried?

0	does not apply	2	no
1	yes, it became worse	3	yes, it improved

12. Did your crying change any further aspect of the situation for the better or did it make the situation worse?

0	does not apply	2	no
1	yes, it became worse	3	yes, it improved

13. If an intimate or familiar person was present while you cried, how did he/she react? (in case more than one intimates were present, limit your answer to the one who you felt is most close to you) **Check all that apply!**

   0 does not apply
   1 he/she ignored my crying
   2 he/she offered comfort with words
   3 he/she comforted me by putting an arm around me
   4 he/she became mad at me
   5 he/she did not know how to behave
   6 he/she was embarassed
   7 he/she expressed understanding
   8 he/she stopped being nasty
   9 he/she became more friendly or warm
   10 otherwise, please specify

   ____

14. If a stranger or less familiar person was present while you cried, how did he/she react? (in case more strangers were present, limit your answer to the one being nearest to you) **Check all that apply!**

   0 does not apply
   1 he/she ignored my crying
   2 he/she offered comfort with words
   3 he/she comforted me by putting an arm around me
   4 he/she became mad at me
   5 he/she did not know how to behave
   6 he/she was embarassed
   7 he/she expressed understanding
   8 he/she stopped being nasty
   9 he/she became more friendly or warm
   10 otherwise, please specify

   ____

15. How did you feel *mentally* after the crying episode in comparison with before?
    1 worse than before
    2 same as before
    3 better than before

16. How did you feel *physically* after the crying episode in comparison with before?
    1 worse than before
    2 same as before
    3 better than before

17. What did you stop crying? (Check all that apply!)
    1 I felt re-stabilized emotionally
    2 other people were watching me
    3 I felt "dried up"
    4 shame/ embarrassment
    5 comforting words and behaviors of others
    6 the situation had changed significantly for the better
    7 other people had forbidden me to cry
    8 I had reached an important goal
    9 I had to concentrate on ongoing activities
    10 I simply had the feeling "it was enough"
    11 continuing would not do my physical health any good (e.g. headache)
    12 I fell asleep
    13 my perception of the situation had changed
    14 I had made my peace with the situation
    15 other reasons, namely ____________________
    ○ ____________________
    ○ ____________________
    ○ ____________________
    ○ ____________________

Part E

Only for women

1. Do you feel that your tendency to cry is dependent of the phase of your menstrual cycle .... yes no

If yes, when are you more willing to cry than usual?
(Please, mark all those days by putting a circle around each of them)
(1a) Days before menstruation 14 13 12 11 10 9 8 7 6 5 4 3 2 1
(1b) Days during menstruation 1 2 3 4 5 6
(1c) Days after menstruation 1 2 3 4 5 6 7 8 9 10 11 12 13 14

2. Do you at present use birth control pills? .... yes no

If you ever have been pregnant

3. Was your tendency to cry greater during (certain months of) pregnancy? .... yes no
   (3a) **If yes,** during which month(s) 1 2 3 4 5 6 7 8 9
   (Please, mark all those months by putting a circle around each of them)

4. Did you have any "crying-days" after the birth of your baby? .... yes no
   (4a) **If yes,** during which postpartal day(s)? 1 2 3 4 5 6 7 8 9 10 11–14 15–19 20–further
   (Please, mark all those days by putting a circle around each of them)

5. Did you breastfeed your baby? .... yes no

If you have more than one child

6. Did you experience any difference in tendency to cry during the first pregnancy and any later pregnancies? .... yes no
   (6a) **If yes,** how?
   (a) I cried more frequently during my first pregnancy
   (b) I cried more frequently during any later pregnancy

7. Did you experience any difference in having experienced any "crying days" after the first delivery compared to any later deliveries? .... yes no
   (7a) **If yes,** how?
   (a) I cried more frequently **after the first** delivery
   (b) I cried more frequently **after any later** deliveries

Index